GW00417606

Geometric Analysis
and
the Calculus of Variations

for

Stefan Hildebrandt

Jürgen Jost, editor

International Press Publications

Mathematical Physics

Quantum Groups: From Coalgebras to Drinfeld Algebras
Steven Shnider and Shlomo Sternberg
75 Years of Radon Transform
edited by Simon Gindikin and Peter Michor
Perspectives in Mathematical Physics
edited by Robert Penner and S.-T. Yau
Essays On Mirror Manifolds
edited by S. T. Yau
Mirror Symmetry II
edited by Brian Greene
XIth International Congress on Mathematical Physics
edited by D. Iagolnitzer
Differential Equations and Mathematical Physics
edited by I. Knowles

Number Theory

Elliptic Curves, Modular Forms, and Fermat's Last Theorem
edited by John Coates and Shing Tung Yau

Geometry and Topology

L^2 Moduli Spaces with 4-Manifolds with Cylindrical Ends
by Clifford Henry Taubes
The L^2 Moduli Space and a Vanishing Theorem for Donaldson Polynomial Invariants
by J. Morgan, T. Mrowka, and D. Ruberman
Algebraic Geometry and Related Topics
edited by J.-H. Yang, Y. Namikawa, and K. Ueno
Lectures on Harmonic Maps
by R. Schoen and S.-T. Yau
Lectures on Differential Geometry
by R. Schoen and S.-T. Yau
Geometry, Topology and Physics for Raoul Bott
edited by S.-T. Yau
Lectures on Low-Dimensional Topology
edited by K. Johannson
Chern, A Great Geometer
edited by S.-T. Yau

Surveys in Differential Geometry
edited by C.C. Hsiung and S.-T. Yau

Analysis

Proceedings of the Conference on Complex Analysis
edited by Lo Yang
Integrals of Cauchy Type on the Ball
by S. Gong
Advances in Geometric Analysis and Continuum Mechanics
edited by P. Concus and K. Lancaster
Lectures on Nonlinear Wave Equations
by C. D. Sogge

Physics

Physics of the Electron Solid
edited by S.-T. Chui
Proceedings of the International Conference on Computational Physics
edited by D.H. Feng and T.-Y. Zhang
Chen Ning Yang, A Great Physicist of the Twentieth Century
edited by S.-T. Yau
Yukawa Couplings and the Origins of Mass
edited by Pierre Ramond

Current Developments in Mathematics, *1995*
Collected and Selected Works

The Collected Works of Freeman Dyson
The Collected Works of C. B. Morrey
The Collected Works of P. Griffiths
V. S. Varadarajan

Journals
Communications in Analysis and Geometry
Mathematical Research Letters
Methods and Applications of Analysis

Library of Congress Catalog Card Number: 96-77901

Jost, Jürgen (Editor)

Geometric Analysis and the Calculus of Variations (for Stefan Hildebrandt)

ISBN 1-57146-037-3

Typeset using LaTeX
Printed on acid-free paper, in the United States of America

Geometric Analysis
and
the Calculus of Variations

for

Stefan Hildebrandt

This volume is dedicated to

Stefan Hildebrandt

on the occasion of his

60th birthday

by his students, colleagues, and friends

Preface.

The present book is dedicated to Stefan Hildebrandt. Stefan Hildebrandt was born in Leipzig on July 13, 1936. He started to study mathematics at the University of Leipzig in 1954. Because of the political circumstances then prevailing in East Germany, he and his girl friend and future wife Brigitte in 1958 fled to West Germany. In Mainz, he remet his former academic teacher Ernst Hölder who had left East Germany as well, and Hildebrandt could continue with a mathematics education that was still substantially based on the Leipzig tradition of mathematical analysis of Leon Lichtenstein, Otto Hölder (the father of Ernst Hölder), Eberhard Hopf and others. In the 60's, he was a frequent visitor at the Courant Institute in New York, at a time when the regularity theory for elliptic systems and minimal surfaces were prominent research topics. At the Courant Institute, he formed lasting friendships with several other brilliant young analysts of his generation, like Paul Rabinowitz, Neil Trudinger, Henry Wente, Kjell-Ove Widman. And, of course, he met the grand masters of his time, like Richard Courant, Hans Lewy, Jürgen Moser, Louis Nirenberg.

In the 30's, Jesse Douglas, Tibor Radó, and Richard Courant had founded the modern theory of minimal surfaces by solving and generalizing Plateau's problem of finding a minimal surface bounded by a prescribed Jordan curve, and connecting minimal surfaces with regularity theory for elliptic partial differential equations and Riemann surfaces. However, there still was a serious gap in the theory: namely, apart from a result of Hans Lewy that applied to real analytic boundary curves only, the question of boundary regularity for minimal surfaces was unsettled. Hildebrandt achieved a complete solution of this difficult problem: he showed that a minimal surface is as regular as the boundary. Hildebrandt's result not only completed the classical theory, but it was also a basis for subsequent new developments, ranging from the index theorems and the global analysis approach of Böhme, Tromba, Tomi to the constructions of minimal surfaces by reflections arguments of Lawson, and more recently of Karcher and others, to the physical theory of strings with boundary that also needs to use Hildebrandt's theorem for a mathematically rigorous development.

This result brought Hildebrandt immediate fame, and he was made a full professor in Mainz in 1967 and in Bonn in 1970. This theorem, however, was only the first one in an impressive series of fundamental results on various geometrically defined variational problems. For example, he found an optimal existence theorem for surfaces of constant mean curvature H with a given boundary curve γ. Again, the result was not only optimal but also one upon which other researchers could build. Using Hildebrandt's result and an important idea of Wente, Struwe and Brézis-Coron succeeded in finding a second solution under Hildebrandt's condition, the so-called large H surface whose existence had already been conjectures by Rellich.

While the existence and regularity for harmonic maps into Riemannian manifolds of nonpositive curvature had been obtained already in 1963 by Eells-Sampson and Al'ber, the corresponding questions for positive target curvature had remained open. In collaboration with Helmut Kaul and Kjell-Ove Widman, he solved the existence and regularity problem for harmonic maps with boundary values in a strictly convex ball. This result is again optimal, and they also showed by an explicit example that regularity fails without that restriction. The example they had discovered was found to be the prototype for singularities of energy minimizing maps in subsequent work of Giaquinta-Giusti and Schoen-Uhlenbeck, and so Hildebrandt's contribution to the theory had again turned out to be fundamental for later research.

Later on, Hildebrandt turned to minimal surfaces with free or partially free boundaries. In collaboration with Johannes Nitsche, he found an example of a cusp singularity at the boundary of such a minimal surface. Most recently, he continued that line of research in collaboration with Friedrich Sauvigny.

The preceding brief outline had to omit several other important mathematical contributions of Stefan Hildebrandt, and, of course, also related results of other researchers.

Stefan Hildebrandt achieved a lasting and formative influence on the geometric calculus of variations not only through his own scientific contributions, but also through his direction of several research projects of the DFG (German Research Foundation) and his systematic education of a younger generation of German analysts. In fact, those research projects with their stimulating seminars and their unique scientific atmosphere made Bonn one of the most popular visiting places for analysts from all over the world. Several of these visits, and the scientific contacts and exchanges fostered by them, like the one of Mariano Giaquinta in the early 80's, had a lasting influence on the development of mathematical research in Germany and abroad.

His broad knowledge of the historical development of the calculus of variations and other areas of mathematics is but one indications of his ability to see individual mathematical results in a wider scientific perspective. This is also testified by the monographs he has written, a beautiful two volume treatise on minimal surfaces written in collaboration with Ulrich Dierkes, Albrecht Küster, and Ortwin Wohlrab, or his encompassing several volume monograph on the calculus of variations written together with Mariano Giaquinta. We should also not omit his longstanding friendship with Anthony Tromba that led not only to mutual mathematical stimulation, but also to beautiful books about the calculus of variations for a general readership.

For those reasons, and from personal gratitude of the editor and many of the authors for his generous support he always provided, the present book is dedicated to him.

Doktoranden von S. Hildebrandt

Ph. D. Students of S. Hildebrandt

1. Bernd Schmidt
2. Klaus Steffen
3. Helmuth Kaul
4. Karl-Heinz Goldhorn
5. Fritz-Peter Harth
6. Claus Gerhardt
7. Klaus Gornik
8. Ederhard Kudszus
9. Emanuel Sperner
10. Josef Bemelmans
11. Michael Meier
12. Jürgen Jost
13. Goswin Eisen
14. Geza Schrauf
15. Alfred Baldes
16. Ulrich Dierkes
17. Albrecht Küster
18. Rugang Ye
19. Ernst Kuwert
20. Christoph Hamburger
21. Leung-Fu Cheung
22. Xiangquing Li
23. Maria Athanassenas
24. Ortwin Wohlrab

Derzeitige Doktoranden in Bonn (present Ph. D. students in Bonn): Heiko von der Mosel, Gudrun Turowski, Katrin Rhode.

Contents.

The N-dimensional Analogue of the Catenary: Prescribed Area

ULRICH DIERKES AND GERHARD HUISKEN

We study "heavy" n-dimensional surfaces which are suspended from some given boundary data φ and have prescribed surface area A. Using a fixed point argument we show existence of a solution provided A is close to the area of the corresponding minimal surface spanned by φ.

The equilibrium condition for a *heavy, inextensible* and *flexible* surface M of constant mass density which is exposed to a vertical gravitational field has been derived by several authors, see Lagrange [L, pp 158–162], Cisa de Gresy [GG, pp 274–276], Jellett [J, p 349–354] and Poisson [P, pp 173–187]. It turns out that there are several model problems available, which are due to different notations of flexibility and inextensibility, and which are all worth to be investigated. Quite generally Poisson [P] considers (flexible-inextensible) surfaces in $\mathbb{R}^3$, which are exposed to an arbitrary force field $F = (X, Y, Z)$, and,—using direct arguments from mechanics—he deduces a system of partial differential equations which, in addition to the unknown function u, also involves two independent "tensions" T and T' which describe the forces inside the surface. Of particular interest is the case where the tensions coincide, i.e. $T = T'$. Then the system of p.d.e.'s reduces to the single equation

$$Z - pX - qY - \frac{T}{k^2}[(1+q^2)u_{xx} - 2pqu_{xy} + (1+p^2)u_{yy}] = 0, \tag{1}$$

where we have set $p = u_x = \frac{\partial u}{\partial x}$, $q = u_y = \frac{\partial u}{\partial y}$, $k^2 = 1+p^2+q^2$ and T satisfies

$$X\,dx + Y\,dy + Z\,dz + dT = 0, \tag{2}$$

that is the external force F must have a potential U and $T = U + c$. From (1) and (2) Poisson deduces:

A) the *minimal surface equation* by taking $X = Y = Z = 0$, $T = \text{const}$;

B) the equation for a *capillary surface* by taking $X = Y = 0$, $Z = \frac{a+bz}{k}$, $T = \text{const}$; as the equilibrium condition of a flexible surface which is covered by a heavy fluid;

C) The equation of a *heavy surface in a gravitational field* by taking $X = Y = 0$, $Z = g\varepsilon$, where g denotes the gravitational constant and ε is the density of the surface. The tension T is then given by $T = -\lambda - g\varepsilon z$, $\lambda \in \mathbb{R}$, and hence (1) implies the condition

$$g\varepsilon - \frac{\lambda + g\varepsilon z}{k^2}\{(1+q^2)u_{xx} - 2pqu_{xy} + (1+p^2)u_{yy}\} = 0. \tag{3}$$

Assuming $g\varepsilon = 1$ we are thus led to the equation

$$\sqrt{1+|Du|^2}\,\operatorname{div}\frac{Du}{\sqrt{1+|Du|^2}} = \frac{1}{(u+\lambda)} \quad \text{in } \Omega \subset \mathbb{R}^2,\ \lambda \in \mathbb{R}, \tag{4}$$

as a model equation for the equilibrium condition of an inextensible, flexible, heavy surface of constant mass density in a gravitational field.

A further application in architecture lends special interest to equation (4), cp. [BHT] and [O]. In fact turning a hanging solution u of (4) upside down gives the optimal shape of a cupola.

Here we are concerned with the Dirichlet problem in $\mathbb{R}^n$ connected with equation (4) and, in addition, we require a solution u to have prescribed area A, i.e.

$$\int_\Omega \sqrt{1+|Du|^2}\,dx = A.$$

In other words we consider the following problem

(P). *Let $\Omega \subset \mathbb{R}^n$ be a bounded domain of class $C^{2,\alpha}$ and suppose $\varphi \in C^{2,\alpha}(\bar{\Omega})$ is given. For some prescribed value $A \in \mathbb{R}^+$ one has to find a function $u \in C^{2,\alpha}(\bar{\Omega})$ and some $\lambda \in \mathbb{R}$ such that*

$$\sqrt{1+|Du|^2}\,\operatorname{div}\frac{Du}{\sqrt{1+|Du|^2}} = \frac{1}{(u+\lambda)} \quad \text{in } \Omega \tag{5}$$

$$u = \varphi \quad \text{on } \partial\Omega, \text{ and}$$

$$A(u) = \int_\Omega \sqrt{1+|Du|^2}\,dx = A. \tag{6}$$

Observe that problem (P) can also be considered as the n-dimensional mathematical analogue of the (one-dimensional) catenary problem: *To find a surface M = graph u of prescribed area A and boundary φ with lowest* *possible center of gravity.* Indeed, since the x_{n+1}-coordinate of the center of gravity is given by the quotient

$$\left(\int_\Omega \sqrt{1+|Du|^2}\,dx\right)^{-1} \int_\Omega u\sqrt{1+|Du|^2}\,dx$$

this amounts to the minimization of the integral

$$\int_\Omega u\sqrt{1+|Du|^2}\,dx$$

subject to the constraints

$$\int_\Omega \sqrt{1+|Du|^2}\,dx = A \quad \text{and} \quad u = \varphi \text{ on } \partial\Omega.$$

Now, introducing a Lagrange multiplier λ one obtains (5) and (6) as the equilibrium condition for this problem.

Nitsche [N, p 146] has shown by way of example that the above variational problem has no solution whatever the value A might be. Thus one has to use more refined techniques from the calculus of variations in order to construct merely relative minima, say. In this paper, however, we tackle equation (5) directly and prove suitable a priori estimates which enable us to apply a fixed point argument.

Clearly, there is an obvious necessary condition on the number A, namely that $A \geq A_0$, A_0 denoting the infimum of area of all graphs bounded by φ.

But, surprisingly, and in contrast to the one-dimensional situation, there is a further necessary condition namely that $A \leq a_1(\varphi)$, $a_1(\varphi)$ denoting some specific number depending on the boundary values φ. In fact, it was pointed out by Nitsche [N] that the Euler equation (5) in the corresponding rotationally symmetric case has no solution, provided $A > a_1(\varphi)$.

In the light of the above remarks the following existence result is natural (and, in a sense, optimal).

Theorem. *Let $\Omega \subset \mathbb{R}^n$ be a bounded, mean-convex domain of class $C^{2,\alpha}$ and suppose $\varphi \in C^{2,\alpha}$. Then there exists some number $A_1 > A_0$ depending only on $n, \Omega, |\varphi|_{2,\alpha}$, such that for all numbers $A \in (A_0, A_1]$ there is some $\lambda \in \mathbb{R}$ and a function $u = u_\lambda \in C^{2,\alpha}(\bar{\Omega})$ which solves problem* (P) *i.e.*

$$\sqrt{1+|Du|^2}\,\operatorname{div}\frac{Du}{\sqrt{1+|Du|^2}} = \frac{1}{(u+\lambda)} \quad \text{in } \Omega \tag{5}$$

$$u = \varphi \quad \text{on } \partial\Omega, \text{ and}$$

$$A(u) = \int_\Omega \sqrt{1+|Du|^2}\,dx = A. \tag{6}$$

Observe that equation (5) is an equation of mean curvature type with (variable) mean curvature $H = H(u, Du) = (u+\lambda)^{-1}(1+|Du|^2)^{-1/2}$ and that $H_u \le 0$, i.e. H is monotone with the "wrong" monotonicity behaviour.

For the proof of the Theorem we use Schauder's fixed point theorem in combination with suitable a priori and monotonicity estimates. We first consider solutions $u = u_{f,\lambda} \in C^{2,\alpha}(\bar\Omega)$ of the related problem

$$\sqrt{1+|Du|^2}\,\mathrm{div}\frac{Du}{\sqrt{1+|Du|^2}} = (f+\lambda)^{-1} \quad \text{in } \Omega \qquad (7)$$
$$u = \varphi \quad \text{on } \partial\Omega,$$

where $f \in C^{1,\alpha}(\bar\Omega)$ is some positive function and $\lambda \in \mathbb{R}$ denotes some positive number. Let $c(n) = n^{-1}\omega_n^{-1/n}$ stand for the isoperimetric constant, $\omega_n = |B_1^n(0)|$ the measure of the unit ball, and put

$$h := \sup_{\partial\Omega}\varphi, \quad k_0 := \inf_{\partial\Omega}\varphi, \quad \text{and} \quad \lambda_0 := (1+\sqrt{2^{n+1}})c(n)|\Omega|^{1/n}.$$

As a first step we establish a priori bounds for $\sup_\Omega u$ and $\inf_\Omega u$.

Lemma 1. *Let $u = u_{f,\lambda} \in C^{2,\alpha}(\bar\Omega)$ denote a solution to the Dirichletproblem (7). If λ, λ_0 and k satisfy*

$$\lambda \ge \lambda_0 = (1+\sqrt{2^{n+1}})c(n)|\Omega|^{1/n} \quad \textit{and}$$
$$k_0 \ge (1+\sqrt{2^{n+1}})^2 c(n)|\Omega|^{1/n} = (1+\sqrt{2^{n+1}})\lambda_0$$

then we have the inequality

$$h \ge u_{f,\lambda} \ge \lambda_0.$$

The proof follows the argument given in [DH]. For completeness we sketch it here.

Since f, λ are positive the first inequality follows from the maximum principle. To prove the second relation we choose $\delta \ge -k_0$ and put $w := \min(u+\delta, 0)$, $A(\delta) := \{x \in \Omega : u < -\delta\}$. Multiplying (7) with w, integrating by parts and using $w|_{\partial\Omega} = 0$, we obtain

$$\int_\Omega \frac{|Dw|^2}{\sqrt{1+|Dw|^2}} = \int_{A(\delta)} \frac{|w|}{(f+\lambda)\sqrt{1+|Du|^2}}, \qquad \text{whence}$$
$$\int_\Omega |Dw| \le |A(\delta)| + \frac{1}{\lambda_0}\int_{A(\delta)} |w|.$$

We use Sobolev's inequality on the left and Hölder's inequality on the right hand and obtain with $c(n) = n^{-1}\omega_n^{-1/n}$ the relation

$$|w|_{n/n-1}\{c^{-1}(n) - \lambda_0^{-1}|\Omega|^{1/n}\} \le |A(\delta)|,$$

where $|w|_{n/n-1}$ stands for the $L_{n/n-1}$-norm of w. Another application of Hölder's inequality yields

$$(\delta_1 - \delta_2)|A(\delta_1)| \le \left\{\frac{c(n)\lambda_0}{\lambda_0 - c(n)|\Omega|^{1/n}}\right\}|A(\delta_2)|^{1+\frac{1}{n}}$$

for all $\delta_1 \ge \delta_2 \ge -k_0$. In view of a well known Lemma due to Stampacchia, [St, Lemma 4.1] this is easily seen to imply

$$\left|A(-k_0 + 2^{n+1}c_1|A(-k_0)|^{1/n})\right| = 0, \qquad \text{where}$$

$$c_1 = \frac{c(n)\lambda_0}{\lambda_0 - c(n)|\Omega|^{1/n}}. \qquad \text{This immediately implies that}$$

$$u \ge k_0 - \frac{2^{n+1}\lambda_0 c(n)|\Omega|^{1/n}}{\lambda_0 - c(n)|\Omega|^{1/n}}.$$

Since $k_0 \ge (1+\sqrt{2^{n+1}})\lambda_0$ and $\lambda_0 = (1+\sqrt{2^{n+1}})c(n)|\Omega|^{1/n}$ we finally obtain $u \ge \lambda_0$ as desired. $\square$

To derive a gradient estimate at the boundary we rewrite (7) into

$$(1+|Du|^2)\Delta u - D_iuD_juD_{ij}u = (f+\lambda)^{-1}(1+|Du|^2). \tag{8}$$

We can apply the results of Serrin [Se, 1], see also [GT, chap. 14.3]. Equation (8) satisfies the structure condition (14.41) and the r.h.s. is of order $O(|Du|^2)$. So we obtain a gradient estimate on the boundary which is independent of $|Df|$:

$$\sup_{\partial\Omega}|Du_{f,\lambda}| \le c_2 = c_2(n,\Omega,h,|\varphi|_{2,\Omega}),$$

provided only that $\partial\Omega$ has non-negative (inward) mean curvature.

It is not possible to derive interior gradient estimates independent of $|Df|$, but we can prove

$$\sup_{\Omega}|Du_{f,\lambda}| \le \max\{2, 1/4\sup|Df|, 2e^{4(h\lambda_0^{-1}-1)}\sup_{\partial\Omega}|Du_{f,\lambda}|\}. \tag{9}$$

Estimate (9) can be obtained from a careful analysis of the structure condition in [GT, chap. 15]. For a selfcontained proof, which uses the geometric nature of equation (7) we refer to [DH].

Having proved the C^1 estimates we now infer from general theory and Schauder-estimates the inequality

$$|u_{f,\lambda}|_{2,\alpha,\Omega} \le C(n, \Omega, \lambda_0, h, |\varphi|_{2,\alpha,\Omega}, |f|_{1,\Omega}) \tag{10}$$

for any solution $u_{f,\lambda}$ of (7) provided

$$\lambda \ge \lambda_0 = (1 + \sqrt{2^{n+1}})c(n)|\Omega|^{1/n},$$

$k_0 = \inf_{\partial\Omega} \varphi \ge (1 + \sqrt{2^{n+1}})\lambda_0$ and $f \ge 0$. Here the constant C only depends on the quantities indicated.

Note that by Arzela–Ascoli this already implies that $u_{f,\lambda} \to u_0$ in $C^2(\Omega)$ as $\lambda \to \infty$, where u_0 denotes the unique (area minimizing) minimal surface spanned by φ. In particular we have $A(u_{f,\lambda}) \to A(u_0)$ as $\lambda \to \infty$ where $A(u) = \int_\Omega \sqrt{1 + |Du|^2}\, dx$ denotes the area of the graph of u.

Later on it will be important to investigate this convergence somewhat more carefully.

For f fixed we now consider the behaviour of solutions $u_{f,\lambda}$ of equation (7) as λ varies. In particular we show that $u_{f,\lambda}$ increases with increasing values of λ. More precisely we have

Lemma 2. *Let u_{f,λ_1} and u_{f,λ_2} denote two solutions of the Dirichletproblem (7) with r.h.s. $(f+\lambda_1)^{-1}$ and $(f+\lambda_2)^{-1}$ respectively, where $\lambda_1 \ge \lambda_2 \ge \lambda_0$ and $0 \le f \le h$. Then we have the inequality*

$$u_{f,\lambda_1}(x) \ge u_{f,\lambda_2}(x) + c_0(2n + 2C \operatorname{diam} \Omega)^{-1} d^2(x, \partial\Omega), \tag{11}$$

where $d(x, \partial\Omega) = \operatorname{dist}(x, \partial\Omega)$ denotes the distant of x to the boundary $\partial\Omega$ and $c_0 = c_0(h, \lambda_1, \lambda_2) = \frac{(\lambda_1 - \lambda_2)}{(h+\lambda_1)^2}$ and $C = C(n, \Omega, \lambda_0, h, |\varphi|_{2,\alpha}, |f|_{1,\Omega})$.

Proof. Put $a_{ij}(p) = \delta_{ij} - \frac{p_i p_j}{1+|p|^2}$, $p \in \mathbb{R}^n$. Then (7) may be rewritten into $a_{ij}(Du)D_{ij}u = (f + \lambda)^{-1}$. Therefore $w := u_{f,\lambda_1} - u_{f,\lambda_2}$ satisfies

$$\begin{aligned} a_{ij}(Du_{f,\lambda_1})D_{ij}w + a_{ij}(Du_{f,\lambda_1})D_{ij}u_{f,\lambda_2} \\ - a_{ij}(Du_{f,\lambda_2})D_{ij}u_{f,\lambda_2} = (f + \lambda_1)^{-1} - (f + \lambda_2)^{-1}, \end{aligned}$$

whence

$$A_{ij}(x)D_{ij}w + B_i(x)D_iw = \frac{\lambda_2 - \lambda_1}{(f+\lambda_1)(f+\lambda_2)} < 0,$$

where

$$A_{ij}(x) := a_{ij}(Du_{f,\lambda_1}(x)) \qquad \text{and}$$
$$B_i(x) := D_{kj}u_{f,\lambda_2}(x)\int_0^1 a_{kj,p_i}(tDu_{f,\lambda_1} + (1-t)Du_{f,\lambda_2})\,dt.$$

(Note that by the Hopf-maximum principle for linear equations this already implies that $u_{f,\lambda_1} \geq u_{f,\lambda_2}$.) Let L denote the linear operator

$$L := A_{ij}(x)D_{ij} + B_i(x)D_i$$

and take some comparison function

$$\varphi(x) := \frac{c_0(h,\lambda_1,\lambda_2)}{2n + 2C\Omega}[|x - x_0|^2 - R^2],$$

where $R = d(x_0, \partial\Omega)$, $x_0 \in \Omega$ and $|B|_{0,\Omega} \leq C = C(n, \Omega, \lambda_0, h, |\varphi|_{2,\alpha,\Omega}, |f|_{1,\Omega})$ denotes a constant depending only on the quantities indicated (cp. (10)). Then we compute

$$\begin{aligned} L\varphi &\leq \frac{c_0}{2n + 2C\operatorname{diam}\Omega}[2\operatorname{trace}(A_{ij}) + 2|B|_{0,\Omega}|x - x_0|] \\ &\leq \frac{c_0}{2n + 2C\operatorname{diam}\Omega}[2n + 2c\operatorname{diam}\Omega] \leq c_0. \end{aligned}$$

Concluding we get

$$L(u_{f,\lambda_1} - u_{f,\lambda_2} + \varphi) \leq \frac{\lambda_2 - \lambda_1}{(f+\lambda_1)(f+\lambda_2)} + \frac{\lambda_1 - \lambda_2}{(h+\lambda_1)^2} \leq 0$$

and $u_{f,\lambda_1} - u_{f,\lambda_2} + \varphi \geq 0$ on the boundary of Ω. Therefore

$$u_{f,\lambda_1}(x) \geq u_{f,\lambda_2}(x) - \varphi(x) = u_{f,\lambda_2}(x) + \frac{c_0}{2n + 2C\operatorname{diam}\Omega}[R^2 - |x - x_0|^2]$$

and, on putting $x = x_0$ we obtain

$$u_{f,\lambda_1}(x_0) \geq u_{f,\lambda_2}(x_0) + \frac{c_0R^2}{2n + 2C\operatorname{diam}\Omega}.$$

□

We now show that the area of solutions $u_{f,\lambda}$ decreases as λ increases; in fact we have the following:

Lemma 3. *Let $\lambda_1 \geq \lambda_2 \geq \lambda_0$ and $0 \leq f \leq h$ be given and denote by $u_{f,\lambda_1}, u_{f,\lambda_2}$ two solutions of (7) with r.h.s. $(f+\lambda_1)^{-1}$ and $(f+\lambda_2)^{-1}$ respectively. Then there is some constant $C > 0$ depending only on n, Ω, λ_0, h, $|\varphi|_{2,\alpha,\Omega}$ and $|f|_{1,\Omega}$ such that the following estimate holds*

$$\int_\Omega \sqrt{1+|Du_{f,\lambda_1}|^2}\,dx + \frac{(\lambda_1-\lambda_2)}{(h+\lambda_1)^3}C \leq \int_\Omega \sqrt{1+|Du_{f,\lambda_2}|^2}\,dx. \tag{12}$$

Proof. Put $u_1 := u_{f,\lambda_1}$ and $u_2 = u_{f,\lambda_2}$. From (7) we obtain for all $\varphi \in C_0^1(\Omega)$

$$-\int_\Omega \frac{Du_1 D\varphi}{\sqrt{1+|Du_1|^2}}dx = \int_\Omega \frac{\varphi}{(f+\lambda_1)\sqrt{1+|Du_1|^2}}dx. \tag{13}$$

We test (13) with $\varphi := u_2 - u_1 \in C_0^1(\Omega)$ and obtain

$$-\int_\Omega \frac{Du_1 D(u_2-u_1)}{\sqrt{1+|Du_1|^2}}dx = \int_\Omega \frac{u_2-u_1}{(f+\lambda_1)\sqrt{1+|Du_1|^2}}dx, \quad \text{whence}$$

$$\int_\Omega \frac{|Du_1|^2}{\sqrt{1+|Du_1|^2}}dx = \int_\Omega \frac{Du_1 Du_2}{\sqrt{1+|Du_1|^2}}dx + \int_\Omega \frac{u_2-u_1}{(f+\lambda_1)\sqrt{1+|Du_1|^2}}dx.$$

We apply Schwarz's inequality and Lemma 2 to obtain

$$\int_\Omega \sqrt{1+|Du_1|^2}\,dx \leq \int_\Omega \sqrt{1+|Du_2|^2}\,dx + \frac{(\lambda_2-\lambda_1)(2n-2C\operatorname{diam}\Omega)^{-1}}{(h+\lambda_1)^3\sqrt{1+|Du_1|^2_{0,\Omega}}}\int_\Omega d^2(x,\partial\Omega)\,dx$$

Concluding we have

$$\int_\Omega \sqrt{1+|Du_1|^2}\,dx \leq \int_\Omega \sqrt{1+|Du_2|^2}\,dx + C\frac{(\lambda_2-\lambda_1)}{(h+\lambda_1)^3},$$

with some constant $C = C(n,\Omega,\lambda_0,h,|\varphi|_{2,\alpha},|f|_{1,\Omega})$. □

It is now desirable to have an explicit bound for the increment of area of the graphs of $u_0 = u_{f,\infty}$ and $u_{f,\lambda}$ respectively. Note that this estimate does not immediately follow from (12) by letting λ_1 tend to infinity.

Lemma 4. *Let $\lambda \geq \lambda_0$, $h \geq f \geq 0$ be given and denote by $u_{f,\lambda}$ and u_0 the unique solution of the Dirichletproblem (7) and the minimal surface spanned* *by φ respectively. Then there exists some constant C depending only on n, Ω, λ_0, h, $|\varphi|_{2,\alpha}$ and $|f|_{1,\Omega}$ such that*

$$\int_\Omega \sqrt{1+|Du_0|^2}\,dx + \frac{C}{(h+\lambda_1)^2} \le \int_\Omega \sqrt{1+|Du_{f,\lambda}|^2}\,dx$$

holds for all $\lambda \ge \lambda_0$.

Proof. Let $a(\lambda) := \displaystyle\int_\Omega \sqrt{1+|Du_{f,\lambda}|^2}\,dx$ denote the area of the graph of $u_{f,\lambda}$. Then (12) implies the inequality

$$\frac{a(\lambda_1)-a(\lambda_2)}{\lambda_1-\lambda_2} \le \frac{-C}{(h+\lambda_1)^3} \quad \text{for } \lambda_1 \ge \lambda_2 \tag{13}$$

and some $C > 0$ independent of λ_1, λ_2. Also, $a(\lambda)$ is monotone decreasing, so $a'(\lambda)$ exists almost everywhere, $a'(\lambda) \le \frac{-C}{(h+\lambda)^3}$ by (13) and

$$a(\infty) \le a(\lambda) + \int_\lambda^\infty a'(\xi)\,d\xi. \tag{14}$$

(Note that from Schauder theory for linear equations we could even infer that $u_{f,\lambda}$, $Du_{f,\lambda}$ depend Lipschitz-continuously on λ, i.e. (14) is in fact an identity.) From (14) we infer

$$\int_\Omega \sqrt{1+|Du_0|^2}\,dx + \frac{C}{(h+\lambda)^2} \le \int_\Omega \sqrt{1+|Du_{f,\lambda}|^2}\,dx$$

with $C > 0$. □

Proof of the Theorem. We define the set $\mathfrak{M}$ by

$$\mathfrak{M} := \{\, f \in C^{1,\alpha}(\bar\Omega) : 0 \le f \le h,\ \sup_\Omega |Df| \le M \,\}.$$

By virtue of our C^1-estimates we may choose $M = M(n, \Omega, h, |\varphi|_{2,\Omega})$ large, so that $u_{f,\lambda} \in \mathfrak{M}$ for all $f \in \mathfrak{M}$, $\lambda \ge \lambda_0$. If f is restricted to $\mathfrak{M}$ then the constant C appearing in Lemma 4 only depends on n, Ω, h, $|\varphi|_{2,\alpha}$ and $M = M(n, \Omega, h, |\varphi|_{2,\Omega})$.

We put $A_1 = A_0 + \frac{C}{(h+\lambda_0)^2}$ where $C = C(n, \Omega, h, |\varphi|_{2,\alpha}, M)$ denotes the constant in Lemma 4, and we also assume for a moment that

$$k_0 := \inf_\Omega \varphi \ge (1+\sqrt{2^{n+1}})\lambda_0.$$

Now fix a value $A \in (A_0, A_1]$. It follows from $a(\lambda) \to A_0$ as $\lambda \to \infty$ and from the monotonicity of $a(\lambda)$ that for $f \in \mathfrak{M}$ given there is precisely one $\lambda = \lambda(A) \geq \lambda_0$ and some unique solution $u_{f,\lambda} \in C^{2,\alpha}(\bar{\Omega})$ of (7) with prescribed area A, i.e. $A(u_{f,\lambda}) = A$.

Consider the operator T_A

$$T_A \colon \mathfrak{M} \to \mathfrak{M}$$
$$f \to u_{f,\lambda(A)}.$$

It follows from the $C^{2,\alpha}$ estimate (10) and Arzela–Ascoli that T_A is compact. Furthermore T_A is continuous. In fact let f_m converge to f in $C^{1,\alpha}(\bar{\Omega})$. Then $\{u_{m,\lambda_m} = T_A f_m\}$ is precompact in $C^2(\bar{\Omega})$ and hence any subsequence in turn has a convergent subsequence. Suppose that $u_{m_j} = u_{m,\lambda_{m_j}} \to u$ in $C^2(\bar{\Omega})$. Then $A(u_{m_j}) = A$ implies $A(u) = A$ and

$$\sqrt{1 + |Du_{m_j}|^2}\,\mathrm{div}\,\frac{Du_{m_j}}{\sqrt{1 + |Du_{m_j}|^2}} = (f_{m_j} + \lambda_{m_j})^{-1}$$

implies

$$\sqrt{1 + |Du|^2}\,\mathrm{div}\,\frac{Du}{\sqrt{1 + |Du|^2}} = (f + \Lambda)^{-1}$$

for some $\Lambda \in \mathbb{R}$. But from $A(u) = A$ and the uniqueness of λ it follows that $\Lambda = \lambda(f)$ and $u = T_A f$. Hence $T_A f_m$ converges to u and we can apply Schauder's fixed point theorem to obtain the existence of a regular $u \in C^{2,\alpha}(\bar{\Omega})$ solving (5) and (6). Now we have to get rid of the additional assumption $k_0 \geq (1 + \sqrt{2^{n+1}})\lambda_0$. To this end we choose some number $\gamma \in \mathbb{R}$ large, so that $\varphi_\gamma := \varphi + \gamma$ satisfies $\inf_{\partial\Omega} \varphi_\gamma \geq (1 + \sqrt{2^{n+1}})\lambda_0$. Then there is some further number $\lambda \in \mathbb{R}$ and a solution $u = u_\gamma \in C^{2,\alpha}(\bar{\Omega})$ satisfying (5) and (6) and $u_\gamma = \varphi_\gamma$ on the boundary of Ω. Therefore the function $u := u_\gamma - \gamma$ has boundary values φ and fulfills the equation

$$\sqrt{1 + |Du|^2}\,\mathrm{div}\,\frac{Du}{\sqrt{1 + |Du|^2}} = \frac{1}{u + (\gamma + \lambda)} \quad \text{in } \Omega, \text{ and}$$
$$A(u) = A.$$

This proves the Theorem.

References.

[BHT] R. Böhme, S. Hildebrandt, E. Tausch, *The two dimensional analogue of the catenary*, Pacific J. Math. **88** (1980), 247–278.

[CG] Cisa de Gresy, *Considération sur l'équilibre des surfaces flexibles et inextensibles*, Mem. Reale. Accad. Sci. Torino (1) **23** (1818), 259–294.

[DH] U. Dierkes, G. Huisken, *The N-dimensional analogue of the catenary: existence and non-existence*, Pacific J. Math. **141** (1990), 47–54.

[GT] D. Gilbarg, N. S. Trudinger, *Elliptic partial differential equations of second order*, Springer, Berlin Heidelberg New York, Grundlehren d. Math. Wiss. 1977.

[J] F. H. Jellet, *Die Grundlehren der Variationsrechnung*, Braunschweig, Verlar der Hofbuchhandlung v. Leibrock, 1860.

[L] J. L. Lagrange, *Mécanique analytique, quatrième édition*, Oeuvre tome onzième.

[N] J. C. C. Nitsche, *A non-existence theorem for the two-dimensional analogue of the catenary*, Analysis 6 (1986), 143–156.

[O] F. Otto (Editor), *Zugbeanspruchte Konstrukionen Bd I und II*, Ullstein Fachverlag, Berlin–Frankfurt/M–Wien, 1962 and 1966.

[P] S. D. Poisson, *Sur les surfaces élastiques.* Mem. Cl. Mathem. Phy. Inst. France, 1812, deux. p., 167–225.

[Se] J. Serrin, *The problem of Dirichlet for quasilinear elliptic differential equations with many independent variables*, Phil. Trans. Roy. Soc. London, Ser A, **264** (1969), 413–496.

[St] G. Stampacchia, *Équations elliptiques du second ordre à coefficients discontinus*, Les Presses de l'Université, Montréal 1966.

Received May 15, 1995.

Mathematischen Institut
Universität Bonn
Beringstrasse 6
53115 Bonn Germany

and

Matematische Fakultät

Universität Tübingen
Auf der Morgenstelle 10
72076 Tübingen Germany

The Plateau Problem for Parametric Surfaces with Prescribed Mean Curvature

FRANK DUZAAR AND KLAUS STEFFEN

Dedicated to S. Hildebrandt

We treat the existence problem for 2 dimensional parametric surfaces with prescribed mean curvature and given boundary curve. In Sections 1 and 2 we describe the contributions of E. Heinz, S. Hildebrandt, R. Gulliver, J. Spruck, H.C. Wente, and the second author to this Plateau problem in Euclidean 3 space. In Section 3 we present some new existence theorems for the case of an oriented Riemannian 3 manifold as ambient space. All the results discussed here are based on the minimization of the energy functional associated with the problem. The emphasis is on geometric conditions for the prescribed mean curvature and the given boundary which are sufficient for the existence of a solution.

0. Introduction.

We consider parametric surfaces $x\colon U \to \mathbb{R}^3$ in Euclidean space which are defined on the unit disc $U = \{w{=}(u,v) \in \mathbb{R}^2 : u^2{+}v^2{<}1\}$. It is convenient to assume *conformal parametrization*, i.e.

$$|x_u|^2 - |x_v|^2 = x_u \cdot x_v = 0 \text{ on } U\,, \tag{0.1}$$

where $x_u \cdot x_v$ denotes the Euclidean inner product of the partial derivatives of x and $|x_u|, |x_v|$ their Euclidean norm.

The advantage of the conformality condition is that we have a simple analytic expression for the mean curvature (the arithmetic mean of the principal curvatures) $h(u,v)$ of x at $(u,v) \in U$. Namely, if x is a conformal C^2 immersion then h is given by $\Delta x = 2h x_u \wedge x_v$, where $\Delta x = x_{uu} + x_{vv}$ is the Laplacean (applied to each component) of x and $\wedge\colon \mathbb{R}^3 \times \mathbb{R}^3 \to \mathbb{R}^3$ denotes the usual skewsymmetric product determined by the Euclidean structure and the orientation of $\mathbb{R}^3$. For a given function $H\colon \mathbb{R}^3 \to \mathbb{R}$ we say that x

has *prescribed mean curvature* H at (u,v) if $h(u,v) = H(x(u,v))$. In the presence of the conformality relations (0.1) the *H-surface-equation*

$$\Delta x = 2(H \circ x)\, x_u \wedge x_v \text{ on } U \tag{0.2}$$

therefore expresses that x has prescribed mean curvature H at each point of maximal rank. By an *H-surface* we mean a nonconstant classical (i.e. C^2) solution x to (0.1) and (0.2). We will always assume that H is continuous and bounded. Then (0.1) and (0.2), interpreted in the sense of distributions, are meaningful for mappings x of Sobolev class $W^{1,2}_{\text{loc}}(U,\mathbb{R}^3)$ and a nonconstant solution in this class is called a *weak H-surface.* (Under mild regularity assumptions on H it is known that weak H-surfaces are of class C^2. Moreover, the H-surfaces obtained in this article are of maximal rank everywhere on U and hence classical immersed surfaces of prescribed mean curvature H.)

We look for H-surfaces with a given oriented closed Jordan curve $\Gamma \subset \mathbb{R}^3$ as boundary. A convenient way to formulate this is the *Plateau boundary condition*

$$x\big|_{\partial U} \text{ is a weakly monotonic parametrization of } \Gamma. \tag{0.3}$$

The notion of a weakly monotonic parametrization of Γ on the unit circle ∂U has the obvious meaning: the uniform limit of a sequence of orientation preserving homeomorphism from ∂U onto Γ. In writing $x|_{\partial U}$ we understand that x is continuous on the closed disc $\overline{U}$ and we take its restriction to the boundary, or x is of class $W^{1,2}(U,\mathbb{R}^3)$ and we take its boundary values in the sense of traces of Sobolev functions. (The H-surfaces x produced below have both properties, and $x\big|_{\partial U}$ is actually a homeomorphism of ∂U onto Γ.) In contrast with a Dirichlet boundary condition where $x\big|_{\partial U}$ is a prescribed mapping, (0.3) is a free boundary condition with one degree of freedom. For Γ we will generally assume that it is at least rectifiable, although sometimes the weaker requirement would suffice that it bounds a parametric surface of finite $W^{1,2}$ norm.

(0.1), (0.2), (0.3) constitute the *Plateau problem* $\mathcal{P}(H,\Gamma)$ with prescribed mean curvature H and given boundary Γ. We will concentrate here on the existence of weak solutions $x \in W^{1,2}(U,\mathbb{R}^3)$ and refer to the literature for the regularity theory leading to classical solutions. The case of vanishing mean curvature $H \equiv 0$ is the classical Plateau problem to find a minimal surface with given boundary curve. This was solved about 1930 by Douglas and by Radó; we refer to the monographies [Ni1], [Ni2], [HDKW]. For nonzero H, however, problem $\mathcal{P}(H,\Gamma)$ may fail to have a solution. This fact was proved

1969 by Heinz [He2] who had already in 1954 obtained the first positive answer to the existence problem for $H \not\equiv 0$ (namely H constant and $|H|$ not too large, [He1]).

It is instructive to consider the example of a Euclidean circle $\Gamma \subset \mathbb{R}^3$ with radius $R > 0$. If H is constant with $0 < |H| \leq R^{-1}$ then we can obtain a solution to $\mathcal{P}(H, \Gamma)$ by parametrizing a spherical cap with radius $|H|^{-1}$ and boundary Γ conformally. However, for $H > R^{-1}$ no such spherical cap exists and one is lead to the conjecture that $\mathcal{P}(H, \Gamma)$ has no solution at all in this case. This was indeed shown in [He2] with the following simple idea: One integrates equation (0.2) over U and uses the Gauss-Green theorem on the left and the constancy of H as well as Stokes theorem on the right to obtain the identity

$$\int_{\partial U} x_r \, d\vartheta = H \int_{\partial U} x \wedge x_\vartheta \, d\vartheta$$

where ϑ is the arc length on ∂U and x_r, x_ϑ denote the radial and angular derivative of x on $\overline{U}$. Since $|x_r| = |x_\vartheta|$ holds on ∂U by (0.1) one concludes the inequality

$$\text{length}(x\big|_{\partial U}) \geq |H| \left| \int_{\partial U} x \wedge x_\vartheta \, d\vartheta \right| ,$$

and evaluating this for a parametrization $x\big|_{\partial U}$ of a circle Γ of radius R one finds $2\pi R \geq |H| 2\pi R^2$, i.e. $|H| \leq R^{-1}$. (The subtelty in the argument is to justify the integration by parts if x is a priori only continuous on $\overline{U}$ and of class C^2 on U.) Gulliver [Gu3], [Gu4] has analyzed and widely generalized this reasoning of Heinz. We should emphasize that it is an open question whether every immersion or even embedding of $\overline{U}$ into $\mathbb{R}^3$ with constant mean curvature $H \neq 0$ and a circle as boundary is in fact a spherical cap (although this has been proved under various additional assumptions on the immersion [EBMS], [Ba], [BE], [BJ], [LM]; note that by [Kap] the answer is negative if surfaces of genus bigger than 2 are admitted).

Since the Plateau problem $\mathcal{P}(H, \Gamma)$ is not solvable in general, one is interested in reasonable conditions on the prescribed mean curvature and the given boundary curve which are sufficient for existence. In view of the example above a natural conjecture is that a solution should exist if H is constant with $|H| \leq R(\Gamma)^{-1}$ or, if one is optimistic, even for variable H with $\sup |H| \leq R(\Gamma)^{-1}$, where $R(\Gamma)$ is the circumscribed radius of Γ, i.e. the radius of the smallest closed ball in $\mathbb{R}^3$ containing Γ. The conjecture has actually been proved by S. Hildebrandt who achieved in 1969 a breakthrough in this problem. His work will be described in Section 1. However, there are curves Γ for which the condition $|H| \leq R(\Gamma)^{-1}$ is far too restrictive.

For example, if Γ is the boundary of a long and narrow rectangle then Γ should bound H-surfaces (resembling pieces of circular cylinders with radius $|H|^{-1}$, if H is constant) for functions H much larger than $R(\Gamma)^{-1}$. Moreover, this should remain true if the long, narrow rectangle and its boundary Γ are twisted and curled through space in a complicated way. In Sections 1 and 2 we will describe the existence theorems aiming in this directrion which have their origin in work of Gulliver & Spruck, Wente and the second author. Finally, we consider in Section 3 the Plateau problem with prescribed mean curvature H and given contractible boundary curve Γ in an oriented Riemannian 3 manifold M instead of $\mathbb{R}^3$. In this situation it is no longer true that a solution exists whenever $\sup|H|$ is smaller than some positive constant determined by the geometry of Γ. For example, a great circle in the sphere S^3 with standard metric does not bound any surface with constant mean curvature $H \neq 0$, [Gu3]. Previous work on the existence problem for H-surfaces in a Riemannian manifold was restricted to situations where one can work in a single normal coordinate system. We will give sufficient criteria for existence which do not assume Γ to be contained in such a coordinate domain on M.

All the solutions of the Plateau problem $\mathcal{P}(H,\Gamma)$ treated in this article are obtained by minimizing in a suitable class $\mathcal{S}$ of surfaces with boundary Γ the energy functional (see Section 1) associated with the prescribed mean curvature function H. Geometric conditions relating H and Γ enter in the first place to secure the existence of energy minimizers in $\mathcal{S}$. If such a minimizer x can be freely varied within the class $\mathcal{S}$ then equations (0.1) and (0.2) follow for x, because these are just the Euler equations of the energy functional for variations of the independent and of the dependent variables respectively. However, in many cases obstacle conditions constraining the range of the admissible surfaces are incorporated in the definition of $\mathcal{S}$ (and must be, to make the direct minimization procedure possible). Then the free variability of a minimizer x in $\mathcal{S}$ is a priori not clear as the surface x might touch the obstacle. To rule out this possibility one needs again certain geometric conditions relating H, Γ, and the boundary mean curvature of the obstacle set, in order that an inclusion principle (a geometric maximum principle) can be applied. We have chosen to restrict the considerations in this article to the simplest formulation of Plateau's problem with prescribed mean curvature and, as we have just pointed out, to the existence theory based on the energy minimization method. As a consequence, we had to leave aside the discussion of various existence results which address generalizations and modifications of the problem or which use other methods.

For instance, the example of a circle $\Gamma \subset \mathbb{R}^3$ with radius R which bounds (in the oriented sense) two distinct spherical caps of constant mean curvature $H \neq 0$ whenever $0 < |H| < R^{-1}$, suggests the conjecture, attributed to Rellich, that there is a fundamental non-uniqueness in the Plateau problem $\mathcal{P}(H, \Gamma)$ for constant $H \neq 0$. Namely, there should exist positive constants $h_-(\Gamma), h_+(\Gamma)$ such that for each constant $H \neq 0$ with $-h_-(\Gamma) < H < h_+(\Gamma)$ one has at least two solutions of $\mathcal{P}(H, \Gamma)$, one energy minimizing and small the other one unstable for the energy functional and large, in a sense. This conjecture was proved with the combined work of Struwe [Str1] and the second author [Ste5] and independently, with the estimate $h_\pm(\Gamma) \geq R(\Gamma)^{-1}$, by Brezis & Coron [BC]. While this estimate cannot be improved for a circle Γ, it is not optimal for many curves Γ, and better results were obtained by Struwe who clarified the situation with [Str2]. We refer to Struwe's notes [Str3], and to [Str4], [Wa], [BR] for recent progress in the non-uniqueness problem for variable H sufficiently close to a constant.

An intuitive idea to produce "large" H-surfaces with constant H is to minimize area in the class of surfaces $x\colon U \to \mathbb{R}^3$ satisfying (0.3) and a volume constraint (i.e. the volume enclosed by x and the cone over Γ is prescribed). The solutions x, whose existence was shown by Wente [Wen2], [Wen3] are solutions to $\mathcal{P}(H, \Gamma)$ with a constant H which is, however, not prescribed but determined by the Lagrange multiplier associated with the volume constraint. Since these surfaces x are "large" if the prescribed volume is big one can infer the existence of "large" H-surfaces with boundary Γ for all values of the constant H which do occur as Lagrange multipliers for big volumes. In [Ste2] is was shown that the set of these values accumulates at 0 from above and below, but it is not clear (and may not be true, in general) that it contains a punctured neighborhood of 0 and, hence, Rellich's conjecture could not be proved in this way.

With regard to unstable H surfaces we mention that there is also an existence theory for "small" unstable surfaces of prescribed mean curvature (see [He5], [Strö], [Str3], [ST]) which extends part of the extensive corresponding theory for parametric minimal surfaces. On the other hand, uniqueness of "small" H-surfaces with constant H was discussed in [GS2], [Ru], and [Sa].

In another direction the existence theory for the Plateau problem has been generalized replacing the unit disc U by a multiply connected domain in $\mathbb{R}^2$ or by an oriented compact surface with boundary. This is called the general Plateau problem or Plateau-Douglas problem, and the principal difficulty is that the conformal structure on the domain cannot be fixed a priori. In the energy minimization process the conformal structure therefore has to be varied and it may degenerate in the limit of a minimizing sequence. (Geometrically speaking the surfaces in the minimizing sequence may break up into a system of surfaces of simpler topological type.) With appropriate assumptions (so-called Douglas conditions) such a behaviour can be excluded, however, and the Plateau-Douglas problem then has a solution. Another variation of the theme is to replace the Plateau boundary condition by a free boundary condition with two degrees of freedom, i.e. the surfaces are required to have (part of) their boundary on a given 2 dimensional supporting manifold. Since the additional difficulties in all these generalizations occur already in the minimal surface case $H \equiv 0$ we refer to the monography [HDKW] of S. Hildebrandt, U. Dierkes and their coauthors which – despite its title – also contains many discussions, hints to the literature, and bibliographical entries related to H-surfaces with H not necessarily zero.

Finally, we note that the Plateau problem for hypersurfaces of prescribed mean curvature has been treated in the setting of geometric measure theory by Fuchs and the present authors, and many of the questions discussed above could be answered satisfactorily in this context (see e.g. [DF1], [DF2], [Du2], [Du3], [DS1]–[DS4]). In fact, the new results in Section 3 below are influenced by our recent work on the Plateau problem for rectifiable integer multiplicity currents of codimension 1 in a Riemannian manifold M [DS4]. For 2 dimensions the geometric measure theory methods produce smooth embedded H-surfaces with given boundary Γ, and indeed smooth embeddings up to the boundary if Γ is smooth. The topological type of these surfaces is, however, not determined a priori. In constrast with this, the 2 dimensional parametric theory gives minimizing smooth immersed H-surfaces with prescribed topology and boundary Γ (they cannot be embedded if, for example, Γ is a knotted curve and the surfaces are of type of the disc). Moreover, one has analytic boundary regularity, i.e. the immersions can be extended as smooth functions to the closure of their domain if the contour Γ is smooth, but the immersed character of the surfaces on the boundary has been established only in quite special situations (e.g. $M = \mathbb{R}^3$, H constant, Γ real analytic). Thus, the *geometric* boundary regularity of the minimizing parametric solutions to Plateau's problem $\mathcal{P}(H, \Gamma)$ is still open – even in the minimal surface case $H \equiv 0$.

1. The method of bounded vector fields.

The H-surface equation (0.2) and the conformality equation (0.1) have a variational structure which is the key to the solution of the Plateau problem and was already used by Heinz [He1] in the first treatment of this problem with prescribed constant mean curvature. The Laplacean appearing on the left of the H-surface equation is the Euler operator associated with *Dirichlet's integral*

$$\mathbf{D}(x) = \tfrac{1}{2} \int_U (|x_u|^2 + |x_v|^2)\, dudv\,, \tag{1.1}$$

and the normal vector field $(H \circ x) x_u \wedge x_v$ appearing on the right-hand side describes the first variation of the volume enclosed by x and a fixed reference surface with the same boundary and measured with respect to H as a weight function. To make this precise we assume in this section that H is represented as the divergence of some vectorfield Z. If Z is continuous on a closed set $A \subset \mathbb{R}^3$ then we define the *volume functional* associated with Z

$$\mathbf{V}_Z(x) = \int_U (Z \circ x) \bullet x_u \wedge x_v\, dudv \tag{1.2}$$

and the corresponding *energy functional*

$$\mathbf{E}_Z(x) = \mathbf{D}(x) + 2\mathbf{V}_Z(x) \tag{1.3}$$

on the set $W^{1,2}(U, A)$ of parametric surfaces $x \in W^{1,2}(U, \mathbb{R}^3)$ which map (almost all of) U into A.

To see the geometric meaning of $\mathbf{V}_Z(x)$ we consider the 1-form ω on $\mathbb{R}^3$ which is dual to Z and note that $\operatorname{div} Z = H$ is equivalent with $d\omega = H\Omega$ where Ω is the Euclidean volume form on $\mathbb{R}^3$. (It may be preferable to work generally with ω instead of Z; we wanted to follow the historical development, however.) Then $\mathbf{V}_Z(x)$ is just the integral $\int_U x^\# \omega$ of ω over the parametric surface x. If x satisfies the Plateau boundary condition (0.3) and Z, ω are of class C^1, then Stokes' theorem tells us that, up to an irrelevant constant depending on Γ only, $\mathbf{V}_Z(x)$ equals the H-weighted volume enclosed by x and the cone over Γ. Note that we had to assume boundedness of Z on the image of X to ensure the existence of the integral (1.2) when x_u, $x_v \in L^2(U, \mathbb{R}^3)$. In the sequel we will need the bound $\sup_A |Z| < \frac{1}{2}$ in order to have coercivity of the energy functional on $W^{1,2}(U, A)$. Note that Z with $\operatorname{div} Z = H$ cannot be bounded on all of $\mathbb{R}^3$ if H is bounded away from zero, for example; this can be seen from Gauss' theorem.

Various constructions have been used to produce bounded vector fields with prescribed bounded divergence H. The simplest method is radial integration (as in the usual proof of Poincaré's Lemma), i.e.

$$Z(a) = \left(\int_0^1 H(ta) t^2\, dt \right) a\,, \tag{1.4}$$

and one has

$$\sup_A |Z| \le \tfrac{1}{3} R \sup_A |H| \tag{1.5}$$

if A is star-shaped with respect to the origin and contained in a ball of radius R. Another possibility is to integrate H in coordinate directions. Z defined by (1.4) may not be of class C^1 if H is merely continuous, but Z is continuous with $\operatorname{div} Z = H$ in the distributional sense and this will be sufficient in the sequel.

It is routine to compute the first variation of $\mathbf{E}_Z$, and one obtains, as expected and well-known:

1.1. Proposition (first variation). *Suppose A is closed in $\mathbb{R}^3$, Z is a C^1 vector field with* $\operatorname{div} Z = H$ *on A, and $x \in W^{1,2}(U, A)$.*

(i) If $\xi \in W_0^{1,2}(U, \mathbb{R}^3)$ is bounded with $x + t\xi \in W^{1,2}(U, A)$ for $0 < t \ll 1$, then

$$\frac{d}{dt}\bigg|_{t=0+} \mathbf{E}_Z(x + t\xi) = \int_U [x_u \cdot \xi_u + x_v \cdot \xi_v + 2(H \circ x)\, \xi \cdot x_u \wedge x_v]\, dudv\,.$$

(ii) If φ_t is the flow of a C^1 vectorfield η on $\overline{U}$ which is tangential to ∂U along ∂U, then

$$\frac{d}{dt}\bigg|_{t=0} \mathbf{E}_Z(x \circ \varphi_t) = \int_U \operatorname{Re}\left[(|x_u|^2 - |x_v|^2 - 2\mathbf{i} x_u \cdot x_v)\bar{\partial}\eta\right]\, dudv$$

where we have used complex notation $\bar{\partial}\eta = \frac{1}{2}(\eta_u + \mathbf{i}\eta_v)$ identifying $\mathbb{R}^2 = \mathbb{C}$.

Proof. (i) Since $\int_U (x_u \cdot \xi_u + x_v \cdot \xi_v) dudv$ is the derivative of $\mathbf{D}(x + t\xi)$ we only have to treat the derivative of the volume. For this we use the identity

$$\mathbf{V}_Z(x + t\xi) - \mathbf{V}_Z(x) = \int_0^t \int_U H \circ (x + s\xi)\, \xi \cdot (x_u + s\xi_u) \wedge (x_v + s\xi_v)\, dudvds \tag{1.6}$$

from which (i) follows by differentiation using the continuity and boundedness of H. For x, ξ smooth on $\overline{U}$ equation (1.6) is a consequence of Stokes' theorem applied to the pull back of ω to $]0, t[\times U$ by the mapping

$(s,w) \mapsto x(w) + s\xi(w)$. For general x, ξ (1.6) follows by approximation, provided Z with $\operatorname{div} Z = H$ has compact support in $\mathbb{R}^3$. The latter restriction is then removed by a further approximation (cf. [Ste3]).

(ii) In view of $\mathbf{V}_Z(x \circ \varphi_t) = \mathbf{V}_Z(x)$ we only have to compute the derivative of $\mathbf{D}(x \circ \varphi_t)$. Using the transformation $w = \varphi_t(\tilde{w})$ we get

$$\int_U |D(x \circ \varphi_t)|^2 \, d\tilde{w} = \int_U |Dx(D\varphi_t) \circ \varphi_t^{-1}|^2 \det(D\varphi_t^{-1}) \, dw \,,$$

and since $D\varphi_t(\tilde{w}) = \mathrm{I} + tD\eta + o(t)$ we obtain by differentiation

$$\left.\frac{d}{dt}\right|_{t=0} \mathbf{D}(x \circ \varphi_t) = \int_U \left[Dx \cdot (DxD\eta) - \tfrac{1}{2}|Dx|^2 \operatorname{trace} D\eta\right] \, dw \,.$$

The claim then follows by appropriately collecting terms in the integrand. □

1.2. Remarks. (1) The integral appearing in (i) is denoted $\delta\mathbf{E}_z(x;\xi)$ and known as the *first variation* of $\mathbf{E}_Z$ at x in the direction of ξ. The vanishing of $\delta\mathbf{E}_Z(x;\xi)$ for all $\xi \in \mathcal{D}(U,\mathbb{R}^3)$ is equivalent with the H-surface equation (0.2) in the weak sense, i.e (0.2) is the Euler equation of the energy functional. For the proof of Proposition 1.1 it is sufficient that Z is continuous with $\operatorname{div} Z = H$ in the distributional sense on a neighborhood of A.

(2) The integral in (ii) is denoted $\partial\mathbf{E}_Z(x;\eta)$ and called the *first variation of independent variables* of $\mathbf{E}_Z$ at x in the direction η. The vanishing of $\partial\mathbf{E}_Z(x;\eta)$ for all $\eta \in \mathcal{D}(U,\mathbb{R}^2)$ is equivalent with the equation $\overline{\partial}\Phi = 0$ in the distributional sense for the function $\Phi = |x_u|^2 - |x_v|^2 - 2\mathrm{i}x_u \cdot x_v$. This equation, sometimes called the second Euler equation of $\mathbf{E}_Z$, means that Φ is a (weakly, hence also classically) holomorphic function on U. It has become customary to call x *stationary* for $\mathbf{E}_Z$ if $\partial\mathbf{E}_Z(x;\eta) = 0$ holds for all vectorfields $\eta \in C^\infty(\overline{U},\mathbb{R}^2)$ which are tangential to ∂U along ∂U. With a formal integration by parts one sees that x is stationary iff $w^2\Phi(w)$ is real on ∂U in the weak sense. This latter condition is thus the natural boundary condition associated with the free Plateau boundary condition. (It can be expressed invariantly by stating the the quadratic holomorphic differential $\Phi(w)(dw)^2$, known as the Hopf differential, is real on the boundary.) But holomorphic functions on U which are real on ∂U in the weak sense are in fact constant, because they can be extended to (weakly) holomorphic functions on a neighborhood of $\overline{U}$. (This is easily verified by transforming U to the upper half plane.) Evaluating at $w = 0$ we deduce that x is

stationary iff the conformality relation (0.1) holds on U. (For more details of the reasoning see [HDKW], [Jo3], [Ni1], [Ni2].) □

In view of the preceding remarks a natural approach to the Plateau problem $\mathcal{P}(H,\Gamma)$ is to minimize the energy $\mathbf{E}_Z$ (with $\operatorname{div} Z = H$) on suitable classes of surfaces. We define the *class of admissible surfaces* $\mathcal{S}(\Gamma)$ to be the set of all $x \in W^{1,2}(U,\mathbb{R}^3)$ which satisfy the Plateau boundary condition (0.3), and for A closed in $\mathbb{R}^3$ we let $\mathcal{S}(\Gamma,A) = \mathcal{S}(\Gamma) \cap W^{1,2}(U,A)$. One then has the fundamental

1.3. Proposition (solution of the variational problem). *Suppose A is closed in $\mathbb{R}^3$, $\Gamma \subset A$, $\mathcal{S}(\Gamma,A) \neq \emptyset$, and Z is a continuous vector field on A with*

$$2\sup_A |Z| = c < 1\,. \tag{1.7}$$

Then the variational problem

$$\mathbf{E}_Z \rightsquigarrow \min \quad on\ \mathcal{S}(\Gamma,A) \tag{1.8}$$

admits a solution.

Proof. From (1.7) we obtain $2\mathbf{V}_Z(x) \leq c\mathbf{D}(x)$ and the *coercivity inequality*

$$\mathbf{E}_Z(x) \geq (1-c)\mathbf{D}(x) \quad \text{for } x \in \mathcal{S}(\Gamma,A), \tag{1.9}$$

from which we deduce that each minimizing sequence x_n for (1.8) is bounded in the norm of $W^{1,2}(U,\mathbb{R}^3)$ (since also the boundary values are uniformly bounded on account of (0.3)). Applying Rellich's theorem and passing to a subsequence we may assume $x_n \to x$ weakly in $W^{1,2}(U,\mathbb{R}^3)$ and almost everywhere on U, hence also x maps U into A. (1.7) also implies that the integrand of the energy functional is convex in the derivatives, and a classical lower semicontinuity theorem of Morrey can be applied to give $\mathbf{E}_Z(x) \leq \liminf_{n\to\infty} \mathbf{E}_Z(x_n)$. Since the integrand is quadratic in the derivatives one can verify this by simply writing $\mathbf{E}_Z(x_n) - \mathbf{E}_Z(x)$ as the sum of the nonnegative integral

$$\int_U \left[\tfrac{1}{2}|(x_n-x)_u|^2 + \tfrac{1}{2}|(x_n-x)_v|^2 + 2(Z\circ x_n)\cdot(x_n-x)_u \wedge (x_n-x)_v\right] dudv$$

and of other integrals whose integrands are products of two factors, one converging L^2 weakly to zero and the other one converging L^2 strongly as $n\to\infty$.

It remains to prove that x satisfies again the Plateau boundary condition (0.3). Actually this may not be true, because due to the action of the noncompact group of conformal automorphisms of U the minimizing sequence can degenerate in the limit. However, it is well-known from the theory of parametric minimal surfaces how to overcome this problem (see [HDKW], [Jo3], [Ni1], [Ni2], [Str3]). One fixes three points w_0, w_1, w_2 on ∂U and p_0, p_1, p_2 on Γ in the order of the orientation and imposes the three-point-condition $x_n(w_i) = p_i$, $0 \leq i \leq 2$. This is no restriction, since $\mathbf{E}_Z$ is invariant under conformal reparametrization of surfaces. Since Γ is a closed Jordan curve it has the following property: For every $\varepsilon > 0$ there exists $\beta > 0$ such that each subarc of Γ with end points at distance $< \beta$ to each other and containing at most one of the points p_i in its interior has diameter less than ε. For $w \in \partial U$ and $0 < \delta < 1$ one can apply a lemma of Courant and Lebesgue to find $\delta \leq \varrho \leq \sqrt{\delta}$ such that the restriction of x_n to $\overline{U} \cap \partial U_\varrho(w)$ is an absolutely continuous curve of length at most $(4\pi M)^{1/2} |\log \delta|^{-1/2}$ where $M = \sup_n \mathbf{D}(x_n)$ is finite by (1.9). (This lemma is readily proved by expressing the Dirichlet integral in polar coordinates centered at w and applying Fubini's theorem and Schwarz' inequality.) Choosing δ so small that $U_{\sqrt{\delta}}(w) \cap \partial \overline{U}$ contains at most one of the points w_i and $(4\pi M)^{1/2} |\log \delta|^{-1/2} < \varepsilon$, we conclude from the Plateau boundary condition that x_n maps the arc $U_\varrho(w) \cap \partial U$ onto a subarc of Γ with diameter less than ε. Thus $x_n\big|_{\partial U}$ is an equicontinuous sequence of weakly monotonic parametrizations of Γ, and the boundary condition for x follows. □

From the preceding proposition we cannot immediately obtain solutions to the Plateau problem with prescribed mean curvature $H = \operatorname{div} Z$, because the solutions x to (1.8) can touch the boundary of A and are not freely variable within $\mathcal{S}(\Gamma, A)$ in such a case. (And we cannot take $A = \mathbb{R}^3$ if, e.g., H is bounded away from zero.) However, we can at least assert the following:

1.4. Remarks. (1) The solutions to the variational problem (1.8) satisfy the conformality relations (0.3). This is immediate from Remark 1.2 and the fact that $\mathcal{S}(\Gamma, A)$ is invariant with respect to reparametrization by oriented self-diffeomorphisms of $\overline{U}$.

(2) If x is a solution to (1.8) then it satisfies the *variational inequality*

$$\delta \mathbf{E}_Z(x;\xi) \geq 0 \quad \text{for } \xi \in W_0^{1,2} \cap L^\infty(U, \mathbb{R}^3) \text{ with } x + t\xi \in W^{1,2}(U, A),\ 0 < t \ll 1. \tag{1.10}$$

(3) Suppose A is the closure of a C^2 domain in $\mathbb{R}^3$, ν is any C^1 extension to $\mathbb{R}^3$ of the inner unit normal field along ∂A, V is a neighborhood of ∂A, x is a solution to (1.8) and $\xi \in W_0^{1,2} \cap L^\infty(U, \mathbb{R}^3)$. Then we have

$$\delta \mathbf{E}_Z(x;\xi) \geq 0 \quad \text{if } \xi \cdot (\nu \circ x) \geq 0 \text{ almost everywhere on } x^{-1}(V). \tag{1.11}$$

To see this we choose $0 \leq \vartheta \in \mathcal{D}(\mathbb{R}^3, \mathbb{R})$ and observe that $(\vartheta \circ x)(\xi + \varepsilon|\xi|\nu \circ x)$ is admissible in (1.10). Letting ε tend to 0 and then ϑ to the constant 1 in appropriate fashion we deduce the assertion. Note that $x^{-1}(V)$ is well defined up to a set of measure zero. □

To obtain a variational equation instead of an inequality for the minimizers x of $\mathbf{E}_Z$ on $\mathcal{S}(\Gamma, A)$ one further ingredient of the theory is needed, namely a *geometric inclusion principle* asserting that x maps into the interior of A. For the case of a ball A of radius R_0 in $\mathbb{R}^3$ centered at the origin such a principle can be deduced from the maximum principle for subharmonic functions. For this one introduces the function $f = |x|^2 \in W^{1,2}(U, \mathbb{R})$ and computes for $0 \leq \eta \in W^{1,2}(U, \mathbb{R})$

$$\begin{aligned} (\eta x)_u \cdot x_u + (\eta x)_v \cdot x_v &= \tfrac{1}{2}\eta_u f_u + \tfrac{1}{2}\eta_v f_v + \eta(|x_u|^2 + |x_v|^2)\,, \\ |2(H \circ x)\, \eta\, x \cdot x_u \wedge x_v| &\leq \eta(|x_u|^2 + |x_v|^2)|x||H \circ x|\,. \end{aligned}$$

Since $\xi = -\eta x$ is admissible in (1.10) whenever $0 \leq \eta \in W_0^{1,2} \cap L^\infty(U, \mathbb{R})$ we find, for such η,

$$\tfrac{1}{2} \int_U (\eta_u f_u + \eta_v f_v)\, dudv \leq 0\,,$$

provided $|x||H \circ x| \leq 1$. This means that f is a (weakly) subharmonic function, and if Γ is contained in a concentric closed ball B of radius $R < R_0$ it follows that $f \leq R^2$ on U, hence x has its image in B. Indeed, using $\eta = \max(f - R^2, 0)$ as test function one finds $D\eta = 0$ and hence $f \leq R^2$ almost everywhere on U. Actually one needs the condition $|x||H \circ x| \leq 1$ only on $x^{-1}(A \setminus B)$, in particular $\sup_{A \setminus B} |H| \leq R_0^{-1}$ will suffice. With the choice (1.4) of Z, Proposition 1.3, Remark 1.4 and the preceding consideration concerning the inclusion principle one has proved

1.5. Theorem (Hildebrandt). *Suppose Γ is contained in a closed ball B of radius R in $\mathbb{R}^3$ and H satisfies*

$$\sup_B |H| < \tfrac{3}{2} R^{-1}\,, \tag{1.12}$$

$$\sup_{\partial B} |H| \leq R^{-1}\,. \tag{1.13}$$

Then there exists a weak solution x with values in B to the Plateau problem $\mathcal{P}(H,\Gamma)$.

More precisely, under the hypotheses of the theorem there exist $\mathbf{E}_Z$ minimizing surfaces x in $\mathcal{S}(\Gamma, B)$ and each such minimizer is a weak H-surface satisfying the Plateau boundary condition (0.3). Moreover, if Γ is contained in a ball $B' \subset B$ of radius $R' < R$ and $|H|$ on $B \setminus B'$ does not exceed the reciprocal distance function to the center of B then x has its values in B'.

The history of this theorem begins in 1954 with Heinz [He1]. He treated constant prescribed mean curvature H and used the vector field $Z(a) = \frac{1}{3}Ha$ to represent $H = \operatorname{div} Z$. His method was to replace the members of a minimizing sequence x_n for (1.8) by solutions y_n to the H-surface equation (0.2) with Dirichlet boundary data specified by x_n. To obtain the y_n he established a priori estimates and applied the method of Leray–Schauder degree. A crucial point was to show that the y_n so constructed were again a minimizing sequence, the a priori estimates then allowed passage to the limit. Heinz had to require $|H| < \frac{1}{8}(\sqrt{17}-1)R^{-1}$ in his proof, but Werner [Wer] could improve the method to allow $|H| < \frac{1}{2}R^{-1}$ and also treated the Plateau–Douglas problem for multiply connected planar domains instead of the unit disc. In a series of papers completed in 1969 Hildebrandt first treated the limit case $|H| = \frac{1}{2}R^{-1}$ in [Hi2], then improved his result in [Hi3] allowing (still constant) H with $|H| \leq R^{-1}$, and finally solved the Plateau problem $\mathcal{P}(H,\Gamma)$ in [Hi4] also for variable H under the condition

$$\sup_B |H| \leq R^{-1}\,. \tag{1.14}$$

That Hildebrandt's arguments actually worked with the weaker assumption (1.12) and (1.13) instead of (1.14) was later pointed out by Gulliver & Spruck [GS3]. Hildebrandt's method in [Hi3], [Hi4] was in essence the variational approach presented above. The geometric inclusion principle for subharmonic functions was already used by Heinz and it was crucial also for Hildebrandt's results. However, at the time it was apparently not clear that the maximum principle could be applied directly to the weakly subharmonic function $f = |x|^2 \in W^{1,2}(U,\mathbb{R})$, and additional considerations were found necessary to justify this, e.g. use of the energy minimizing property of x as in [Hi3] or reduction to the situation where x is smooth as in [He1], [Hi4].

By the non-existence result of Heinz [He2] the sufficient condition (1.14) is sharp in the sense that $\mathcal{P}(H,\Gamma)$ does not have solutions for all $\Gamma \subset B$ if H is constant with $|H| > R^{-1}$. However, it was soon observed that (1.14) is unnecessarily restrictive in many cases. For example Gulliver & Spruck

[GS1] (see also [Hi5]) considered 1971 the case of a cylinder $A = \{a \in \mathbb{R}^3 : |Pa| \leq R_0\}$ where $P\colon \mathbb{R}^3 \to \mathbb{R}^2 \times \{0\}$ denotes the projection. For $\mathbf{E}_Z$ minimizing x in $\mathcal{S}(\Gamma, A)$ it is natural to consider $f = |P \circ x|^2$. Now, given $0 \leq \eta \in W_0^{1,2} \cap L^\infty(U, \mathbb{R})$ the vector field $\xi = -\eta(P \circ x)$ is admissible in (1.10), and one computes

$$-\tfrac{1}{2}(\eta_u f_u + \eta_v f_v) = \xi_u \cdot x_u + \xi_v \cdot x_v + \eta(|Px_u|^2 + |Px_v|^2)\,. \tag{1.15}$$

On the other hand, the conformality of x (cf. Remark 1.4 (1)) implies

$$\begin{aligned} |2(H \circ x)\,\xi \cdot x_u \wedge x_v| &\leq 2|P \circ x||H \circ x|\eta|P(x_u \wedge x_v)| \\ &\leq 2|P \circ x||H \circ x|\eta(|Px_u|^2 + |Px_v|^2) \end{aligned} \tag{1.16}$$

since $1 \geq |P(a \wedge b)|^2 = 2 - |Pa|^2 - |Pb|^2$ and hence

$$|P(a \wedge b)|^2 = |P^\perp a|^2 + |P^\perp b|^2 \leq 1 \leq |Pa|^2 + |Pb|^2 \tag{1.17}$$

holds for a, b orthonormal in $\mathbb{R}^3$. From (1.15) – (1.17) we deduce that f is weakly subharmonic whenever $|P \circ x||H \circ x| \leq 1$, and if Γ is contained in a concentric cylinder C of radius $R < R_0$ we can conclude that also x has its image in C, provided $\sup_{A \setminus C} |H| \leq \frac{1}{2} R_0^{-1}$ or, at least, $|P \circ x||H \circ x| \leq \frac{1}{2}$ on $x^{-1}(A \setminus C)$. Choosing the vectorfield

$$Z(a) = \left(\int_0^1 H(ta_1, ta_2, a_3)t\, dt \right) Pa \tag{1.18}$$

with $\operatorname{div} Z = H$ and $|Z(a)| \leq \frac{1}{2}|Pa| \sup_C |H|$ we may now repeat the reson-ing leading to Theorem 1.5 to obtain

1.6. Theorem (Gulliver & Spruck). *Suppose Γ is contained in a rotationally symmetric cylinder C of radius R in $\mathbb{R}^3$ and H satisfies*

$$\sup_C |H| < R^{-1}\,, \quad \sup_{\partial C} |H| \leq \tfrac{1}{2} R^{-1}\,. \tag{1.19}$$

Then $\mathcal{P}(H, \Gamma)$ has a weak solution $x \in W^{1,2}(U, C)$.

For constant H condition (1.19) reduces to $|H| \leq \frac{1}{2} R^{-1}$ and is again sharp (see [GS1]). We note that the conformality of x was not used in the proof of the inclusion principle leading to Theorem 1.5, hence this theorem of Hildebrandt is valid in an analogous formulation also for the Dirichlet problem where one minimizes $\mathbf{E}_Z$ on $W^{1,2}(U,A)$ subject to Dirichlet boundary conditions. This assertion is not true, however, with regard to Theorem 1.6 above.

A subsequent analysis by Hildebrandt [Hi6], [Hi7], Kaul [Kau], and Gulliver & Spruck [GS3] made it evident that the comparison of the prescribed mean curvature H with the *boundary mean curvature* $H_{\partial A}$ of the prospective inclusion domain A is decisive for the validity of an inclusion principle. Namely, assuming that A is the closure of a C^2 domain, such a principle holds for conformal solutions to the variational inequality (1.10) if (and essentially also only if, cf. [GS3]) $|H|$ does nowhere exceed on ∂A the mean curvature $H_{\partial A}$ of ∂A. (Here we define $H_{\partial A}$ with respect to the inner unit normal vector field, of course, so that $H_{\partial A}$ is nonnegative for convex A.) Gulliver & Spruck used the energy minimality and continuity of x in their proof of the inclusion principle, whereas Hildebrandt assumed the variational inequality and a certain amount of smoothness (which he established for minimizers in [Hi8]). The following strong version is valid for all conformal solutions $x \in W^{1,2}(U,A)$ of the variational inequality (1.10). The proof uses arguments from [Du1] and [DS4]; for details we refer to [DS5].

1.7. Proposition. *Suppose A is the closure of a C^2 domain in $\mathbb{R}^3$, ν is the inner unit normal on ∂A, the mean curvature $H_{\partial A}$ of ∂A with respect to ν is bounded from below, and $x \in W^{1,2}(U,A)$ is a conformal solution to the variational inequality*

$$\delta\mathbf{E}_Z(x;\xi) = \int_U [x_u \cdot \xi_u + x_v \cdot \xi_v + 2(H\circ x)\,\xi \cdot x_u \wedge x_v]\, dudv \geq 0$$

for all $\xi \in W_0^{1,2} \cap L^\infty(U,\mathbb{R}^3)$ with $x + t\xi \in W^{1,2}(U,A)$, $0 < t \ll 1$. Then the following assertions hold:

(i) There exists a nonnegative Radon measure λ on U which is absolutely continuous with respect to Lebesgue measure $\mathcal{L}^2$ and concentrated on the coincidence set $x^{-1}(\partial A)$ such that

$$\delta\mathbf{E}_Z(x;\xi) = \int_{x^{-1}(\partial A)} \xi \cdot (\nu\circ x)\, d\lambda \qquad \text{for all } \xi \in W_0^{1,2} \cap L^\infty(U,\mathbb{R}^3).$$

(ii) The estimate

$$\lambda \leq \mathcal{L}^2 \llcorner \left[(|x_u|^2 + |x_v|^2)(|H| - H_{\partial A})_+ \circ x\right] \qquad \text{on } x^{-1}(A)$$

holds; in particular we have $\lambda = 0$ and x is a weak H-surface, if $|H| \leq H_{\partial A}$ is true pointwise on ∂A.

(iii) If $|H(a)| < H_{\partial A}(a)$ *holds at some point* $a \in \partial A$ *and the boundary trace* $x\big|_{\partial U}$ *does not meet a neigborhood of this point then also the image of* x *omits a neighborhood of this point.*

Here the *coincidence set* $x^{-1}(\partial A)$ is the set of $w \in U$ (defined up to a set of $\mathcal{L}^2$ measure zero) such that $x(w) \in \partial A$, i.e. x touches the obstacle ∂A at w. Similarly, the assertion (iii) means that $x^{-1}(V)$ is a set of $\mathcal{L}^2$ measure zero for some neighborhood V of a. For the vectorfield Z we understand that $\operatorname{div} Z = H$ holds. However, Z is irrelevant here and one may simply interprete $\delta\mathbf{E}_Z(x;\xi)$ as an abbreviation for the integral appearing in the variational inequality. It can be seen from the proof that x actually has its image in an interior parallel set $A_\varepsilon = \{a \in A : \operatorname{dist}(a, \partial A) \geq \varepsilon\}$ if the following conditions are satisfied: A has bounded principal curvatures, smooth global inner parallel surfaces to ∂A exist up to distance ε, the conformal solution $x \in W^{1,2}(U, A)$ to the variational inequality has boundary trace $x\big|_{\partial U}$ with values in A_ε and $|H| \leq H_A$ holds pointwise on $A \setminus A_\varepsilon$, where $H_A(a)$ denotes the mean curvature of the parallel surface to ∂A at $a \in A \setminus A_\varepsilon$. If the boundary mean curvature $H_{\partial A}$ is replaced by the minimum of the principal curvatures in all the statements of Proposition 1.7 then the conformality assumption for x may be dropped.

From Proposition 1.3, Remark 1.4 and Proposition 1.7 we now immediately obtain the following general existence theorem of Gulliver & Spruck [GS3] (with weaker assumptions on A and ∂A here):

1.8. Theorem (Gulliver & Spruck). *Suppose A is the closure of a C^2 domain in $\mathbb{R}^3$, the prescribed mean curvature H and the boundary mean curvature $H_{\partial A}$ of A satisfy*

$$|H| \leq H_{\partial A} \quad \textit{pointwise on } \partial A, \tag{1.20}$$

and there exists a continuous vector field Z with $\operatorname{div} Z = H$ on a neighborhood of A in the distributional sense such that

$$\sup_A |Z| < \tfrac{1}{2}\,. \tag{1.21}$$

Then, for every Jordan curve $\Gamma \subset A$ which is contractible in A the Plateau problem $\mathcal{P}(H, \Gamma)$ has a weak solution in $W^{1,2}(U, A)$. Moreover, if $|H(a)| < H_{\partial A}(a)$ holds at some point $a \in (\partial A) \setminus \Gamma$, then each solution surface omits a neighborhood of this point.

The contractibility hypothesis on Γ is needed here to ensure $\mathcal{S}(\Gamma, A) \neq \emptyset$. This fact is not implied by other assumptions in the theorem as can be seen from the example of a solid torus A in $\mathbb{R}^3$ with positive boundary mean curvature. An analogous theorem, with $H_{\partial A}$ replaced by the minimum principle curvature on ∂A, is valid for the Dirichlet problem.

Choosing A as a ball or a rotationally symmetric cylinder and Z as in (1.4) or (1.8) respectively we recover Theorem 1.5 and 1.6 as special cases of the preceding general theorem. There are many other concrete applications, e.g. one may consider solid ellipsoids or rotationally symmetric bodies bounded by Delaunay surfaces as in [Hi6], [GS3] and use

$$Z(a) = \left(\int_0^1 H(ta_1, a_2, a_3)\, dt \right) (a_1, 0, 0) \tag{1.22}$$

so that (1.21) holds if A is contained in a slab $[-R, R] \times \mathbb{R}^2$ and $\sup |H| < \frac{1}{2} R^{-1}$ on this slab. If A is an ellipsoid with axes $R_1 \leq R_2 \leq R_3$, then (1.21) is satisfied whenever

$$\sup_A |H| < \max \left\{ \tfrac{3}{2} R_3^{-1}, R_2^{-1}, \tfrac{1}{2} R_1^{-1} \right\} ,$$

as can be seen with the choices (1.4), (1.18), and (1.22) for Z, while (1.20) is implied by

$$\sup_{\partial A} |H| \leq \tfrac{1}{2} R_1 (R_2^{-1} + R_3^{-1})$$

as one checks by calculating the minimum of the boundary mean curvature of A.

Gulliver & Spruck [GS3] have also observed that one can use solutions to the *nonparametric mean curvature equation*

$$\operatorname{div} \frac{\nabla f}{\sqrt{1 + |\nabla f|^2}} = 2H \quad \text{on } D \subset \mathbb{R}^3, \tag{1.23}$$

in order to obtain on $A = \overline{D}$ a C^1 vectorfield

$$Z = \tfrac{1}{2} \frac{\nabla f}{\sqrt{1 + |\nabla f|^2}} \tag{1.24}$$

satisfying $\operatorname{div} Z = H$ on A and $\sup_A |Z| < \frac{1}{2}$, provided f has bounded gradient ∇f on A. The geometric meaning of (1.23) is that the graph of the scalar function f has mean curvature $H(a)$ at the point $(a, f(a))$ for each $a \in D$. The nonparametric mean curvature equation was solved (in general dimensions) with arbitrary continuous Dirichlet boundary data by Serrin

[Se] for bounded C^2 domains D and bounded C^1 functions H satisfying the inequality (1.20) on ∂A as well as

$$|H| \le (1 - \kappa d_A) H_A + \tfrac{1}{2}\kappa \quad \text{on } A \text{ for some } 0 \le \kappa \le r_A^{-1}. \tag{1.25}$$

Here $d_A(a) = \operatorname{dist}(a, \partial A)$ for $a \in A$, $r_A = \sup_A d_A$ is the *inner radius* of A (assumed to be finite) and $H_A(a)$ denotes the mean curvature of the parallel surface to ∂A through a whenever this is defined (i.e. whenever a has a unique nearest point in ∂A and the principal curvatures of ∂A there are smaller than $d_A(a)^{-1}$), while we set $H_A(a) = \infty$ otherwise. We call H_A the *parallel mean curvature function* of A.

Serrin's result was extended to unbounded domains D with bounded principal curvatures and with global inner parallel surfaces by Gulliver & Spruck [GS3]. Assuming a uniform bound for the C^1 norm of H on A they also established that the solution f of (1.23) with zero Dirichlet boundary conditions has a bounded C^1 gradient on $A = \overline{D}$ so that $\sup_A |Z| < \frac{1}{2}$ is valid for Z from (1.24) and Theorem 1.8 is applicable. In fact, we will give in Section 2 a different proof of the resulting existence theorem for the Plateau problem $\mathcal{P}(H, \Gamma)$ which does not use (1.23) and clarifies the meaning of condition (1.25). This proof also uses weaker hypotheses for ∂A and for H (bounded and continuous, by our general assumption in the introduction) and gives the following strengthened version of the result in [GS3]:

1.9. Theorem (Gulliver & Spruck). *Suppose A is the closure of a C^2 domain in $\mathbb{R}^3$ with finite inner radius r_A, and H satisfies (1.25) on A and (1.20) on ∂A. If A is unbounded assume further that ∂A has global exterior parallel surfaces and in the case $\kappa = r_A^{-1}$ also $2 \sup_{V \cap A}(|H| - H_A) < r_A^{-1}$ for some uniform neighborhood V of ∂A. Then the conclusions of Theorem 1.8 are valid.*

We conclude this section with some comments on the *regularity theory* that has been developed in order to prove that the weak solutions to the Plateau problem $\mathcal{P}(H, \Gamma)$ produced in the theorems above are in fact classical H-surfaces.

Grüter [Grü1] has proved that weak H-surfaces $x \in W^{1,2}(U, \mathbb{R}^3)$, i.e. conformal solutions to the H-surface equation (0.2), are continuous on U and hence, by a regularity theorem of Tomi [Tom1], also of class $C^{1,\alpha}$ on U for $0 < \alpha < 1$. (In fact, Grüter only assumes (0.2) with $H \circ x$ replaced by some measurable bounded function h on U.) It follows then from potential theory that x is of class $C^{k+2,\beta}$ on U if the prescribed mean curvature H

is $C^{k,\beta}$. It is conjectured that also without the conformality assumption weak solutions to the H-surface equation (which then looses its geometric meaning) are regular, if H is bounded and continuous. With additional assumptions on H (constant, resp. Lipschitz with some decay at infinity, resp. Lipschitz with a certain decay of $|\nabla H|$) Wente [Wen1], Tomi [Tom2], and Heinz [He6], [He7] had proved this some time ago, and Bethuel [Be] could settle the conjecture for H with $|\nabla H|$ bounded in $\mathbb{R}^3$. It was also shown by Bethuel & Ghidaglia ([BG], and a recent preprint) that the conjecture is true if H depends only on two coordinates or if $|\partial_e H(a)| \leq \mathrm{const}(1+|a \cdot e|)^{-1}$ for some direction e in $\mathbb{R}^3$.

With regard to boundary regularity no results are known to us in the literature which are valid for general weak solutions. However, the situation is better for parametric surfaces x which minimize an energy functional $\mathbf{E}_Z$ (where $\operatorname{div} Z = H$, as usual) in a class $W^{1,2}(U, A)$ subject to Dirichlet boundary conditions. Namely, if A has a certain uniform local convexity property called *quasiregularity* in [HK], [Hi8], then one can apply a classical method of Morrey to deduce that x is Hölder continuous on U and also continuous on the closed disc $\overline{U}$, provided the boundary trace $x|_{\partial U}$ is continuous. Quasiregularity of $A \subset \mathbb{R}^3$ means that A is uniformly locally biLipschitz equivalent to a convex set in the sense that there exists $\Lambda < \infty$ and for each $a \in A$ a biLipschitz map Φ from a convex set $K \subset \mathbb{R}^3$ into A such that $\operatorname{Lip}\Phi \leq \Lambda$, $\operatorname{Lip}\Phi^{-1} \leq \Lambda$, and the image of Φ contains the intersection of A with the ball of radius Λ^{-1} and center a. In order to include a proof of the important fact that energy minimizing solutions to the Plateau problem are continuous up to the boundary we now briefly repeat the reasoning of Hildebrandt & Kaul [HK]. Given a disc $U' = U_r(w)$ in U such that the restriction of x to $\partial U'$ is an absolutely continuous curve of length $< (2\Lambda)^{-1}$ in A we choose $a \in A$ on the trace of this curve and Φ as above, define $h \in W^{1,2}(U', \mathbb{R}^3)$ as the harmonic function with boundary trace $\Phi^{-1} \circ x|_{\partial U'}$ and use $\tilde{x} = \Phi \circ h$ on U', $\tilde{x} = x$ on $U \setminus U'$ as comparison surface. The $\mathbf{E}_Z$ minimality of x and the coercivity inequality (1.9) on U' as well as the reversed inequality (valid with $1+c$ instead of $1-c$ where c is the constant from (1.7)) now imply

$$\int_{U_r(w)} |Dx|^2 \, dudv \leq \frac{1+c}{1-c} \int_{U_r(w)} \Lambda^2 |Dh|^2 \, dudv \,. \tag{1.26}$$

Since the Dirichlet integral of a harmonic function on the unit disc is dominated by the Dirichlet integal of its boundary trace (as can be seen from expanding into a Fourier series), one infers from (1.26)

$$\int_{U_r(w)} |Dx|^2 \, dudv \le Lr \int_{\partial U_r(w)} |Dx|^2 \, ds \tag{1.27}$$

with $L = \frac{1+c}{1-c}\Lambda^4$. On the other hand, (1.27) is trivially valid with $L = \pi\Lambda^2\mathbf{D}(x)$ in the case where $x\big|_{\partial U'}$ is absolutely continuous with length $\ge 2\Lambda^{-1}$. Choosing the maximum value for L it follows that (1.27) is valid for almost all radii $0 < r < 1 - |w|$, and since (1.27) is a differential inequality for the absolutely continuous function appearing on the left, one obtains by integration

$$\int_{U_r(w)} |Dx|^2 \, dudv \le \left(\frac{r}{R}\right)^{2\alpha} \int_{U_R(w)} |Dx|^2 \, dudv \quad \text{for } 0 < r \le R \le 1 - |w| \tag{1.28}$$

where $\alpha = (2L)^{-1}$. By Morrey's well known Dirichlet growth theorem, (1.28) implies Hölder continuity with exponent α for x on U and if $x\big|_{\partial U}$ is continuous, also continuity of x on $\overline{U}$ (see [HK] for a proof of the last assertion).

We emphasize that the arguments in the preceding paragraph are valid for minimizers to the obstacle problem and do not use the H-surface equation for x. Hildebrandt has shown in [Hi8], [Hi9] interior $C^{1,\alpha} \cap W^{2,p}$ regularity for the solutions to more general obstacle problems under appropriate smoothness assumptions on the data.

Boundary regularity of higher order for parametric H-surfaces has first been proved for constant H by Hildebrandt [Hi2] who extended his fundamental work [Hi1] on the boundary regularity of parametric minimal surfaces with Plateau boundary conditions. Then Heinz [He3], [He4], using earlier work of Heinz & Tomi [HT], showed $C^{1,\alpha}$ regularity for $0 < \alpha < 1$ and in the case $H \in C^{0,\beta}(\mathbb{R}^3)$ also $C^{2,\beta}$ regularity of x up to the boundary, if $x \in W^{1,2}(U, \mathbb{R}^3)$ is a classical (i.e. C^2 on U and continuous on $\overline{U}$) H-surface satisfying the Plateau boundary condition (0.3) with a Jordan curve of class C^2 or $C^{2,\beta}$ respectively. Jäger [Jä] still improved this result requiring only class $C^{1,\alpha}$ for Γ (and finiteness of the Dirichlet energy of x). The method of Heinz is also presented in chapter 7 of [HDKW].

The question of analytic regularity being settled by the above we now turn to the problem of *geometric regularity* which is to prove the immersed character of the solutions x to the Plateau problem $\mathcal{P}(H, \Gamma)$. (One cannot expect embeddings, in general, even if Γ is unknotted; see [GS2], [Sa], however.) By the conformality relations (0.2), we must have $x_u(w_0) = 0 = x_v(w_0)$ at a point $w_0 \in \overline{U}$ where x does not have maximal rank. A device of Hartman & Winter [HW] can be applied as in [HT], [He3] to demonstrate, for H-surfaces with sufficient degree of smoothness on $\overline{U}$,

$$x_u(w) - \mathbf{i}x_v(w) = (w - w_0)^k a + o(|w - w_0|^k) \quad \text{as } \overline{U} \ni w \to w_0 \tag{1.29}$$

with an integer $k \geq 1$ and a vector $a \in \mathbb{C}^3$, $a \neq 0$ (since x is not constant). (See also [HDKW], Chapter 8 for a thorough discussion.) In view of the expansion (1.29) w_0 is called a *branch point* of the H-surface x and k its branching order.

Geometric regularity of x then is equivalent to the absence of branch points. Clearly, (1.29) implies that branch points w_0 are isolated and that the tangent plane $\mathbb{R}x_u(w) + \mathbb{R}x_v(w)$ has a limit position as $w \to w_0$. Heinz & Hildebrandt [HH1] have given estimates on the number of branch points in terms of geometric quantities associated with the boundary curve Γ. From (1.29) one also infers that x cannot be constant on an arc $\gamma \subset \partial U$. (Otherwise x would be smooth on $U \cup \gamma$, if γ is relatively open in ∂U, with $x_u = x_v = 0$ on γ contradicting the isolatedness of branch points; see [He3].) Therefore, the solutions to the Plateau problem $\mathcal{P}(H, \Gamma)$ have boundary values which are not only weakly monotonic parametrizations of Γ but really homeomorphisms from ∂U onto Γ.

In 1969 Ossermann [Os] could prove the non-existence of true interior branch points for minimizing solutions to the classical Plateau problem. These are branch points $w_0 \in U$ accompanied by arcs of transversal selfintersection of x emanating from w_0. Osserman considered parametric minimal surfaces minimizing Dirichlet's integral under Plateau boundary conditions, but his arguments are valid for $\mathbf{E}_Z$ minimizing H-surfaces with H Lipschitz as well. He did not rule out false branch points, i.e. branch points $w_0 \in U$ near which x is a branched covering of a smoothly embedded surface in $\mathbb{R}^3$. This problem was solved, using unique continuation arguments, topological considerations and the minimizing property of x, by Alt [Al1], [Al2] and Gulliver [Gu2] (see also [GOR], [Gu5], [Gu6], [SW]). Consequently, energy minimizing solutions to a Plateau problem $\mathcal{P}(H, \Gamma)$ with Lipschitz mean curvature H are free of interior branch points.

The results are less complete with regard to boundary branch points $w_0 \in \partial U$. These must be of even order, as can be seen from (1.29) and the monotonicity of $x\big|_{\partial U}$. The only general result that excludes boundary branch points for energy minimizing solutions x to a Plateau problem $\mathcal{P}(H, \Gamma)$ is due to Gulliver & Lesley [GL] and limited to constant mean curvature H and analytic Jordan curves Γ in $\mathbb{R}^3$. The reason is that a reflection argument is used in the proof. Also, if A is the closure of a smooth domain, Γ a smooth curve on the boundary ∂A, the H-surface x has boundary Γ and values in A, and if the boundary mean curvature of A satisfies $H_{\partial A} \geq |H|$ pointwise on ∂A with strict inequality somewhere on the component of ∂A containing Γ, then one can conclude from the representation (1.29) and from the maximum principle (boundary point lemma) that boundary branch points do not exist. False boundary branch points can be excluded using the energy minimzing property of x ([Gu5], [Gu6]), for true boundary branch points this is not known, however; the subtelty of the matter is illustrated in [Gu7]. Thus the question of the geometric boundary regularity of the energy minimizing solutions has not yet found a final answer.

2. The method of isoperimetric inequalities.

There are situations where all the existence theorems of Section 1 for the Plateau problem $\mathcal{P}(H,\Gamma)$ impose stronger restrictions on the prescribed mean curvature H than necessary. For example, if Γ is the boundary of a very long and narrow strip of surface which is curled and knotted quite densley in a large region of space, then Γ will not be contained in a domain with large mean boundary curvature so that the results of Section 1 are not applicable for large constant H, but one would conjecture that Γ nevertheless bounds H-surfaces for quite big values H. A new approach to the Plateau problem, which confirms such a conjecture and eventually lead to much more geometric insight and to several new existence theorems, was proposed by Wente [Wen1]. It is based on the isoperimetric inequality in $\mathbb{R}^3$.

Wente considered constant prescribed mean curvature H and used the energy functional $\mathbf{E}_H = \mathbf{D} + 2H\mathbf{V}$, where

$$\mathbf{V}(x) = \tfrac{1}{3} \int_U x \cdot x_u \wedge x_v \, dudv \tag{2.1}$$

is the volume enclosed by the parametric surface $x \in W^{1,2}(U, \mathbb{R}^3)$ and the cone over the boundary trace of x. To ensure the existence of the integral (2.1) one needs to know that x is bounded. Wente did not want to assume this, however, because then he would eventually have to work in a bounded domain of $\mathbb{R}^3$ and could allow only values of $|H|$ not exceeding its mean boundary curvature, in order to have the inclusion principle available. He observed that the volume $\mathbf{V}(x)$ could be well defined by continuous extension for all surfaces $x \in W^{1,2}(U,\mathbb{R}^3)$ with bounded trace $x|_{\partial U}$, although the integral representation (2.1), valid in the case $x \in L^\infty(U,\mathbb{R}^3)$, will in general not hold for unbounded x.

To define this extension one decomposes $x = y + z$ into its bounded harmonic part y and its $W_0^{1,2}$ part z. (y is the minimizer for Dirichlet's integral on $x + W_0^{1,2}(U,\mathbb{R}^3)$.) Then z is a closed surface with area not exceeding $\mathbf{D}(z)$, and the isoperimetric inequality implies

$$|\mathbf{V}(z)| \le \frac{1}{\sqrt{36\pi}} \mathbf{D}(z)^{3/2} . \tag{2.2}$$

This holds for smooth z and proves, since $\mathbf{V}(z)$ is a cubic form in z, that $\mathbf{V}$ can be extended continuously to $W_0^{1,2}(U,\mathbb{R}^3)$ with the same inequality. Furthermore, with an integration by parts we verify

$$\mathbf{V}(x) = \int_U y \cdot \left(y_u \wedge y_v + \tfrac{1}{2} y_u \wedge z_v + \tfrac{1}{2} z_u \wedge y_v + z_u \wedge z_v\right) dudv + \mathbf{V}(z) \tag{2.3}$$

for smooth $x = y + z$, and we deduce that $\mathbf{V}$ can be extended to the space of $x \in W^{1,2}(U,\mathbb{R}^3)$ with $x|_{\partial U} \in L^\infty(\partial U,\mathbb{R}^3)$ as a functional that is continuous with respect to $W^{1,2}$ convergence and simultaneous L^∞ convergence of boundary traces (i.e. of the harmonic parts).

With regard to weak convergence one verifies with (2.2), (2.3), integration by parts, and Rellich's theorem:

$$\begin{aligned} &\mathbf{V}(y + z_n) - \mathbf{V}(y + z) - \mathbf{V}(z_n - z) \to 0 \\ &\qquad \text{as } z_n \to z \text{ weakly in } W_0^{1,2}(U,\mathbb{R}^3), \end{aligned} \tag{2.4}$$

$$\begin{aligned} &\mathbf{V}(y_n) \to \mathbf{V}(y) \\ &\qquad \text{as } y_n \to y \text{ uniformly and } W^{1,2} \text{ weakly in } \mathcal{S}(\Gamma). \end{aligned} \tag{2.5}$$

(See [Wen1], [Ste1] for details. Instead of $y_n \in \mathcal{S}(\Gamma)$ with rectifiable Γ it would suffice to assume a bound on the length of the continuous curves $y_n|_{\partial U}$, in order to control the boundary terms.)

Following Wente we now attempt to minimize $\mathbf{E_H} = \mathbf{D} + 2H\mathbf{V}$ on

$$\mathcal{S}(\Gamma;\sigma) = \{x \in \mathcal{S}(\Gamma) : \mathbf{D}(x) \le \sigma a_\Gamma\} \tag{2.6}$$

where $1 < \sigma < \infty$ and a_Γ is the *least spanning area of* Γ, i.e. the area $a_\Gamma = \mathbf{D}(y_\Gamma)$ of a parametric surface y_Γ minimizing $\mathbf{D}$ on $\mathcal{S}(\Gamma)$ (which exists by the solution of the classical Plateau problem, i.e. Proposition 1.3 with $H = 0$

and $A = \mathbb{R}^3$). Choosing a minimizing sequence $x_n = y_n + z_n$, decomposed as above, we may assume $x_n \to x = y + z \in \mathcal{S}(\Gamma;\sigma)$ weakly in $W^{1,2}(U,\mathbb{R}^3)$ with $x_n\big|_{\partial U} \to x\big|_{\partial U}$ uniformly, hence also $y_n \to y$ uniformly on $\overline{U}$ (cf. the proof of Proposition 1.3). We put $\tilde{x}_n = y + z_n$ and assert $\mathbf{V}(x_n) - \mathbf{V}(\tilde{x}_n) \to 0$ as $n \to \infty$. This means that the integral in (2.3) is the limit of the corresponding integral with y, z replaced by y_n, z_n, and is readily verified with (2.5) and integration by parts. It follows that $\liminf_{n\to\infty} \mathbf{E}_H(\tilde{x}_n)$ does not exceed $\liminf_{n\to\infty} \mathbf{E}_H(x_n)$. Using also $\mathbf{D}(\tilde{x}_n) - \mathbf{D}(z_n - z) \to \mathbf{D}(x)$ and (2.4) we arrive at

$$\mathbf{E}_H(x) \leq \liminf_{n\to\infty} \mathbf{E}_H(x_n) + \limsup_{n\to\infty} \left[2|H||\mathbf{V}(z_n - z) - \mathbf{D}(z_n - z)\right] ,$$

and with (2.2) we infer

$$\mathbf{E}_H(x) \leq \liminf_{n\to\infty} \mathbf{E}_H(x_n) , \tag{2.7}$$

provided

$$\limsup_{n\to\infty} \frac{2|H|}{\sqrt{36\pi}} \mathbf{D}(z_n - z)^{1/2} \leq 1 .$$

Taking into account $\limsup_{n\to\infty} \mathbf{D}(z_n - z) \leq \limsup_{n\to\infty} \mathbf{D}(x_n)$ and (2.6) we see that the last inequality is satisfied if

$$|H| \leq \frac{3\sqrt{\pi}}{\sqrt{\sigma a_\Gamma}} , \tag{2.8}$$

and with this condition on H we have verified by (2.7) that x is a minimizer for $\mathbf{E}_H$ in $\mathcal{S}(\Gamma;\sigma)$.

To show that x is a solution to $\mathcal{P}(H,\Gamma)$ it remains to prove $\mathbf{D}(x) < \sigma a_\Gamma$ so that x can be freely varied and the variational formulas of Proposition 1.1 (which are easily checked for the present definition of volume using an expansion similar to (2.3)) together with Remarks 1.2 can be applied. Following Wente again, we use the inequality $\mathbf{E}_H(x) < \mathbf{E}_H(y_\Gamma)$ for this purpose which is valid in the case $H \neq 0$, because $y_\Gamma \in \mathcal{S}(\Gamma;\sigma)$ then cannot be a minimizer for $\mathbf{E}_H$. Consequently, we have

$$\mathbf{D}(x) = \mathbf{E}_H(x) - 2H\mathbf{V}(x) < \mathbf{E}_H(y_\Gamma) - 2H\mathbf{V}(x) = a_\Gamma + 2H(\mathbf{V}(y_\Gamma) - \mathbf{V}(x)) .$$

Now, x and y_Γ together form a closed surface to which the isoperimetric inequality in $\mathbb{R}^3$ can be applied, and similarly to (2.2) we get

$$|\mathbf{V}(y_\Gamma) - \mathbf{V}(x)| \leq \frac{1}{\sqrt{36\pi}} (\mathbf{D}(y_\Gamma) + \mathbf{D}(x))^{3/2} ,$$

and hence, recalling $\mathbf{D}(y_\Gamma) = a_\Gamma$ and $\mathbf{D}(x) \leq \sigma a_\Gamma$,

$$\mathbf{D}(x) < a_\Gamma \left(1 + \frac{2|H|}{\sqrt{36\pi}}(1+\sigma)^{3/2} {a_\Gamma}^{1/2}\right). \tag{2.9}$$

The desired conclusion follows if the right-hand side does not exceed σa_Γ, and it is readily checked that this is true if we make the optimal choice $\sigma = 5$ and require

$$|H| \leq \sqrt{\frac{2}{3}}\sqrt{\frac{\pi}{a_\Gamma}}. \tag{2.10}$$

Note that (2.10) also implies the restriction (2.8) imposed earlier on H.

For constant H we have thus proved the following theorem which was originally formulated by Wente [Wen1] with constant $c \approx \frac{2}{5}$ in (2.10), then sharpened by the second author to $c \approx 0.52$ in [Ste1], and later further improved in [Ste4] to the above constant $c = \sqrt{2/3}$, allowing also variable H.

2.1. Theorem (Wente, Steffen). *Suppose*

$$\sup_{\mathbb{R}^3} |H| \leq \sqrt{\frac{2}{3}}\sqrt{\frac{\pi}{a_\Gamma}},$$

where a_Γ is the least spanning area of Γ. Then the Plateau problem $\mathcal{P}(H,\Gamma)$ has a weak solution.

It is clear that this result is better than the existence theorems of Section 1 for curves with a shape described at the beginning of this section. The constant $c = \sqrt{2/3}$ is, however, probably not optimal. Considering the example of a circle Γ one is lead to the conjecture that it can be replaced by $c = 1$. This would be best possible by the non-existence theorem of Heinz [He2]. It has been observed by Struwe [Str3, III.3] that $\sqrt{2/3}$ can be replaced by a larger constant depending on Γ, if only constant H are considered. This also follows from the proof above, because $\mathbf{E}_H$ is uniformly close to $\mathbf{E}_{\tilde{H}}$ on $\mathcal{S}(\Gamma;\sigma)$ if $|H - \tilde{H}|$ is small.

It turns out that Wente's method can be generalized and made more transparent if one treats the volume functional in a more geometric fashion. To do this we note that integration

$$J_x(\beta) = \int_U x^\# \beta = \int_U \langle \beta \circ x, x_u \wedge x_v \rangle \, du dv \tag{2.11}$$

of 2-forms $\beta \in \mathcal{D}^2(\mathbb{R}^3)$ (i.e. smooth and with compact support) associates a 2-current J_x of finite mass

$$\mathbf{M}(J_x) \le \mathbf{D}(x) \tag{2.12}$$

with a given parametric surface $x \in W^{1,2}(U, \mathbb{R}^3)$. (A 2-current is a continuous linear functional T on $\mathcal{D}^2(\mathbb{R}^3)$ and its mass is the supremum of values $T(\beta)$ on 2-forms $\beta \in \mathcal{D}^2(\mathbb{R}^3)$ with $|\beta| \le 1$ on $\mathbb{R}^3$. The standard references are [Fe], [Si].) Using approximation by smooth mappings, in the sense of $W^{1,2}$ norm or in the sense of a Lusin type theorem ([EG, §6.6]), one sees that J_x is an *integer multiplicity rectifiable 2-current* on $\mathbb{R}^3$, i.e. J_x can be described by $\mathcal{H}^2$-integration of 2-forms over a locally 2-rectifiable set with $\mathcal{H}^2$-measurable orientation and $\mathcal{H}^2$-summable integer valued multiplicity function. (In the language of [Fe] this would be called a locally rectifiable 2-current of finite mass. $\mathcal{H}^2$ is the 2-dimensional Hausdorff measure. For details of the approximation arguments needed here and in the sequel we refer to [Ste3].) If $y \in W^{1,2}(U, \mathbb{R}^3)$ is another surface, then $(J_x - J_y)(\beta)$ is given by integration over the set G of $w \in U$ with $x(w) \ne y(w)$ because $Dx = Dy$ holds almost everywhere on $U \setminus G$. Hence (2.12) can be generalized to

$$\mathbf{M}(J_x - J_y) \le \mathbf{D}_G(x) + \mathbf{D}_G(y) \quad \text{if } x = y \text{ on } U \setminus G, \tag{2.13}$$

where $\mathbf{D}_G(x) = \frac{1}{2}\int_G |Dx|^2\, dudv$ for measurable $G \subset U$.

The boundary ∂T of a current T is generally defined by $\partial T(\alpha) = T(d\alpha)$. For J_x we find that ∂J_x is given by integration of 1-forms over the boundary $x\big|_{\partial U}$, if this curve is continuous and rectifiable. In particular, we have

$$\partial(J_x - J_y) = 0 \quad \text{if } x,\, y \in \mathcal{S}(\Gamma), \tag{2.14}$$

which is just a precise way of expressing that x and y form a closed surface in $\mathbb{R}^3$.

Now, every closed 2-current on $\mathbb{R}^3$ is the boundary of a 3-current which is unique up to a constant. For $J_x - J_y$ from (2.14) one can use the homotopy formula [Fe, 4.1.9] to obtain such a 3-current $I_{x,y}$ as follows:

$$I_{x,y}(\gamma) = \int_{[0,1]\times U} h^{\#}\gamma = \int_U \int_0^1 \langle \gamma \circ h, h_t \wedge h_u \wedge h_v \rangle\, dtdudv \tag{2.15}$$

where $h(t, w) = tx(w) + (1-t)y(w)$ and $\gamma \in \mathcal{D}^3(\mathbb{R}^3)$. The equation

$$\partial I_{x,y} = J_x - J_y \tag{2.16}$$

is just a version of Stokes' theorem if x, y are smooth on $\overline{U}$, and it follows in general by approximation. We note that (2.15) implies the mass estimate

$$\mathbf{M}(I_{x,y}) \leq \|x - y\|_{L^\infty}(\mathbf{D}_G(x) + \mathbf{D}_G(y)) \quad \text{if } x = y \text{ on } U \setminus G. \tag{2.17}$$

$I_{x,y}$ is an integer multiplicity locally rectifiable current in the top dimension and hence representable by a locally integrable integer valued function $i_{x,y}$ on $\mathbb{R}^3$ in the form

$$I_{x,y}(\gamma) = \int_{\mathbb{R}^3} i_{x,y}\gamma = \int_{\mathbb{R}^3} \langle \gamma, \vec{e}\,\rangle i_{x,y}\, d\mathcal{L}^2 \tag{2.18}$$

where $\vec{e} \in \bigwedge_3 \mathbb{R}^3$ is the orientation of $\mathbb{R}^3$. Geometrically speaking, $i_{x,y}$ is a *set with integer multiplicities*, and (2.16), (2.18) just say that this set has the closed surface composed by x and y as its boundary. From the isoperimetric inequality [Fe, 4.5.9(31)] one infers

$$\mathbf{M}(I_{x,y}) = \int_{\mathbb{R}^3} |i_{x,y}|\, d\mathcal{L}^2 \leq \int_{\mathbb{R}^3} |i_{x,y}|^{3/2}\, d\mathcal{L}^2 \leq \frac{1}{\sqrt{36\pi}} \mathbf{M}(J_x - J_y)^{3/2}, \tag{2.19}$$

proving in particular that the mass of $I_{x,y}$ is finite. We note that $I_{x,y}$ is uniquely determined by (2.16) and by the condition $\mathbf{M}(I_{x,y}) < \infty$, and we deduce

$$I_{x,y} + I_{y,z} = I_{x,z} \quad \text{for } x,\ y,\ z \in \mathcal{S}(\Gamma). \tag{2.20}$$

After this geometric-measure-theoretic discussion it is evident how one should define the *H-volume functional* associated with a prescribed mean curvature function H, namely

$$\mathbf{V}_H(x, y) = \int_{\mathbb{R}^3} H i_{x,y}\, d\mathcal{L}^2 \quad \text{for } x,\ y \in \mathcal{S}(\Gamma). \tag{2.21}$$

The integral exists on account of (2.19), because we generally assume that H is bounded and continuous. (The H-volume can be defined under much weaker conditions, see [Ste3].) Approximating $H\Omega$ suitably by 3-forms $\gamma \in \mathcal{D}^3(\mathbb{R}^3)$ (recall that Ω is the volume form of $\mathbb{R}^3$) we obtain from (2.17), (2.18), (2.21)

$$|\mathbf{V}_H(x, y)| \leq \|H\|_{L^\infty} \|x - y\|_{L^\infty} (\mathbf{D}(x) + \mathbf{D}(y)), \tag{2.22}$$

and for $\xi \in W_0^{1,2}(U,\mathbb{R}^3) \cap L^\infty(U,\mathbb{R}^3)$ we see that $\mathbf{V}_H(x+t\xi,x)$ has the integral representation (1.6). Taking (2.20) into account, i.e.

$$\mathbf{V}_H(x,y) + \mathbf{V}_H(y,z) = \mathbf{V}_H(x,z) \quad \text{for } x,\ y,\ z \in \mathcal{S}(\Gamma), \tag{2.23}$$

we deduce for the functional $x \mapsto \mathbf{V}_H(x,y)$ exactly the same first variation formula as in the proof of Proposition 1.1. Moreover, J_x, J_y and hence also $I_{x,y}$, $\mathbf{V}_H(x,y)$ are invariant with respect to orientation preserving self-diffeomorphisms of $\overline{U}$. It follows that Propositions 1.1, 1.7 and Remarks 1.2, 1.4 are valid if we replace $\mathbf{E}_Z$ by $\mathbf{E}_H$ there and define the *H-energy* now by

$$\mathbf{E}_H(x) = \mathbf{D}(x) + 2\mathbf{V}_H(x,y_\Gamma) \quad \text{for } x \in \mathcal{S}(\Gamma), \tag{2.24}$$

where y_Γ is a fixed surface in the class $\mathcal{S}(\Gamma)$.

In order to minimize $\mathbf{E}_H$ on $\mathcal{S}(\Gamma)$ or on suitable subclasses we next turn to the question of lower semicontinuity of $\mathbf{E}_H$ on a $W^{1,2}$ weakly convergent sequence $x_n \to x$ with $x_n\big|_{\partial U} \to x\big|_{\partial U}$ uniformly. The volume $\mathbf{V}_H(x_n,y_\Gamma)$ is not, in general, continuous for this convergence, because a "bubble" can split off x_n in the limit carrying away a certain amount of volume that is not enclosed by the limit surface x. The idea is then that such a bubble must be parametrized over an arbitrarily small part of U and its boundary mass must be dominated by the jump $\limsup_{n\to\infty}(\mathbf{D}(x_n) - \mathbf{D}(x))$ of Dirichlet's integral. On the other hand, we can dominate the jump of H-volume $\limsup_{n\to\infty} 2|\mathbf{V}_H(x_n,x)|$ by the boundary mass of the bubble, using the isoperimetric inequality and suitable hypotheses for H and the x_n. The conclusion will then be $\mathbf{E}_H(x) \le \liminf_{n\to\infty} \mathbf{E}_H(x_n)$ as desired. To make this idea precise we need the following technical

2.2. Lemma. *Suppose x_n, $x \in \mathcal{S}(\Gamma)$, $x_n \to x$ weakly in $W^{1,2}(U,\mathbb{R}^3)$, and $x_n\big|_{\partial U} \to x\big|_{\partial U}$ uniformly on ∂U. Then for every $\varepsilon > 0$ there exist $R > 0$, a measurable set $G \subset U$, and surfaces $\tilde{x}_n \in \mathcal{S}(\Gamma)$ such that the following assertions hold after passage to subsequences (still denoted $\tilde{x}_n$, x_n):*

(i) $\tilde{x}_n = x_n$ *on* $U \setminus G$ *and* $\mathcal{L}^2(G) < \varepsilon$;

(ii) $\lim_{n\to\infty} \|\tilde{x}_n - x_n\|_{L^\infty} = 0$ *and* $\tilde{x}_n(w) = x(w)$ *if* $|x_n(w)| \ge R$;

(iii) $\limsup_{n\to\infty}[\mathbf{D}_G(\tilde{x}_n) + \mathbf{D}_G(x)] \le \varepsilon + \liminf_{n\to\infty}[\mathbf{D}(x_n) - \mathbf{D}(x)]$;

(iv) if the x_n have values in the closure A of a C^2 domain in $\mathbb{R}^3$, then the $\tilde{x}_n$ may be chosen to have values in A, too.

Proof. Using the theorems of Rellich and Egoroff we can find $R > 3$, $\frac{1}{2} \geq \delta_n \to 0$ and $G \subset U$ with $\mathcal{L}^2(G) + \mathbf{D}_G(x)$ arbitrarily small, such that $\sup_{U \setminus G} |x| \leq \frac{1}{3}R$ and, after passage to a subsequence, $\sup_{U \setminus G} |x_n - x| \leq \delta_n$ for all n. We choose $\eta \in \mathcal{D}(\mathbb{R})$ with $\eta \equiv 1$ on $[0, \frac{1}{3}R]$, $\operatorname{spt} \eta \subset]-\infty, \frac{2}{3}R]$, $0 \geq \eta' \geq -4R$, and we put $\vartheta_n(t) = 1$ for $0 \leq t \leq \delta_n$, $\vartheta_n(t) = \delta_n(1-\delta_n)^{-1}(t^{-1}-1)$ for $\delta_n \leq t \leq 1$, and $\vartheta_n(t) = 0$ for $t \geq 1$. Defining

$$\tilde{x}_n = x_n + (\eta \circ |x|)(\vartheta_n \circ |x_n - x|)(x - x_n)$$

one immediately has (i), (ii), and one verifies (iii) by a direct calculation. (iv) is then achieved by applying a C^1 neighborhood retraction of A to the $\tilde{x}_n$. (For more details we refer to [DS5]. See also [Ste3], where another construction has been used to define the $\tilde{x}_n$ which gives a somewhat weaker result.) □

To pursue the idea above we must now estimate $2\mathbf{V}_H(\tilde{x}_n, x)$ in terms of $\mathbf{D}_G(\tilde{x}_n) + \mathbf{D}_G(x)$. The following definition is useful for this (cf. [Ste3], [DS4]):

2.3. Definition. Given $0 \leq c < \infty$, $0 < s \leq \infty$ we say that H satisfies an *isoperimetric condition* of type c, s on A if

$$\left| \int_E 2H \, d\mathcal{L}^2 \right| \leq c\mathbf{P}(E) \tag{2.25}$$

holds whenever $E \subset A$ is a set with finite perimeter $\mathbf{P}(E) \leq s$. □

Here, by a *set with finite perimeter* $E \subset \mathbb{R}^3$ we mean a Borel set of finite $\mathcal{L}^2$-measure such that its characteristic function has distributional gradient of finite total variation. This total variation is then the perimeter $\mathbf{P}(E) < \infty$, it equals the boundary mass $\mathbf{M}(\partial[\![E]\!])$ of the 3-current given by integration of 3-forms over E, and its geometric meaning is the boundary area of E. For smooth bounded domains E one has $\mathbf{P}(E) = \mathcal{H}^2(\partial E)$, and it is sufficient to verify (2.25) for such domains $E \subset A$ when A is the closure of a smooth domain. A standard decomposition theorem from geometric measure theory [Fe, 4.5.17] implies that every integer multiplicity rectifiable 3-current I on $\mathbb{R}^3$ can be decomposed into a L^1-convergent sum of (suitably oriented) sets E_k, $k \in \mathbb{Z}$, with finite perimeter such that $\sum_{k \in \mathbb{Z}} \mathbf{P}(E_k) = \mathbf{M}(\partial I)$. Applying this to our currents $I_{x,y}$ above and recalling (2.16), (2.21) we obtain

$$2|\mathbf{V}_H(x,y)| \leq c\mathbf{M}(J_x - J_y) \quad \text{for } x,\, y \in \mathcal{S}(\Gamma) \text{ with } \mathbf{M}(J_x - J_y) \leq s, \tag{2.26}$$

whenever an isoperimetric condition of type c, s is valid on $\mathbb{R}^3$. If we know this condition only on $A \subset \mathbb{R}^3$ we can draw the same conclusion for x, $y \in \mathcal{S}(\Gamma, A)$, provided $I_{x,y}$ has support in A. This will be the case, for instance, if $\mathbb{R}^3 \setminus A$ has no components of finite measure, because $\partial I_{x,y} = J_x - J_y$ has support in A, hence $I_{x,y}$ must be constant on each component of $\mathbb{R}^3 \setminus A$, and $\mathbf{M}(I_{x,y}) < \infty$ by (2.19). (For a more thorough discussion of Definition 2.3 and its consequence (2.26) see [Ste3], [DS4].)

We are now prepared to prove the following general existence theorem for the Plateau problem $\mathcal{P}(H, \Gamma)$, which is a strengthening of the results of the second author [Ste3].

2.4. Theorem. *Suppose A is closed in $\mathbb{R}^3$, $\mathbb{R}^3 \setminus A$ has no component of finite measure, H satisfies an isoperimetric condition of type c, s on A, $\Gamma \subset A$, and $y_\Gamma \in \mathcal{S}(\Gamma, A)$ with $(1+\sigma)\mathbf{D}(y_\Gamma) \leq s$ for some $1 < \sigma \leq \infty$. Set $\mathcal{S}(\Gamma, A; \sigma) = \{x \in \mathcal{S}(\Gamma, A) : \mathbf{D}(x) \leq \sigma \mathbf{D}(y_\Gamma)\}$. Then the following assertions hold:*

(i) If $\sigma < \infty$ and $c \leq 1$ or $\sigma = \infty$ and $c < 1$, then the variational problem

$$\mathbf{E}_H(x) = \mathbf{D}(x) + 2\mathbf{V}_H(x, y_\Gamma) \rightsquigarrow \min \quad \text{on } \mathcal{S}(\Gamma, A; \sigma) \tag{2.27}$$

has a solution.

(ii) If

$$c \leq \frac{\sigma - 1}{\sigma + 1}, \tag{2.28}$$

then (2.27) posesses a solution x with $\mathbf{D}(x) < \sigma \mathbf{D}(y_\Gamma)$; if $\sigma = \infty$ or strict inequality holds in (2.28), then this is true for each solution x of (2.27).

(iii) If A is the closure of a C^2 domain and

$$|H| \leq H_{\partial A} \quad \text{holds pointwise on } \partial A, \tag{2.29}$$

where $H_{\partial A}$ denotes the (inward) boundary mean curvature of A, then each minimizer x of (2.27) with $\mathbf{D}(x) < \sigma \mathbf{D}(y_\Gamma)$ is a weak solution to the Plateau problem $\mathcal{P}(H, \Gamma)$. Moreover, if $|H(a)| < H_{\partial A}(a)$ holds at some point $a \in (\partial A) \setminus \Gamma$, then x does not meet a neighborhood of this point.

Proof. (i) We choose a minimizing sequence x_n for (2.27). Then

$$\sup_n \mathbf{D}(x_n) < \infty,$$

because in the case $\sigma = \infty$ we have $c < 1$ and

$$|\mathbf{D}(x_n) - \mathbf{E}_H(x_n)| = 2|\mathbf{V}_H(x_n, y_\Gamma)| \leq c(\mathbf{D}(x_n) + \mathbf{D}(y_\Gamma)), \tag{2.30}$$

by (2.26), (2.13), the definition of $\mathcal{S}(\Gamma, A; s)$, and the assumption $(1 + \sigma)\mathbf{D}(y_\Gamma) \leq s$. As in the proof of Proposition 1.3 we can assume $x_n \to x \in \mathcal{S}(\Gamma, A; \sigma)$ weakly in $W^{1,2}(U, \mathbb{R}^3)$ and uniformly on ∂U. We then apply Lemma 2.2 with a given $\varepsilon > 0$ to obtain, after passage to a subsequence, surfaces $\tilde{x}_n \in \mathcal{S}(\Gamma, A)$ with the properties described there. From (2.22) and parts (i)–(iii) of the Lemma we see $\mathbf{V}_H(\tilde{x}_n, x_n) \to 0$. Choosing $2\varepsilon < \mathbf{D}(x)$ we also infer from (2.13) and parts (i), (iii) of Lemma 2.2 that

$$\mathbf{M}(J_{\tilde{x}_n} - J_x) \leq \mathbf{D}_G(\tilde{x}_n) + \mathbf{D}_G(x) \leq 2\varepsilon + \mathbf{D}(x_n) - \mathbf{D}(x) < \sigma\mathbf{D}(y_\Gamma) < s \tag{2.31}$$

for large n. Therefore, (2.26) and $c \leq 1$ imply

$$2|\mathbf{V}_H(\tilde{x}_n, x)| \leq 2\varepsilon + \mathbf{D}(x_n) - \mathbf{D}(x),$$

and we conclude with (2.23)

$$\begin{aligned}
\mathbf{E}_H(x_n) &= \mathbf{D}(x_n) + 2\mathbf{V}_H(x_n, y_\Gamma) = \mathbf{E}_H(x) + \mathbf{D}(x_n) - \mathbf{D}(x) + 2\mathbf{V}_H(x_n, x) \\
&= \mathbf{E}_H(x) + \mathbf{D}(x_n) - \mathbf{D}(x) + 2\mathbf{V}_H(\tilde{x}_n, x) - 2\mathbf{V}_H(\tilde{x}_n, x_n) \\
&\geq \mathbf{E}_H(x) - 3\varepsilon
\end{aligned}$$

for sufficiently large n.

(ii) If y_Γ is not a solution to (2.27), then we have

$$\begin{aligned}
\mathbf{D}(x) &= \mathbf{E}_H(x) - 2\mathbf{V}_H(x, y_\Gamma) < \mathbf{E}_H(y_\Gamma) - 2\mathbf{V}_H(x, y_\Gamma) \\
&= \mathbf{D}(y_\Gamma) - 2\mathbf{V}_H(x, y_\Gamma) \leq \mathbf{D}(y_\Gamma) + c(\mathbf{D}(x) + \mathbf{D}(y_\Gamma)) \\
&\leq \mathbf{D}(y_\Gamma)(1 + c(\sigma + 1)),
\end{aligned}$$

by (2.30) (applied with x instead of x_n) and the definition of $\mathcal{S}(\Gamma, A; \sigma)$. Consequently, with the assumption $1 + c(\sigma + 1) \leq \sigma$ we obtain $\mathbf{D}(x) < \sigma\mathbf{D}(y_\Gamma)$. If y_Γ is a solution, then the same inequality holds for $x = y_\Gamma$, because $\sigma > 1$.

(iii) Here we can repeat the proof of the corresponding assertion in Theorem 1.8, since we have already noted in connection with the definition of H-energy (2.24) that all the steps are valid in the present context. □

To give concrete *applications* of the preceding theorem we have to verify the isoperimetric condition (2.25) or (2.26). For example, we can use the isoperimetric inequality for sets of finite perimeter to obtain

$$\left|\int_E 2H\,d\mathcal{L}^2\right| \le 2\sup_{\mathbb{R}^3}|H|\mathcal{L}^2(E) \le 2\sqrt{\frac{s}{36\pi}}\sup_{\mathbb{R}^3}|H|\mathbf{P}(E)\,,$$

if $\mathbf{P}(E) \le s < \infty$. (Or apply (2.19) directly in (2.21) to see that (2.26) holds with $c = 2s^{1/2}(36\pi)^{-1/2}\sup_{\mathbb{R}^3}|H|$.) Choosing $s = (\sigma+1)\mathbf{D}(y_\Gamma)$ the condition (2.28) on c reduces to

$$2(\sigma+1)^{1/2}\sqrt{\frac{\mathbf{D}(y_\Gamma)}{36\pi}}\sup_{\mathbb{R}^3}|H| \le \frac{\sigma-1}{\sigma+1}\,,$$

and it is readily verified that σ=5 is the optimal choice and then

$$\sup_{\mathbb{R}^3}|H| \le (\frac{3}{2\pi}\mathbf{D}(y_\Gamma))^{-1/2}$$

is the resulting condition on H. Of course, we choose here y_Γ as a minimal surface of least area $\mathbf{D}(y_\Gamma) = a_\Gamma$ in $\mathcal{S}(\Gamma)$ and we have proved Theorem 2.1 with the additional inclusion statement from part (iii) of the preceding theorem if $y_\Gamma \in \mathcal{S}(\Gamma, A)$. We note that Dierkes [Di1], [Di2] has stated inclusion principles of a different nature which require that $\sup_{\partial A}|H_{\partial A}|$ and $\sup_A |H|$ are (quite) small compared to ${a_\Gamma}^{-1/2}$, but allow negative values of $H_{\partial A}$.

Another application results when we first use Hölder's inequality in (2.25) or (2.26) and then apply the isoperimetric inequality:

$$\begin{aligned}\left|\int_E 2H\,d\mathcal{L}^2\right| &\le 2\left[\int_E |H|^3\,d\mathcal{L}^2\right]^{1/3}\mathcal{L}^2(E)^{2/3}\\ &\le 2\left[\int_E |H|^3\,d\mathcal{L}^2\right]^{1/3}\frac{1}{\sqrt[3]{36\pi}}\mathbf{P}(E)\,.\end{aligned}$$

In fact, the same inequality holds with $|H|_{L^3}$ replaced by the maximum of $|H_+|_{L^3}$ and $|H_-|_{L^3}$, where $H = H_+ - H_-$ is the decomposition of H into its positive and negative part. Using Theorem 2.4 with $\sigma = s = \infty$ and $A = \mathbb{R}^3$

we obtain the following theorem proved by the second author [Ste4] (and a corresponding inclusion version in the case $A \neq \mathbb{R}^3$ is also valid):

2.5. Theorem (Steffen). *The Plateau problem $\mathcal{P}(H,\Gamma)$ is solvable whenever*

$$\int_{\mathbb{R}^3} |H_+|^3 \, d\mathcal{L}^2 < \frac{9\pi}{2} \quad \textit{and} \quad \int_{\mathbb{R}^3} |H_-|^3 \, d\mathcal{L}^2 < \frac{9\pi}{2} . \tag{2.32}$$

It is interesting that no condition on the boundary curve Γ is required here. There are many other possibilities to derive an isoperimetric condition. For example we may apply Fubini's theorem to the integral (2.25) and use Schwarz' inequality and the isoperimetric inequality for parametric 2-dimensional domains to prove a cylinder version of the preceding Theorem where the inequalities (2.32) are replaced by

$$\sup_{t\in\mathbb{R}} \int_{\mathbb{R}^2} |H_\pm(r,s,t)|^2 \, drds < \pi .$$

The interested reader can find in [Ste4] and [DS4] further examples where an isoperimetric condition can be verified and a corresponding existence theorem for the Plateau problem follows.

If Z is a bounded continuous vector field with $\operatorname{div} Z = H$ on a neighborhood of A, then we obtain an isoperimetric condition with $c = 2\sup_A |Z|$ from Gauss' theorem (applied to bounded smooth domains $E \subset A$, or using [Fe, 4.5.6]):

$$\left| \int_E 2H \, d\mathcal{L}^2 \right| \leq \int_{\partial E} 2|Z| \, d\mathcal{H}^2 \leq 2 \sup_A |Z| \, \mathbf{P}(E) . \tag{2.33}$$

In particular, we have $c < 1$ if $\sup_A |Z| < \frac{1}{2}$, and we now recover all the existence results of Section 1 as special cases of Theorem 2.4. Moreover, for (2.33) we do not need $\operatorname{div} Z = H$ but only $\operatorname{div} Z \geq |H|$ in the distributional sense, which allows much more freedom in the construction of Z.

For example, if A is the closure of a proper C^2 domain in $\mathbb{R}^3$,

$$d_A \colon A \to [0, \infty[$$

denotes the boundary distance and $2Z = -\nabla d_A$ its gradient, then one has $|Z| = \frac{1}{2}$ and $\operatorname{div} Z = H_A$ on $A \setminus C$ where H_A is the parallel mean curvature function of A introduced before Theorem 1.9 and $C \subset A$ is the $\mathcal{L}^2$ null set consisting of cut points and focal points of ∂A (i.e. points $a \in A$ with non-unique nearest point b in ∂A or with $|a-b|^{-1}$ equal to the maximum of the principal curvatures of ∂A at b). Moreover, it is not difficult to see that $\operatorname{div} Z \geq H_A$ holds in the sense of distributions on A (see [DS4]), and we deduce an isoperimetric condition of type c, ∞ for H on A whenever $|H| \leq cH_A$ holds on A. This gives immediately a version of Theorem 1.9.

To obtain a *proof of Theorem* 1.9 as formulated in Section 1 one uses variants of the definition of Z above. For instance, assuming that the inner radius $r_A = \sup_A d_A$ is finite we define

$$2Z = -\frac{e^{\lambda(r_A - d_A)}}{\sqrt{1+e^{2\lambda(r_A-d_A)}}}\nabla d_A \quad \text{with } \lambda = 2\sup_A |H|$$

and verify $\operatorname{div} Z \geq |H|$ when $|H| \leq H_A$ on A. An isoperimetric condition with $c = e^{\lambda r_A}(1+e^{2\lambda r_A})^{-1/2} < 1$ follows for H. In fact we can allow $|H| \leq \sqrt{1+\varepsilon^2} H_A$ on A with $\varepsilon > 0$ so small that $\lambda r_A < \sqrt{1+\varepsilon^2}\log|\varepsilon|$, and we still obtain an isoperimetric condition with constant $c < 1$. With the choice

$$2Z = -\left(1 - \kappa d_A - \varepsilon e^{(\kappa+\lambda)d_A/(1-\kappa r_A)}\right)\nabla d_A$$

we can calculate $\operatorname{div} Z \geq |H|$ whenever $0 < \kappa < r_A^{-1}$ and

$$2|H| \leq 2(1-\kappa d_A)H_A + \kappa$$

holds on A. $\varepsilon > 0$ here is small enough to make the coefficient $(\dots)$ in front of ∇d_A positive, and an isoperimetric condition with constant $c = 1-\varepsilon$ follows for H on A. To include the case $\kappa = r_A^{-1}$ we assume in addition that $2|H| \leq 2H_A + (1-2\varepsilon)\kappa$ holds on $A \cap d_A^{-1}[0,\delta]$ for some $0 < \delta, \varepsilon \ll 1$ with $\delta\lambda \leq \varepsilon$. We can then check that $\operatorname{div} Z \geq |H|$ and $2|Z| \leq 1 - \varepsilon\kappa\delta$ hold for

$$2Z = \min(1-\kappa d_A, 1-\kappa\delta + (1-\varepsilon)\kappa(\delta - d_A))\nabla d_A\,.$$

Consequently we have an isoperimetric condition for H with $c = 1-\varepsilon\kappa\delta < 1$. To apply Theorem 2.4 we must also know that $\mathbb{R}^3 \setminus A$ has no components of finite measure. It follows from $H_{\partial A} \geq 0$ that none of these components is bounded. Unbounded components of finite measure can be excluded with suitable uniformity assumptions on A at infinity, e.g. the existence of global exterior parallel surfaces for ∂A. With these observations and Theorem 2.4 we have proved all the assertions made in Theorem 1.9 (and, in fact, stronger statements).

To conclude this Section we briefly indicate how to prove the *regularity of weak solutions* $x \in \mathcal{S}(\Gamma, A)$ to the Plateau problem $\mathcal{P}(\Gamma, H)$ which are obtained from Theorem 2.4. The main point is to show continuity of x on $\overline{U}$, because we then use a representation $H = \operatorname{div} Z$ with Z bounded on a neighborhood V of the compact set $x(\overline{U})$ and we have, with a constant V_0 depending on Γ, Z only,

$$\mathbf{V}_H(\tilde{x}, y_\Gamma) = \int_U (Z \circ \tilde{x}) \cdot \tilde{x}_u \wedge \tilde{x}_v \, du dv - V_0 = \mathbf{V}_Z(\tilde{x}) - V_0$$

for $\tilde{x} \in \mathcal{S}(\Gamma, A)$ with image in V. (The constant is just $V_0 = \mathbf{V}_Z(y_\Gamma)$.) Moreover, for a given $w_0 \in \overline{U}$ we can make $|Z|$ as small as we like near $x(w_0)$. It follows that all the results on analytic and geometric regularity mentioned at the end of Section 1 can be applied to the present situation. (There is one exception related to the exclusion of false branch points. In this connection various authors have used the global condition $\sup |Z| < \frac{1}{2}$. An inspection of the proofs in [Al1], [Al2], [Gu2], [GOR] reveals, however, that an isoperimetric condition for H with constant $c < 1$ is actually sufficient, and this was also shown in [SW].) Now, to verify continuity of x we employ the energy minimizing property and repeat the argument leading to (1.26). The only property of the energy functional needed there was the inequality, with a constant $0 \leq c < 1$,

$$\mathbf{D}_G(x) \leq \frac{1+c}{1-c} \mathbf{D}_G(\tilde{x}) \tag{2.34}$$

for admissible comparison surfaces $\tilde{x}$ with $x = \tilde{x}$ on $U \setminus G$. The inequality follows here from the isoperimetric condition (2.26) and from (2.13) which imply

$$\begin{aligned} \mathbf{D}_G(x) - \mathbf{D}_G(\tilde{x}) &= \mathbf{E}_H(x) - \mathbf{E}_H(\tilde{x}) - 2\mathbf{V}_H(x, \tilde{x}) \\ &\leq 2|\mathbf{V}_H(x, \tilde{x})| \leq c(\mathbf{D}_G(x) + \mathbf{D}_G(\tilde{x})) \,. \end{aligned}$$

(Note that we have $c < 1$ by assumption or by (2.28), if Theorem 2.4 is applicable to produce an energy minimizing solution of $\mathcal{P}(H, \Gamma)$. Also note that we need only consider comparison surfaces with $\mathbf{D}_G(x) + \mathbf{D}_G(\tilde{x})$ small so that the application of (2.26) is justified.) Therefore, if A is quasiconvex we can use the device of Morrey, described after (1.26), in order to prove Hölder continuity of x on U and continuity on $\overline{U}$.

3. H-Surfaces in a Riemannian manifold.

In this section we consider the Plateau problem with prescribed mean curvature in a Riemannian manifold (M, g). We assume that M is of dimension 3, connected, oriented, complete, without boundary, and sufficiently smooth. Riemannian inner products and norms will be denoted $(\tau,\tau')_M$, $|\tau|_M$ (and it is understood that these are taken in the tangent space T_aM with respect to its inner product $g(a)$ if τ, $\tau' \in T_aM$). The Riemannian metric and the orientation determine a volume form Ω on M. Furthermore, since $\dim M = 3$ we have the exterior vector product $\tau\wedge_M\tau'$ on each tangent space T_aM which is characterized by $(\tau\wedge_M\tau',\tau'')_M = \langle\Omega(a),\tau\wedge\tau'\wedge\tau''\rangle$ for $a \in M$ and $\tau,\tau',\tau'' \in T_aM$.

To obtain the Riemannian analogue for the H-surface-equation (0.2) we interprete x_u, x_v as tangential vector fields along the parametric surface $x\colon U \to M$ and replace $x_u\wedge x_v$ in (0.2) by the normal field $x_u\wedge_M x_v$ along x. For the Laplacean appearing in (0.2) we have to substitute the Euler operator Δ^M associated with *Dirichlet's integral* which now is, of course,

$$\mathbf{D}(x) = \tfrac{1}{2}\int_U \left(|x_u|_M^2 + |x_v|_M^2\right) dudv\,. \tag{3.1}$$

The in general nonlinear operator Δ^M is well known from the theory of harmonic mappings (cf. [EL], [Jo], [Jo2], [Jo3]) and called the *tension field operator* for mappings from the Euclidean disc U to the Riemannian manifold M. Note that $\Delta^M x$ is a tangent vector field of M along the mapping x, i.e. $\Delta^M x(w) \in T_{x(w)}M$ for $w \in U$ (assuming x is of class C^2).

The *Plateau problem $\mathcal{P}_M(H,\Gamma)$ in the Riemannian manifold M* with prescribed mean curvature $H\colon M \to \mathbb{R}$ (always assumed bounded and continuous) and given closed oriented Jordan boundary curve $\Gamma \subset M$ (always assumed rectifiable and contractible in M or in the subdomain of M where we want to find a solution) then has the following formulation:

$$\Delta^M x = 2(H\circ x)x_u\wedge_M x_v \quad \text{on } U, \tag{3.2}$$

$$|x_u|_M^2 - |x_v|_M^2 = (x_u,x_v)_M = 0 \quad \text{on } U, \tag{3.3}$$

$$x\big|_{\partial U} \text{ is a weakly monotonic parametrization of } \Gamma. \tag{3.4}$$

Here (3.2) is again called the *H-surface-equation*, (3.3) is the condition of *conformal parametrization* in the present Riemannian context, and the *Plateau boundary condition* (3.4) is the same as (0.3). C^2 solutions $x\colon U \to M$ to (3.3), (3.2) will be called *H-surfaces* in M. To see the geometric meaning of (3.2), (3.3) we note that $\int_U(\xi,x_u\wedge_M x_v)_M dudv$ is the initial rate of change of volume swept out by a deformation x_t of $x = x_0$ with initial vector field $\xi = \frac{d}{dt}\big|_0 x_t$, while $\int_U(\xi,\Delta^M x)_M dudv$ describes the initial change of the *area* $\mathbf{A}(x) = \int_U |x_u\wedge_M x_v|_M dudv$, because we have $\mathbf{A}(x_t) \le \mathbf{D}(x_t)$ with equality at $t = 0$ and hence, assuming $x_u\wedge_M x_v \ne 0$, the equation $\frac{d}{dt}\big|_0\mathbf{A}(x_t) = \frac{d}{dt}\big|_0\mathbf{D}(x_t)$ holds. Since the mean curvature is (half) the ratio of the initial rates of change of area and volume under a normal deformation we recognize that (3.2), (3.3) express that the conformal parametric surface x has prescribed mean curvature H wherever it is of maximal rank. Of course, one can also readily verify this fact using the classical definition of mean curvature, i.e. $2H \circ x$ is the trace of the second fundamental form of x relative to M.

Introducing local coordinates on M we can write equations (3.2), (3.3), on each subset of U which is mapped into the coordinate domain by x, as follows (see eg. [HK], [Gu1], [Jo3]):

$$\begin{aligned} x^l_{uu} + x^l_{vv} &+ (\Gamma^l_{ij} \circ x)(x^i_u x^j_u + x^i_v x^j_v) \\ &= 2(H \circ x)\sqrt{\gamma \circ x}\, \varepsilon_{ijk}\,(g^{lk} \circ x)(x^i_u x^j_u - x^i_v x^j_v)\,, \end{aligned} \tag{3.5}$$

$$\begin{aligned} (g_{ij} \circ x)&(x^i_u x^j_u - x^i_v x^j_v) \\ &= (g_{ij} \circ x) x^i_u x^j_v = 0\,, \end{aligned} \tag{3.6}$$

where x^l, g_{ij}, Γ^l_{ij} are the coordinate components of x, g and the Levi-Civita connection (Christoffel symbols of g), we have set $(g^{lk}) = (g_{ij})^{-1}$, $\gamma = \det(g_{ij})$, $\varepsilon_{ijk} = \frac{1}{2}(i-j)\,(j-k)(k-i)$, and summation over repeated indices from 1 to 3 is understood.

As we want to solve $\mathcal{P}_M(H,\Gamma)$ with a variational method we need a weak formulation of the H-surface-equation. This is no problem if we restrict our considerations to surfaces x in the domain of a fixed coordinate domain in M, because we can use (3.5). Such a point of view, which is equivalent with studying parametric surfaces in $\mathbb{R}^3$ equipped with a general Riemannian metric, was adapted by Hildebrandt & Kaul [HK] and Gulliver [Gu1] in their first treatment of the Plateau problem with prescribed mean curvature in a Riemannian manifold. However, here we do not want to assume that our surfaces are contained in a coordinate domain on M, because we are aiming for an analogue of Theorem 2.1 in the Riemannian situation which is valid for *all* curves Γ in M bounding a surface of sufficiently small area. We then have the problem to define the Sobolev space $W^{1,2}(U,M)$ and to give a meaning to the H-surface-equation for $W^{1,2}$ mappings $x\colon U \to M$. Following the usual procedure in the theory of harmonic mappings we therefore assume that M is isometrically embedded in some Euclidean space $\mathbb{R}^m$ as closed subset. This is not really necessary, but it is convenient and no essential loss of generality, by the well-known embedding theorems of Nash and Gromov & Rohlin.

We then define $W^{1,2}(U,M)$ as the closed subclass of $W^{1,2}(U,\mathbb{R}^m)$ consisting of surfaces x with values in M (almost everywhere on U). Interpreting

$x: U \to M$ as a mapping into $\mathbb{R}^m$ we can replace Riemannian inner products and norms in M by Euclidean inner products and norms in $\mathbb{R}^m$, the Riemannian Dirichlet integral (3.1) becomes the Euclidean one, and the same is true for the conformality relations (3.3). Moreover, by a general principle from the calculus of variations, the Euler operator Δ^M associated with Dirichlet's integral and the constraint $x(w) \in M$ is just the tangential projection of the Euler operator for the unconstrained problem which here is the Euclidean Laplacean $\Delta x = x_{uu} + x_{vv}$ (applied to each of the m components of x). Denoting by Π^M the field of orthogonal projections $\Pi^M(a): \mathbb{R}^m \to T_aM$ and by $A^M(a): T_aM \times T_aM \to (T_aM)^\perp$ the second fundamental form of M we have

$$\Delta^M x = (\Pi^M \circ x)\Delta x = \Delta x - (A^M \circ x)(x_u, x_u) - (A^M \circ x)(x_v, x_v), \tag{3.7}$$

and (3.2) in this setting becomes

$$(\Pi^M \circ x)\Delta x = 2(H \circ x) x_u \wedge_M x_v \quad \text{on } U. \tag{3.8}$$

A weak formulation of (3.8), meaningful for $x \in W^{1,2}(U, M)$, is then the following:

$$\int_U \left[x_u \cdot ((\Pi^M \circ x)\xi)_u + x_v \cdot ((\Pi^M \circ x)\xi)_v + 2(H \circ x)\, \xi \cdot x_u \wedge_M x_v\right] dudv = 0$$
$$\text{for all vector fields } \xi \in W_0^{1,2}(U, \mathbb{R}^m) \cap L^\infty(U, \mathbb{R}^m), \tag{3.9}$$

or equivalently,

$$\int_U \left[x_u \cdot \nabla_u^M \zeta + x_v \cdot \nabla_v^M \zeta + 2(H \circ x)\, \zeta \cdot x_u \wedge_M x_v\right] dudv = 0 \tag{3.10}$$
$$\text{for all } \zeta \in W_0^{1,2}(U, \mathbb{R}^m) \cap L^\infty(U, \mathbb{R}^m) \text{ tangential to } M \text{ along } x,$$

the latter condition meaning that $\zeta(w) \in T_{x(w)}M$ for (almost all) $w \in U$. We call (3.9), (3.10) the *weak H-surface-equation* in M and any conformal solution $x \in W^{1,2}(U, M)$ a *weak H-surface* in M, or a *weak solution to the Plateau problem $\mathcal{P}_M(H, \Gamma)$* if also (3.4) is valid.

The covariant derivatives ∇^M appearing in (3.10) are defined by $\nabla_u^M \zeta = (\Pi^M \circ x)\zeta_u$. Since x_u, x_v are tangential to M along x we could, of course, simply write ζ_u, ζ_v in (3.10) instead of $\nabla_u^M \zeta$, $\nabla_v^M \zeta$. However, we prefer the latter expression, because it has an intrinsic meaning (ignoring the fact that we have used the extrinsic embedding in $\mathbb{R}^m$, in order to define tangential vector fields of class $W^{1,2}$ along x). Note that in (3.9) we can equivalently use smooth $\xi \in \mathcal{D}(U, \mathbb{R}^m)$ while in (3.10) we cannot, in general, assume that ζ is smooth (because $\zeta = (\Pi^M \circ x)\zeta$ cannot have a higher degree of regularity than $x \in W^{1,2}(U, M)$). The tension field operator may be expressed in terms of covariant derivatives simply by $\Delta^M x = \nabla_u^M x_u + \nabla_v^M x_v$, and in local coordinates on M as above we have the representation

$$(\nabla_u^M \zeta)^l = \zeta_u^l + (\Gamma_{ij}^l \circ x) x_u^i \zeta^j \tag{3.11}$$

at almost all points $w \in U$ with $x(w)$ in the coordinate domain.

To set up a variational functional with (3.10) as variational equation we need to define the H-volume functional with $x \mapsto (H \circ x)\, x_u \wedge_M x_v$ as its Euler operator. The method of Section 1 works well also in the Riemannian context, i.e. we consider bounded continuous vector fields Z on M with $\mathrm{div}_M Z = H$ or, preferably, the dual 2-form ω on M with $d\omega = H\Omega$, and we define the *volume functional* associated with Z, Ω

$$\mathbf{V}_Z(x) = \int_U (Z \circ x, x_u \wedge_M x_v)_M \, du dv = \int_U x^\# \omega \tag{3.12}$$

for $x \in W^{1,2}(U, M)$. For variations x_t of $x = x_0$ we then have formally, by Stokes' theorem,

$$\begin{aligned} \mathbf{V}_Z(x_t) - \mathbf{V}_Z(x) &= \int_{[0,t] \times U} h^\#(H\Omega) \\ &= \int_U \int_0^t H \circ h \langle \Omega \circ h, h_s \wedge h_u \wedge h_v \rangle \, ds du dv \,, \end{aligned} \tag{3.13}$$

where $h(s, u, v) = x_s(u, v)$. Taking here the derivative and using also $\langle \Omega \circ x, \zeta \wedge x_u \wedge x_v \rangle = (\zeta, x_u \wedge_M x_v)_M$ for $\zeta = \frac{d}{dt}\big|_0 x_t$ we see that $\mathbf{V}_Z$ formally has the desired Euler equation. (We will be more rigorous later on.) The *energy functional* associated with Z, ω is then

$$\mathbf{E}_Z(x) = \mathbf{D}(x) + 2\mathbf{V}_Z(x) \,, \tag{3.14}$$

and it has the H-surface equation as its formal Euler equation. This definition of the energy was used in [HK], [Gu1] and also in recent work of Toda [Tod1] who treated closed H-surfaces in 3-manifolds. It has been noted [Grü2], [Jo3] that (3.14) is, in a sense, the most general conformally invariant variational functional for parametric surfaces in a 3-dimensional space.

Noting that $2|(Z\circ x, x_u\wedge_M x_v)| \le \frac{c}{2}(|x_u|^2+|x_v|^2)$ if $2\sup|Z|\le c$, we can repeat the proof of Proposition 1.3 verbatim to obtain

3.1. Proposition. *If A is closed in M, $\sup_A|Z|<\frac{1}{2}$, and $\mathcal{S}(\Gamma,A)\ne\emptyset$, then $\mathbf{E}_Z$ attains a minimum on $\mathcal{S}(\Gamma,A)$.*

Here $\mathcal{S}(\Gamma,A)$ is defined as before, i.e. the set of surfaces $x\in W^{1,2}(U,M)$ with values in $A\subset M$ (almost everywhere on U). In order to derive the weak H-surface-equation for $\mathbf{E}_Z$ minimizers $x\in\mathcal{S}(\Gamma,A)$ we have to consider now variations x_t of $x=x_0$ within the manifold M. Various constructions have been used for this in the theory of harmonic mappings, e.g.

$$x_t(w)=\exp^M_{x(w)}[t\eta(w)\vartheta(x(w))\xi(w)] \tag{3.15}$$

with $\eta\in\mathcal{D}(U)$, $\vartheta\in\mathcal{D}(M)$, and ξ a bounded tangential vector field along x of class $W^{1,2}$, or

$$x_t(w)=\Psi(t\eta(w),x(w))\,, \tag{3.16}$$

where $\eta\in\mathcal{D}(U)$ and Ψ is the flow of a compactly supported smooth vector field Y on M, or, using the extrinsic space $\mathbb{R}^m$,

$$x_t(w)=\pi^M\,[x(w)+t\vartheta(x(w))\xi(w)] \tag{3.17}$$

where $\vartheta\in\mathcal{D}(M)$, $\xi\in\mathcal{D}(U,\mathbb{R}^m)$, and π^M denotes the nearest point retraction of some neighborhood of M in $\mathbb{R}^m$ onto M.

For such variations (3.13) can be verified by approximation whenever $\operatorname{div}_M Z=H$, i.e. $d\omega=H\Omega$, holds in the sense of distributions on a neighborhood of A. (One first considers a smooth 2-form ω on $\mathbb{R}^m$ and smooth variations of smooth surfaces $\overline{U}\to M$. Then equation (3.13) holds, with $d\omega$ instead of $H\Omega$, by Stokes' theorem. It is next verified by approximation for variations x_t as above, and finally for continuous ω with a further approximation.) Furthermore, differentiation of the integral in (3.13) is justified for the variations considered here, and

$$\begin{aligned}\frac{d}{dt}\Big|_0\mathbf{V}_Z(x_t)&=\int_U H\circ x\,\langle\Omega\circ x,\zeta\wedge x_u\wedge x_v\rangle\,dudv\\&=\int_U(H\circ x)(\zeta,x_u\wedge_M x_v)_M\,dudv\end{aligned} \tag{3.18}$$

follows for the initial vector field $\zeta=\frac{d}{dt}\big|_0 x_t$. The variation of Dirichlet's integral is computed using the rules of covariant differentiation or the Euclidean

Dirichlet integral for $\mathbb{R}^m$ valued mappings, and the result is

$$\frac{d}{dt}\Big|_0 \mathbf{D}(x_t) = \int_U \left(x_u \bullet \nabla_u^M \zeta + x_v \bullet \nabla_v^M \zeta\right) \, dudv \,. \tag{3.19}$$

From (3.18), (3.19) we see that (3.10) holds for ζ if $\frac{d}{dt}\big|_o \mathbf{E}_Z(x_t) = 0$. If (3.10) is satified for all variations of one of the types above we can deduce the weak H-surface-equation for x, because we get enough initial fields ζ to approximate suitably the general vector fields allowed in (3.10) by linear combinations ($\zeta = \eta(\vartheta \circ x)\xi$ in the case (3.15), $\zeta = \eta(Y \circ x)$ in the case (3.16), $\zeta = (\vartheta \circ x)(\Pi^M \circ x)\xi$ in the case (3.17)).

With regard to variations of the independent variables we note that

$$\mathbf{V}_Z(x) = \int_U x^\# \omega$$

is invariant with respect to orientation preserving self-diffeomorphisms of U, and the proof of Proposition 1.1 (i) remains valid, i.e. we have conformality of $\mathbf{E}_Z$ minimizers in $\mathcal{S}(\Gamma, A)$. Furthermore, if A is the closure of a C^2 domain in M, then the statements of Proposition 1.7 are valid in the present situation if as variational inequality we require $\frac{d}{dt}\big|_o \mathbf{E}_Z(x_t) \geq 0$ whenever $x_t \in \mathcal{S}(\Gamma, A)$ for $0 < t \ll 1$ and the variation is of the type above. (We may use any of the types (3.15)–(3.17); it seems, however, that (3.16) is most convenient to deal with the variational inequality, cf. [Du1] where the Riemannian Dirichlet integral is treated, and [DS4] where surfaces in M are represented by integer multiplicity rectifiable currents and $H \not\equiv 0$ is allowed.) We thus have the following existence theorem analogous to Theorem 1.8:

3.2. Theorem. *Suppose A is the closure of a C^2 domain in M, the prescribed mean curvature H and the (inward) boundary mean curvature $H_{\partial A}$ of ∂A relative to M satisfy $|H| \leq H_{\partial A}$ along ∂A, and there exists a continuous vector field Z on M with $\sup_A |Z|_M < \frac{1}{2}$ and $\operatorname{div}_M Z = H$ in the distributional sense on a neighborhood of A. Then, for every Jordan curve Γ which is contractible in A the Plateau problem $\mathcal{P}_M(H, \Gamma)$ has a weak solution $x \in W^{1,2}(U, M)$ with values in A. Moreover, if $|H(a)| < H_{\partial A}(a)$ holds at a point $a \in (\partial A) \setminus \Gamma$, then each such solution omits a neighborhood of this point.*

For concrete applications of this theorem it remains to construct vector fields Z with $\operatorname{div}_M Z = H$ and $\sup_A |Z|_M < \frac{1}{2}$. The simplest way to do this is to consider a *geodesically star shaped domain* D, by which we mean the diffeomorphic image under the exponential map of a domain in some tangent space T_aM with the origin as star point. One then defines Z by integration along rays analogous to (1.4). This is what Hildebrandt & Kaul [HK] and Gulliver [Gu1] did. The pull back $\overline{Z}$ of Z by $\exp_a^M$ is given by the formula

$$\overline{Z}(\tau) = \left[\int_0^1 \overline{H}(s\tau)J(s\tau)\,ds\right]\tau \quad \text{for } \tau \in T_aM, \tag{3.20}$$

where $\overline{H} = H \circ \exp_a^M$ and $J(s\tau)$ is the Jacobian of $\exp_a^M$ at $s\tau$ (with respect to the constant metric $g(a)$ on T_aM). For the dual 2-form ω of Z equation (3.20) reads $\overline{\omega}(\tau) = \int_0^1 s^2\overline{H}(s\tau)\overline{\Omega}(s\tau) \llcorner \tau\, ds$, and the familiar Poincaré Lemma tells us $d\overline{\omega} = \overline{H}\,\overline{\Omega}$, i.e. $\operatorname{div}_M Z = H$ (note $\overline{\Omega}(s\tau) = J(s\tau)\Omega(a)$).

If the sectional curvature of M satisfies $\operatorname{Sec}(M) \leq \kappa^2$, where κ is imaginary or real, then we can apply the comparison theorem for Jacobians [Gü], [GHL, 3.101] which tells us that $J(s\tau)^{-1}\frac{d}{ds}J(s\tau)$ does not exceed the corresponding quantity for the space M_κ of constant sectional curvature κ^2, i.e. the number $2\kappa|\tau|_M \cot(\kappa s|\tau|_M)$. (Here we assume $|\kappa|s|\tau|_M < \frac{\pi}{2}$ in the case $\kappa \in \mathbb{R}$.) One may integrate this as in [Gu1] to obtain a bound for $J(s\tau)$ and to deduce $\sup_D |Z|_M < \frac{1}{2}$, whenever $\sup_D |H| \leq \kappa \cot \kappa R$ and the geodesics in D emanating from a have at most length R. Since $J(s\tau)$ is also the Jacobian of the restriction of $\exp_a^M$ to the sphere of radius $s|\tau|_M$ in T_aM, we see that the quantity $J(s\tau)^{-1}\frac{d}{ds}J(s\tau)$ has a nice geometric interpretation: Up to a factor $|\tau|_M$ it describes the ratio of initial area change and volume change of an infinitesimal piece of a geodesic sphere with center a under a radial deformation, i.e. (twice) the mean curvature of such a sphere at the point $\exp_a^M(s\tau)$. The comparison theorem says that this mean curvature is not smaller than the mean curvature of a sphere of radius $s|\tau|_M$ in M_κ, which is $\kappa\cot(\kappa s|\tau|_M)$. With these observations and with Propositions 3.1, 3.2 we have proved

3.3. Theorem (Gulliver, Hildebrandt & Kaul). *Suppose D is a geodesically star shaped domain in M with respect to the point $a \in M$, A is the closure of a C^2 domain in D, geodesic arcs connecting a in D to a point in A have length $\leq R$,* $\operatorname{Sec}(M) \leq \kappa^2$, *and H satisfies the two conditions*

$$\sup_A |H| \leq \kappa \cot \kappa R, \tag{3.21}$$

$$|H| \leq H_{\partial A} \quad \textit{pointwise on } \partial A. \tag{3.22}$$

Then the conclusions of Proposition 3.2 concerning the solvability of the Plateau problem $\mathcal{P}_M(H,\Gamma)$ in A are valid. Moreover, condition (3.22) *on H is redundant, if A itself is a closed geodesic ball in M with center a.*

This was proved by Gulliver [Gu1] for the particular case where A is a geodesic ball. Gulliver's arguments show in fact that $\kappa \cot \kappa R$ in (3.21) can be replaced by a computable larger constant. Hildebrandt & Kaul [HK] have considered sets A which are star shaped with respect to $a \in M$ and they have proved the theorem under conditions on $\sup_A |H|$ somewhat more restrictive than (3.21). In the case $\kappa = 0$ we must interprete $\kappa \cot \kappa R = R^{-1}$, of course, and we recover Hildebrandt's theorem (cf. Theorem 1.5). The injectivity assumption on the exponential map, which is implicit in the hypotheses of the preceding theorem, can be dropped by pulling back the metric to the tangent space $T_a M$ (cf. [Gu1, §6]). Note that in the case $\kappa^2 < 0$ the theorem implies existence of an H-surface with boundary Γ in M whenever $\sup_M |H| \leq |\kappa|$; no condition on Γ is needed for this (except contractibility)!

The reasoning so far is limited to boundary curves Γ contained in a normal coordinate domain in M, and in general it will not be possible to construct Z with $\operatorname{div}_M Z = H$ and $|Z|_M < \frac{1}{2}$ on M under a smallness condition on $|H|$. This is so even when M is compact, because then $\operatorname{div}_M Z = H$ is impossible for constant $H \neq 0$, by Stokes' theorem. We therefore attempt to extend the method of Section 2 to the Riemannian case. There is no problem with the definition of the 2-current J_x associated with $x \in W^{1,2}(U, M)$ by $J_x(\beta) = \int_U x^\# \beta$ for $\beta \in \mathcal{D}^2(M)$. J_x is an (integer multiplicity) rectifiable 2-current on M. Using the isometric embedding $M \subset \mathbb{R}^m$ we may also consider J_x as a rectifiable 2-current on $\mathbb{R}^m$ which is supported by a 2-rectifiable set contained in M. For x, $y \in \mathcal{S}(\Gamma, M)$ we have again $\partial(J_x - J_y) = 0$, but difficulties arise when we try to find the 3-current $I_{x,y}$ on M with $\partial I_{x,y} = J_x - J_y$, which is needed for the definition of the H-volume $\mathbf{V}_H(x, y)$. In the first place, such a current need not exist at all, i.e. $J_x - J_y$ could represent a nontrivial 2-dimensional integral homology class of M (see [Fe, §4.4]). It would not help much to restrict ourselves to surfaces $x \in \mathcal{S}(\Gamma, M)$ which are homologous to a fixed surface $y_\Gamma \in \mathcal{S}(\Gamma, M)$, because the "splitting off of bubbles" in the limit of a minimizing sequence, described before Lemma 2.2, may carry away not only volume but also homology. (Simple examples show that this phenomenon can actually occur for weakly convergent sequences in $W^{1,2}(U, M)$.) Secondly, if $I_{x,y}$ does exist, there will be a uniqueness problem, since we can add integer multiples of $[\![M]\!]$, the 3-current defined by integration of 3-forms over M (assuming that M has finite volume). This indeterminacy does not matter, if H has mean value zero on M, but the latter condition excludes constant $H \neq 0$ which we find undesirable. (Toda [Tod2] has shown that the volume can be defined without ambiguity on a homotopy class of closed surfaces, provided M has nonpositive curvature. In our case we can then simply lift the contractible curve Γ to the universal cover $\tilde{M}$ of M and work in $\tilde{M}$ where all the problems have disappeared.)

To proceed we will therefore impose a bound $\frac{1}{2}s \leq \infty$ on the Dirichlet integal, i.e. we work with $x \in \mathcal{S}(\Gamma, A)$ satisfying $\mathbf{D}(x) \leq \frac{1}{2}s$, and we henceforth make the *assumption* that each closed rectifiable 2-current T on M with support in A and mass $\mathbf{M}(T) \leq s$ is the boundary $T = \partial Q$ of some rectifiable 3-current Q on M with support in A. (When saying "rectifiable" we imply integer multiplicity and finite mass, but not necessarily compact support.) The assumption can be verified [DS4, §2] for small $s > 0$, if A is a quasiregular closed subset of M as defined in Section 1, i.e. uniformly locally biLipschitz equivalent to convex sets in $\mathbb{R}^3$. For the case $M = A$ noncompact, this uniformity condition is also known as *homogeneous regularity* of M in the sense of Morrey. For a Cartan-Hadamard manifold M (i.e. simply connected with sectional curvature $\mathrm{Sec}(M) \leq 0$, hence M diffeomorphic with $\mathbb{R}^3$) the assumption is true with $s = \infty$.

We now distinguish two cases: The *closed case* is $M = A$ compact and will be treated later. In the *non-closed case* to be considered next, we assume that the rectifiable 3-current Q above is unique. This is true, for example, if M has infinite volume $|M|$, because two such 3-currents differ only by an integer multiple of $[\![M]\!]$, and the mass $\mathbf{M}([\![M]\!]) = |M|$ is infinite now. It is also true when A is a proper closed subset of M, whether or not M is compact. In the non-closed case we thus have a unique rectifiable 3-current $I_{x,y}$ on M with support in A and $\partial I_{x,y} = J_x - J_y$, whenever x, $y \in \mathcal{S}(\Gamma, A)$ with $\mathbf{M}(J_x - J_y) \leq s$. The latter inequality is implied by $\mathbf{D}(x) + \mathbf{D}(y) \leq s$, because (2.12) and (2.13) are valid. $I_{x,y}$ may be viewed as the "set with integer multiplicities" in A whose boundary is formed by x and y, and we have the representation by an integer valued function $i_{x,y} \in L^1(M)$ with support in A and with distributional gradient of total variation $\mathbf{M}(J_x - J_y)$. We then define the *H-volume*

$$\mathbf{V}_H(x,y) = \int_A i_{x,y} H\Omega \quad \text{for } x,\, y \in \mathcal{S}(\Gamma, A),\ \mathbf{M}(J_x - J_y) \leq s, \tag{3.23}$$

and, fixing y_Γ in $\mathcal{S}(\Gamma, A)$ with minimal Dirichlet energy $a_\Gamma = \mathbf{D}(y_\Gamma)$ (existence of y_Γ follows from Proposition 3.1), also the H-energy

$$\mathbf{E}_H(x) = \mathbf{D}(x) + 2\mathbf{V}_H(x, y_\Gamma) \quad \text{for } x \in \mathcal{S}(\Gamma, A),\ \mathbf{M}(J_x - J_{y_\Gamma}) \leq s. \tag{3.24}$$

Assuming $a_\Gamma < \frac{1}{2}s$ we see that $\mathbf{E}_H(x)$ is defined for $x \in \mathcal{S}(\Gamma, A)$ with $\mathbf{D}(x) \leq \sigma a_\Gamma$ whenever $\sigma > 1$ satisfies $(1 + \sigma)a_\Gamma \leq s$.

The definition of an *isoperimetric condition* of type c, s for H on A in the present situation is the same as in Definition 2.3, except that we integrate with respect to the Riemannian measure and we additionally require our general assumption above. For the H-volume we then have, as before in (2.26),

$$2|\mathbf{V}_H(x,y)| \le c\mathbf{M}(J_x - J_y) \quad \text{for } x,\, y \in \mathcal{S}(\Gamma, A),\ \mathbf{M}(J_x - J_y) \le s\ . \tag{3.25}$$

The proof of Theorem 2.4 now can be repeated in the present situation almost verbatim. For the existence we only have to observe that Lemma 2.2 holds for $x_n \in W^{1,2}(U, \mathbb{R}^m)$, and the $\tilde{x}_n$ obtained there can be deformed to surfaces with values in $A \subset M$ by first applying a C^1 neighborhood retraction in $\mathbb{R}^m$ onto M, then another C^1 neighborhood retraction in M onto A. For the first variation of the H-volume we verify that (3.13) now holds for $\mathbf{V}_H(x_t, x)$ instead of $\mathbf{V}_Z(x_t) - \mathbf{V}_Z(x)$. In fact, it suffices to check this for variations of the type (3.15), (3.16), or (3.17) which are chosen to deform the values of x only in a normal coordinate domain on M, and then we can choose Z with $\operatorname{div}_M Z = H$ on this domain and verify $\mathbf{V}_H(x_t, x) = \mathbf{V}_Z(x_t) + \text{const}$. With these considerations one proves (see [DS5]):

3.4. Theorem. *Suppose A is the closure of a C^2 domain in M with infinite volume or $A \ne M$, $0 < s \le \infty$, every closed rectifiable 2-current in A of mass $\le s$ is homologous to zero in A, $a_\Gamma < \infty$ is the least area spanned by Γ in A, and $a_\Gamma < \frac{1}{2}s$. Assume further that $|H|$ does nowhere on ∂A exceed the inward mean boundary curvature $H_{\partial A}$ and that an isoperimetric condition of type c, s is valid for H on A with*

$$s = \infty\,, \quad c < 1 \quad \text{or} \quad s < \infty\,, \quad c < \frac{s - 2a_\Gamma}{s}\,. \tag{3.26}$$

Then there exists a weak solution x to the Plateau problem $\mathcal{P}_M(H, \Gamma)$ in A and the conclusions of Proposition 3.2 hold for x.

When writing $a_\Gamma < \infty$ we understand that $\mathcal{S}(\Gamma, A)$ is non empty and $a_\Gamma = \mathbf{D}(y_\Gamma)$ for some minimizer y_Γ of Dirichlet's integral on $\mathcal{S}(\Gamma, A)$. In [DS4] we have studied isoperimetric conditions in Riemannian manifolds. One result is that if A is compact, $A \ne M$, then for every s a linear isoperimetric inequality $|E| \le c_A(s)\mathbf{P}(E)$ is valid for all $E \subset A$ of finite perimeter $\mathbf{P}(E) \le s$, with a finite constant $c_A(s)$. (Here $|E| = \left|\int_E \Omega\right|$ is the volume of E.) This inequality implies an isoperimetric condition for H with $c = 2c_A(s) \sup_A |H|$, hence the

3.5. Corollary. *Suppose $A \subset M$ is the closure of a bounded proper C^2 subdomain of M with nonnegative inward boundary mean curvature $H_{\partial A}$, $0 < s < \infty$, each closed rectifiable 2-current on A with mass $\leq s$ is homologous to zero in A, and $\Gamma \subset A$ has least spanning area $a_\Gamma < \frac{1}{2}s$. Then the Plateau problem $\mathcal{P}_M(H;\Gamma)$ has a weak solution in A, whenever $|H| \leq H_{\partial A}$ holds pointwise on ∂A and*

$$\sup_A |H| \leq \frac{s - 2a_\Gamma}{2sc_A(s)},$$

where $c_A(s)$ is the isoperimetric constant described above.

Assuming $\mathrm{Sec}(M) \leq \kappa^2$ we have also studied in [DS4] isoperimetric conditions on geodesically star shaped domains as in Theorem 3.3. Combining the results of [DS4] with the preceding theorem we find that the constant appearing in (3.21) can be replaced by a larger number. (In the case $\kappa = 0$ this amounts to the replacement of R^{-1} by $\frac{3}{2}R^{-1}$.) We may even add a function to H with not too large L^3-norm, thereby simultaneously improving Theorem 2.5 and extending it to Riemannian manifolds:

3.6. Corollary. *The conclusions of Theorem 3.3 remain valid if (3.21) is replaced by the weaker requirement*

$$\sup_A |H| \leq \frac{\sin^2 \kappa R}{R - \frac{1}{2\kappa}\sin 2\kappa R} =: c(\kappa, R), \tag{3.27}$$

or, in the case $\mathrm{Sec}(M) \leq \kappa^2 \leq 0$, by the still weaker condition

$$\int_{\{a \in A:\, |H_\pm(a)| > c(\kappa, r)\}} |H_\pm|^3 \, d\mathcal{H}^3 < \frac{9\pi}{2}, \tag{3.28}$$

where H_+, H_- are the positive and negative parts of H.

Note $c(\kappa, R) \to |\kappa|$ as $R \to \infty$ in the case $\kappa^2 \leq 0$. Hence the domain of integration in (3.28) can be replaced by $\{a \in A: |H(a)| > |\kappa|\}$ to obtain a condition which is sufficient for all curves Γ, independent of the diameter. With Theorem 3.4 we also obtain a Riemannian analogue of Theorem 1.9; the arguments used in Section 2 to prove this result of Gulliver & Spruck are valid in the Riemannian context as well (cf. [DS4]):

3.7. Corollary. *Theorem 1.9 also holds if $\mathbb{R}^3$ is replaced by the Riemannian manifold M, and we assume in addition that every closed rectifiable 2-current in A is homologous to zero in A.*

Still another result from [DS4] is that if M is compact and quasiregular (homogeneously regular in Morrey's sense), then there exist $s_0 > 0$, $\gamma < \infty$ such that a nonlinear isoperimetric inequality $|E| \le \gamma \mathbf{P}(E)^{3/2}$ is valid for all $E \subset M$ with perimeter $\mathbf{P}(E) \le s_0$. Decreasing s_0, if necessary, we also know that rectifiable closed 2-currents in M of mass $\le s_0$ are boundaries of rectifiable 3-currents, and an isoperimetric condition of type $2\gamma s^{1/2} \sup_M |H|$, s follows for H on M when $0 < s \le s_0$. If we now apply Theorem 3.4, then the second condition for c in (3.26) becomes

$$2\gamma s^{1/2} \sup_M |H| < \frac{s - 2a_\Gamma}{s}.$$

With the optimal choice $s = 6a_\Gamma$ we obtain from this the next corollary which generalizes Wente's theorem to non-closed manifolds. The constant γ can be explicitely determined in many cases. For example, if M is simply connected with $\mathrm{Sec}(M) \le 0$, then γ is the isoperimetric constant $(36\pi)^{-1/2}$ of $\mathbb{R}^3$. (See [Kl] and the discussion in [DS4].)

3.8. Corollary. *If M is non-compact and homogeneously regular, s_0 and γ are as above, and Γ has least spanning area $a_\Gamma \le \frac{1}{6}s_0$, then the Plateau problem $\mathcal{P}_M(H,\Gamma)$ is weakly solvable for all prescribed mean curvature functions H with*

$$\sup_M |H| < \frac{1}{3\gamma\sqrt{6a_\Gamma}}. \tag{3.29}$$

We finally turn to the *closed case*, i.e. M is a compact manifold (and $A = M$). For the definition of the H-volume $\mathbf{V}_H(x,y)$ correspodimg to x, $y \in \mathcal{S}(\Gamma, M)$ with $\mathbf{M}(J_x - J_y) \le s$ it is no longer enough to know that $J_x - J_y$ is the boundary of some rectifiable 3-current on M, because this 3-current will not be unique. Of course, we could circumvent this difficulty by requiring that the mean value of H on M vanishes, and we would then carry out the proof of Theorem 3.4 and its Corollaries exactly as before. However, we do not want to exclude constant $H \ne 0$, hence we must select a particular $I_{x,y}$ in a reasonable way. For this, we note that there can be at most one rectifiable 3-current with given boundary in M which has mass $<\frac{1}{2}|M|$, because any two such currents differ by an integer multiple of $[\![M]\!]$. Therefore, if we now make the *assumption* that every closed rectifiable 2-current of mass $\le s$ in M is the boundary of some 3-current with mass $<\frac{1}{2}|M|$, then we can define $I_{x,y}$ for x, $y \in \mathcal{S}(\Gamma, M)$ with $\mathbf{M}(J_x - J_y) \le s$ by the conditions $\partial I_{x,y} = J_x - J_y$, $\mathbf{M}(I_{x,y}) < \frac{1}{2}|M|$ uniquely. As before, we have the representation of $I_{x,y}$ by an integer valued summable function, and we can use (3.23), (3.24) to define $\mathbf{V}_H(x,y)$ and $\mathbf{E}_H(x)$.

In addition, we will require the following *linear isoperimetric inequality* for sets $E \subset M$ with finite perimeter:

$$|E| \leq c_M(s)\mathbf{P}(E) \quad \text{when } \mathbf{P}(E) \leq s,\ |E| < \tfrac{1}{2}|M| \tag{3.30}$$

with a smallest possible constant $c_M(s)$ satisfying also

$$c_M(s)s < \tfrac{1}{2}|M|\,. \tag{3.31}$$

By the decomposition theorem, such an inequality implies

$$\mathbf{M}(I_{x,y}) \leq c_M(s)\mathbf{M}(J_x - J_y) \leq c_M(s)s < \tfrac{1}{2}|M|\,, \quad \text{if } \mathbf{M}(J_x - J_y) \leq s\ . \tag{3.32}$$

On a compact manifold M the inequality $|E| \leq c_M(\infty)\mathbf{P}(E)$ always holds, with a finite constant $c_M(\infty)$, for all E satisfying $|E| < \frac{1}{2}|M|$. For the standard 3-sphere $M = S^3$ the isoperimetric function $c_M(s)$ is known explicitely and determined by the isoperimetric property of geodesic balls in M. Various estimates are available for $c_M(s)$ in terms of other geometric quantities. We refer to [DS4, §4] for a discussion.

By an *isoperimetric condition* of type c, s for H on a compact manifold M we mean again the condition of Definition 2.3, except that the sets E admitted in (2.25) are now restricted to have volume $|E| < \frac{1}{2}|M|$ and we also require the assumption made above in order to define the currents $I_{x,y}$. It follows from the decomposition theorem that (3.25) is valid, and we can repeat the proof of Theorem 2.4. However, in part (i) of that proof we now encounter a problem, because in deriving the lower semicontinuity of $\mathbf{E}_H$ on the minimizing sequence $x_n \to x$ in $\{y \in \mathcal{S}(\Gamma, M) \colon \mathbf{D}(y) \leq \sigma a_\Gamma\}$ we have used additivity of the volume in the following form:

$$\mathbf{V}_H(\tilde{x}_n, x_n) + \mathbf{V}_H(x_n, y_\Gamma) = \mathbf{V}_H(\tilde{x}_n, x) + \mathbf{V}_H(x, y_\Gamma)\,, \tag{3.33}$$

where the $\tilde{x}_n$ are from Lemma 2.2. The volume additivity does not hold unrestrictedly in the closed case, however, because we can have $\mathbf{M}(I_{x,y}) < \frac{1}{2}|M|$, $\mathbf{M}(I_{y,z}) < \frac{1}{2}|M|$, but $\mathbf{M}(I_{x,y} + I_{y,z}) > \frac{1}{2}|M|$, so that $I_{x,y} + I_{y,z} \neq I_{x,z}$. With regard to the left-hand side of (3.33) we observe $\mathbf{M}(I_{\tilde{x}_n,x_n}) \to 0$ as $n \to \infty$, since $\|\tilde{x}_n - x_n\|_{L^\infty} \to 0$ (cf. the estimate (2.17); here we can use a suitable homotopy from x_n to $\tilde{x}_n$ in M). With (3.32) we infer

$$\mathbf{M}(I_{\tilde{x}_n,x_n} + I_{x_n,y_\Gamma}) \leq \mathbf{M}(I_{\tilde{x}_n,x_n}) + c_M(s)\mathbf{M}(J_{x_n} - J_{y_\Gamma}) < \tfrac{1}{2}|M| \quad \text{for } n \gg 1\,,$$

hence $I_{\tilde{x}_n,x_n} + I_{x_n,y_\Gamma} = I_{\tilde{x}_n,y_\Gamma}$. On the other side we estimate similarly, using also Lemma 2.2:

$$\begin{aligned}\mathbf{M}(I_{\tilde{x}_n,x} + I_{x,y_\Gamma}) &\leq \mathbf{M}(I_{\tilde{x}_n,x}) + \mathbf{M}(I_{x,y_\Gamma})\\ &\leq c_M(s)\mathbf{M}(J_{\tilde{x}_n} - J_x) + c_M(s)\mathbf{M}(J_x - J_{y_\Gamma})\\ &\leq c_M(s)\,[2\varepsilon + \mathbf{D}(x_n) - \mathbf{D}(x) + \mathbf{D}(x) + \mathbf{D}(y_\Gamma)]\\ &\leq c_M(s)(4\varepsilon + s) < \tfrac{1}{2}|M| \quad \text{for } n \gg 1\,,\end{aligned}$$

provided we have chosen ε in Lemma 2.2 small enough. We conclude $I_{\tilde{x}_n,x} + I_{x,y_\Gamma} = I_{\tilde{x}_n,y_\Gamma}$ and hence (3.33) holds for large n. (Of course, when using (3.32) on $I_{y,z}$ we must verify $\mathbf{M}(J_y - J_z) \leq s$ first. This can be done in all the instances where we have applied (3.32) in the preceding calculation, using $\mathbf{D}(y_\Gamma) = a_\Gamma$, $\mathbf{D}(x_n) \leq \sigma a_\Gamma$, $\mathbf{D}(x) \leq \sigma a_\Gamma$, $\mathbf{D}_G(\tilde{x}_n) + \mathbf{D}_G(x) \leq 2\varepsilon + (\sigma-1)a_\Gamma$, and $(1+\sigma)a_\Gamma \leq s$. These inequalities are obtained in the proof of Theorem 2.4 and in Lemma 2.2.)

The rest of the proof is the same as in Section 2. For the variational equation we note that $\mathbf{V}_H(x_t, y_\Gamma) = \mathbf{V}_H(x_t, x) + \mathbf{V}_H(x, y_\Gamma)$ holds for $|t| \ll 1$, since x_t tends to x uniformly and hence $\mathbf{M}(I_{x_t,x}) \to 0$ as $t \to 0$, and for $\mathbf{V}_H(x_t, x)$ the representation (3.13) is again valid. In this way we have proved the following theorem (see also [DS5]):

3.9. Theorem. *Suppose M is compact $0 < s < \infty$, the linear isoperimetric inequality (3.30) holds in M with*

$$c_M(s)s < \tfrac{1}{2}|M|\,, \tag{3.34}$$

every rectifiable closed 2-current of mass $\leq s$ on M is homologous to zero, and Γ has least spanning area $a_\Gamma < \frac{1}{2}s$. Then, if H satisfies an isoperimetric condition of type c, s in M with

$$c < \frac{s - 2a_\Gamma}{s}\,, \tag{3.35}$$

the Plateau problem $\mathcal{P}_M(H, \Gamma)$ has a weak solution.

One possibility to verify an isoperimetric condition for H is, of course, to use the isoperimetric inequality (3.32). This gives $c = 2c_M(s)\sup_M |H|$, and hence the

3.10. Corollary. *The conclusion of Theorem 3.9 is valid whenever H satisfies*

$$\sup_M |H| < \frac{s - 2a_\Gamma}{2sc_M(s)}\,. \tag{3.36}$$

We can also see from Theorem 3.9 that Γ bounds H-surfaces in M for large $|H|$ if a_Γ is small. For this we note that all the assumptions on M, s are satisfied for $s \le s_0 \ll 1$, and that $\lim_{s\downarrow 0} c_M(s) = 0$. Actually a nonlinear isoperimetric inequality $|E| \le \gamma \mathbf{P}(E)^{3/2}$ is valid with a finite constant γ for all $E \subset M$ with $|E| < \frac{1}{2}|M|$, so that $c_M(s) \le \gamma s^{1/2}$ follows (see [DS4]). Using this we get an isoperimetric condition for H with $c = 2\gamma s^{1/2} \sup_M |H|$, and we thus have obtained another

3.11. Corollary. *The assertions in Corollary 3.8 are valid also when M is a compact manifold.*

As an *example* we consider the standard sphere $M = S^3$. Then the hypothesis (3.34) in Theorem 3.9 is equivalent with $s < 4\pi$, and the condition $a_\Gamma < \frac{1}{2}s$ says that the least spanning area of Γ should be smaller than the area of a half great sphere in S^3. If this is satisfied, then Corollary 3.10 shows that we can solve $\mathcal{P}_M(H,\Gamma)$ for $\sup_M |H|$ sufficiently small. On the other hand, Gulliver [Gu3] has proved that $\mathcal{P}_M(H,\Gamma)$ has no solution for constant $H \ne 0$ if Γ is a great circle in S^3. Hence Corollary 3.10 is sharp with respect to the conditions imposed on Γ to imply solvability of $\mathcal{P}_M(H,\Gamma)$ for sufficiently small $|H|$. Hovever, the Corollary is most likely not optimal with respect to the bound required for $\sup_M |H|$. If one optimizes (3.36) using the explicitely known form of $c_M(s)$ for $M = S^3$, one finds the following sufficient condition for existence ([DS4]):

$$\sup_{S^3} |H| < \cot r\,, \tag{3.37}$$

where r is the solution of the equation

$$r \cot r = 1 + \tfrac{1}{2\pi} a_\Gamma\,, \quad \arcsin \sqrt{\tfrac{1}{2\pi} a_\Gamma} < r < \tfrac{\pi}{2}\,.$$

One can see that a stronger condition than (3.37) is

$$\sup_{S^3} |H| < \tfrac{1}{\pi^3}\left(1 - \tfrac{1}{2\pi} a_\Gamma\right), \tag{3.38}$$

but for a_Γ small this is much more restrictive than necessary.

We conclude with some brief comments on the *regularity* of the energy minimizing weak solutions x in $\mathcal{S}(\Gamma, A) \subset W^{1,2}(U,M)$ to the Plateau problem $\mathcal{P}_M(H,\Gamma)$. First of all, Morrey's method can be applied exactly as in Section 2 to prove Hölder continuity of x on U and continuity on $\overline{U}$, provided A is quasiregular. With this, the problem of higher regularity can be localized in the manifold M, and one can work with coordinates and the equations (3.5) (in the distributional sense) and (3.6). Interior $C^{k+2,\beta}$ regularity was proved for x in [HK], [Hi8], assuming H of class $C^{k,\beta}$; the simplest method is [Grü3]. Boundary regularity was estabilished by Heinz & Hildebrandt [HH2]. They showed how to deal with the conformality relations in the Riemannian form (3.6) using the earlier regularity theorems of Heinz mentioned at the end of Section 1. Their reasoning can also be applied to H-surfaces in a 3-manifold, because the extra terms appearing in (3.5) when $H \neq 0$ are of the same quadratic nature in x_u, x_v as the terms which come from the Levi-Civita connection and are already present in the case $H \equiv 0$ (see also [HDKW, Chap.7]).

Hildebrandt & Heinz [HH2] have also given an asymptotic expansion near a branch point of x in $\overline{U}$ (a point where x_u, x_v vanish), from which it follows that branch points are isolated and that $x\big|_{\partial U}$ is a homeomorphism of ∂U onto Γ. Interior branch points were ruled out for energy minimizing H-surfaces in Riemannian manifolds by Gulliver [Gu2]. (He used the energy functional $\mathbf{E}_Z$ with $\sup_M |Z| < \frac{1}{2}$, but the isoperimetric conditions above are in fact sufficient; cf. [SW]). Except for the case of minimal surfaces in an analytic Riemannian 3-manifold [GL] it is not known, however, whether there can exist a (true) boundary branch point in an energy minimizing solution to the Plateau problem $\mathcal{P}_M(H, \Gamma)$.

References.

[Al1] Alt, H.W., *Verzweigungspunkte von H–Flächen, I*, Math. Z. **127** (1972), 333–362.

[Al2] Alt, H.W., *Verzweigungspunkte von H–Flächen, II*, Math. Ann. **201** (1973), 33–55.

[Ba] Barbosa, J.L., *Constant mean curvature surfaces bounded by a planar curve*, Matematica Contemporanea **1** (1991), 3–15.

[BJ] Barbosa, J.L., Jorge, L.P., *Stable H-surfaces spanning $S^1(1)$*, Ann. Acad. Bras. Ci. **66** (1994).

[Be] Bethuel, F., *Un résultat de régularité pour les solutions d l'équation des surfaces à courbure moyenne prescrite*, C.R. Acad. Sci. Paris **314** (1992), 1003–1007.

[BG] Bethuel, F., Ghidaglia, J.M., *Improved regularity of solutions to elliptic equations involving Jacobians and applications*, J. Math. Pures et Appliquées **72** (1993), 441–474.

[BR] Bethuel, F., Rey, O., *Multiple solutions to the Plateau problem for nonconstant mean curvature*, Duke Math. J. **73** (1994), 593–646.

[BC] Brézis, H.R., Coron, M., *Multiple solutions of H–systems and Rellich's conjecture*, Commun. Pure Appl. Math. **37** (1984), 149–187.

[BE] Brito, F., Earp, R., *Geometric configurations of constant mean curvature surfaces with planar boundary*, An. Acad. Bras. Ci. **63** (1991), 5–19.

[Di1] Dierkes, U., *Plateau's problem for surfaces of prescribed mean curvature in given regions*, Manuscr. Math. **56** (1986), 313–331.

[Di2] Dierkes, U., *A geometric maximum principle for surfaces of prescribed mean curvature in Riemannian manifolds*, Z. Anal. Anwend. **8** (2) (1989), 97–102.

[Du1] Duzaar, F., *Variational inequalities and harmonic mappings*, J. Reine Angew. Math. **374** (1987), 39–60.

[Du2] Duzaar, F., *On the existence of surfaces with prescribed mean curvature and boundary in higher dimensions*, Ann. Inst. Henri Poincaré (Anal. Non Lineaire) **10** (1993), 191–214.

[Du3] Duzaar, F., *Boundary regularity for area minimizing currents with prescribed volume*, To appear in J. Geometric Analysis 1996.

[DF1] Duzaar, F., Fuchs, M., *On the existence of integral currents with prescribed mean curvature vector*, Manuscr. Math. **67** (1990), 41–67.

[DF2] Duzaar, F., Fuchs, M., *A general existence theorem for integral currents with prescribed mean curvature form*, Bolletino U.M.I. (7) **6**–B (1992), 901–912.

[DS1] Duzaar, F., Steffen, K., *Area minimizing hypersurfaces with prescribed volume and boundary*, Math. Z. **209** (1992), 581–618.

[DS2] Duzaar, F., Steffen, K., *λ minimizing currents*, Manuscr. Math. **80** (1993), 403–447.

[DS3] Duzaar, F., Steffen, K., *Boundary regularity for minimizing currents with prescribed mean curvature*, Calc. Var. **1** (1993), 355–406.

[DS4] Duzaar, F., Steffen, K., *Existence of hypersurfaces with prescribed mean curvature in Riemannian mannifolds*, Preprint, SFB 256 Bonn, 1995.

[DS5] Duzaar, F., Steffen, K., *Parametric surfaces of least H-energy in a Riemannian manifold*, In preparation.

[EBMS] Earp, R., Brito, F., Meeks III, W.H., Rosenberg, H., *Structure theorems for constant mean curvature spaces bounded by a planar curve*, Indiana Univ. Math. J. **40** (1991), 333–343.

[EL] Eells, J., Lemaire, L., *A report on harmonic maps*, Bull. London Math. Soc. **10** (1978), 1–68. Another report on harmonic maps. Bull. London Math. Soc. **20** (1988), 385–542.

[EG] Evans, L.C., Gariepy, L.F., *Measure theory and fine properties of functions*, CRC Press, Boca Raton Ann Arbor London, 1992.

[Fe] Federer, H., *Geometric measure theory*, Springer–Verlag, Berlin Heidelberg New York, 1969.

[GHL] Gallot, S., Hulin, D., Lafontaine, J., *Riemannian geometry*, 2nd edition. Springer–Verlag, Berlin Heidelberg New York, 1990.

[Grü1] Grüter, M., *Regularity of weak H–surfaces*, J. Reine Angew. Math. **329** (1981), 1–15.

[Grü2] Grüter, M., *Conformally invariant variational integrals and the removability of isolated singularities*, Manuscr. Math. **47** (1984), 85–104.

[Grü3] Grüter, M., *Eine Bemerkung zur Regularität stationärer Punkte von konform invarianten Variationsintegralen*, Manuscr. Math. **55** (1986), 451–453.

[Gü] Günther, P., *Einige Vergleichssätze über das Volumenelement eines Riemannschen Raumes*, Publ. Math. Debrecen **7** (1960), 258–287.

[Gu1] Gulliver, R., *The Plateau problem for surfaces of prescribed mean curvature in a Riemannian manifold*, J. Differ. Geom. **8** (1973), 317–330.

[Gu2] Gulliver, R., *Regularity of minimizing surfaces of prescribed mean curvature*, Ann. Math. **97** (1973), 275–305.

[Gu3] Gulliver, R., *On the non–existence of a hypersurface of prescribed mean curvature with a given boundary*, Manuscr. Math. **11** (1974),15–39.

[Gu4] Gulliver, R., *Necessary conditions for submanifolds and currents with prescribed mean curvature vector*, Seminar on minimal submanifolds, ed. E. Bombieri, Princeton, 1983.

[Gu5] Gulliver, R., *Branched immersions of surfaces and reduction of topological type*, I. Math. Z. **145** (1975), 267–288.

[Gu6] Gulliver, R., *Branched immersions of surfaces and reduction of topological type*, II. Math. Ann. **230** (1977), 25–48.

[Gu7] Gulliver, R., *A minimal surface with an atypical boundary branch point*, Preprint CMA- R38-88, ANU Canberra (1988), 0–19.

[GL] Gulliver, R., Lesley, F.D., *On boundary branch points of minimizing surfaces*, Arch. Ration. Mech. Anal. **52** (1973), 20–25.

[GOR] Gulliver, R., Osserman, R., Royden, H.L., *A theory of branched immersions of surfaces*, Am. J. Math. **95** (1973), 750–812.

[GS1] Gulliver, R., Spruck, J., *The Plateau problem for surfaces of prescribed mean curvature in a cylinder*, Invent. math. **13** (1971), 169–178.

[GS2] Gulliver, R., Spruck, J., *Surfaces of constant mean curvature which have a simple projection*, Math. Z. **129** (1972), 95–107.

[GS3] Gulliver, R., Spruck, J., *Existence theorems for parametric surfaces of prescribed mean curvature*, Indiana Univ. Math. J. **22** (1972), 445–472.

[HW] Hartmann, P., Winter, A., *On the local behaviour of solutions of non-parabolic partial differential equations*, Amer. J. Math. **75** (1953), 449-476.

[He1] Heinz, E., *Über die Existenz einer Fläche konstanter mittlerer Krümmung mit gegebener Berandung*, Math. Ann. **127** (1954), 258–287.

[He2] Heinz, E., *On the non-existence of a surface of constant mean curvature with finite area and prescribed rectifiable boundary*, Arch. Rat. Mech. Anal. **35** (1969), 249–252.

[He3] Heinz, E., *Ein Regularitätssatz für Flächen beschränkter mittlerer Krümmung*, Nachr. Akad. Wiss. Gött., II. Math.-Phys. Kl. (1969), 107–118.

[He4] Heinz, E., *Über das Randverhalten quasilinearer elliptischer Systeme mit isothermen Parametern*, Math. Z. **113** (1970), 99–105.

[He5] Heinz, E., *Unstable surfaces of constant mean curvature*, Arch. Ration. Mech. Anal. **38** (1970), 257–267.

[He6] Heinz, E., *Ein Regularitätssatz für schwache Lösungen nichtlinearer elliptischer Systeme*, Nachr. Akad. Wiss. Gött., II. Math.-Phys. Kl. (1975), 1–13.

[He7] Heinz., E., *Über die Regularität schwacher Lösungen nichtlinarer elliptischer Systeme*, Nachr. Akad. Wiss. Gött. II, Math.-Phys. Kl. (1985), 1–15.

[HH1] Heinz, E., Hildebrandt, S., *Some remarks on minimal surfaces in Riemannian manifolds*, Commun. Pure Appl. Math. **23** (1970), 371–377.

[HH2] Heinz, E., Hildebrandt, S., *On the number of branch points of surfaces of bounded mean curvature*, J. Differ. Geom. **4** (1970), 227–235.

[HT] Heinz, E., Tomi, F., *Zu einem Satz von S*, Hildebrandt über das Randverhalten von Minimalflächen. Math. Z. **111** (1969), 372–386.

[Hi1] Hildebrandt, S., *Boundary behavior of minimal surfaces*, Arch. Ration. Mech. Anal. **35** (1969), 47–82.

[Hi2] Hildebrandt, S., *Über Flächen konstanter mittlerer Krümmung*, Math. Z. **112** (1969), 107–144.

[Hi3] Hildebrandt, S., *On the Plateau problem for surfaces of prescribed mean curvature*, Commun. Pure Appl. Math. **23** (1970), 97–114.

[Hi4] Hildebrandt, S., *Randwertprobleme für Flächen mit vorgeschriebener mittlerer Krümmung und Anwendungen auf die Kapillaritätstheorie I*, Math. Z. **112** (1969), 205–213.

[Hi5] Hildebrandt, S., *Über einen neuen Existenzsatz für Flächen vorgeschriebener mittlerer Krümmung*, Math. Z. **119** (1971), 267–272.

[Hi6] Hildebrandt, S., *Einige Bemerkungen über Flächen beschränkter mittlerer Krümmung*, Math. Z. **115** (1970), 169–178.

[Hi7] Hildebrandt, S., *Maximum principles for minimal surfaces and for surfaces of continuous mean curvature*, Math. Z. **128** (1972), 253–269.

[Hi8] Hildebrandt, S., *On the regularity of solutions of two-dimensional variational problems with obstructions*, Commun. Pure Appl. Math. **25** (1972), 479–496.

[Hi9] Hildebrandt, S., *Interior $C^{1+\alpha}$-regularity of solutions of two-dimensional variational problems with obstacles*, Math. Z. **131** (1973), 233–240.

[HDKW] Dierkes, U., Hildebrandt, S., Küster, A., Wohlrab, O., *Minimal surfaces vol. 1, vol.2*, Grundlehren math. Wiss. 295, 296. Springer, Berlin Heidelbeg New York, 1992.

[HK] Hildebrandt S., Kaul, H., *Two-dimensional variational problems with obstructions, and Plateau's problem for H-surfaces in a Riemannian manifold*, Commun. Pure Appl. Math. **25** (1972), 187–223.

[Jä] Jäger, W., *Das Randverhalten von Flächen beschränkter mittlerer Krümmung bei $C^{1,\alpha}$-Rändern*, Nachr. Akad. Wiss. Gött., II. Math.-Phys. Kl. (1977), 45–54.

[Jo] Jost, J., *Harmonic mappings between Riemannian manifolds*, Proc. CMA, Vol. 4, ANU-Press, Canberra, 1984.

[Jo2] Jost, J., *Lectures on harmonic maps (with applications to conformal mappings and minimal surfaces)*, Lect. Notes Math. **1161**, Springer, Berlin Heidelberg New York (1985), 118–192.

[Jo3] Jost, J., *Two-dimensional geometric variational problems*, Wiley–Interscience, Chichester New York, 1991.

[Kap] Kapouleas, N., *Compact constant mean curvature surfaces in Euclidean three-space*, J. Differ. Geom. **33** (1991), 683–715.

[Kau] Kaul, H., *Ein Einschließungssatz für H–Flächen in Riemannschen Mannigfaltigkeiten*, Manuscr. Math. **5** (1971), 103–112.

[Kl] Kleiner, B., *An isoperimetric comparison theorem*, Invent. math. **108** (1992), 37–47.

[LM] López, S., Montiel, S., *Constant mean curvature discs with bounded area*, Proc. Amer. Math. Soc. **123** (1995), 1555–1558.

[Ni1] Nitsche, J.C.C., *Vorlesungen über Minimalflächen*, Grundlehren math. Wiss., vol. 199. Springer, Berlin Heidelberg New York, 1975.

[Ni2] Nitsche, J.C.C., *Lectures on minimal surfaces, vol 1: Introduction, fundamentals, geometry and basic boundary problems*, Cambridge Univ. Press, 1989.

[Os] Osserman, R., *A proof of the regularity everywhere of the classical solution to Plateau's problem*, Ann. Math. **91** (1970), 550–569.

[Ru] Ruchert, H., *Ein Eindeutigkeitssatz für Flächen konstanter mittlerer Krümmung*, Arch. Math. **33** (1979), 91–104.

[Sa] Sauvigny, F., *Flächen vorgeschriebener mittlerer Krümmung mit eindeutiger Projektion auf eine Ebene*, Math. Z. **180** (1982), 41–67.

[ST] Schüffler, K., Tomi, F., *Ein Indexsatz für Flächen konstanter mittlerer Krümmung*, Math. Z. **182** (1983), 245–258.

[Se] Serrin, J., *The problem of Dirichlet for quasilinear elliptic differential equations in many independent variables*, Phil. Trans. Royal Soc. London **264** (1969), 413–419.

[Si] Simon, L., *Lectures on geometric measure theory*, Proc. CMA, Vol. 3, ANU Canberra, 1983.

[Ste1] Steffen, K., *Flächen konstanter mittlerer Krümmung mit vorgegebenem Volumen oder Flächeninhalt*, Arch. Ration. Mech. Anal. **49** (1972), 99–128.

[Ste2] Steffen, K., *Ein verbesserter Existenzsatz für Flächen konstanter mittlerer Krümmung*, Manuscr. Math. **6** (1972), 105–139.

[Ste3] Steffen, K., *Isoperimetric inequalities and the problem of Plateau*, Math. Ann. **222** (1976), 97–144.

[Ste4] Steffen, K., *On the existence of surfaces with prescribed mean curvature and boundary*, Math. Z. **146** (1976), 113–135.

[Ste5] Steffen, K., *On the nonuniqueness of surfaces with prescribed constant mean curvature spanning a given contour*, Arch. Ration. Mech. Anal. **94** (1986), 101–122.

[SW] Steffen, K., Wente, H., *The non-existence of branch points in solutions to certain classes of Plateau type variational problems*, Math. Z. **163** (1978), 211–238.

[Strö] Ströhmer, G., *Instabile Flächen vorgeschriebener mittlerer Krümmung*, Math. Z. **174** (1980), 119–133.

[Str1] Struwe, M., *Nonuniqueness in the Plateau problem for surfaces of constant mean curvature*, Arch. Ration. Mech. Anal. **93** (1986), 135–157.

[Str2] Struwe, M., *Large H-surfaces via the mountain-pass-lemma*, Math. Ann. **270** (1985), 441–459.

[Str3] Struwe, M., *Plateau's problem and the calculus of variations*, Mathematical notes 35, Princeton University Press, Princeton, New Jersey, 1988.

[Str4] Struwe, M., *Multiple solutions to the Dirichlet problem for the equation of prescribed mean curvature*, Moser-Festschrift, Academic Press, 1990.

[Tod1] Toda, M., *On the existence of H–surfaces into Riemannian manifolds*, Preprint, University of Tokyo, 1994.

[Tod2] Toda, M., *Existence and non–existence results of H–surfaces into 3–dimensional Riemannian maninfolds*, Preprint, University of Tokyo, 1995.

[Tom1] Tomi, F., *Ein einfacher Beweis eines Regularitätssatzes für schwache Lösungen gewisser elliptischer Systeme*, Math. Z. **112** (1969), 214–218.

[Tom2] Tomi, F., *Bemerkungen zum Regularitätsproblem der Gleichung vorgeschriebener mittlerer Krümmung*, Math. Z. **132** (1973), 323–326.

[Wa] Wang, G., *The Dirichlet problem for the equation of prescribed mean curvature*, Ann. Inst. Henri Poincaré (Anal. Non Linéaire) **9** (1992), 643–655.

[Wen1] Wente, H., *An existence theorem for surfaces of constant mean curvature*, J. Math. Anal. Appl. **26** (1969), 318–344.

[Wen2] Wente, H., *A general existence theorem for surfaces of constant mean curvature*, Math. Z. **120** (1971), 277–288.

[Wen3] Wente, H., *An existence theorem for surfaces in equilibrium satisfying a volume constraint*, Arch. Ration. Mech. Anal. **50** (1973), 139–158.

[Wer] Werner, H., *Das Problem von Douglas für Flächen konstanter mittlerer Krümmung*, Math. Ann. **133** (1957), 303–319.

RECEIVED NOVEMBER 10, 1995.

MATHEMATISCHES INSTITUT DER HUMBOLDT-UNIVERSITÄT ZU BERLIN,
UNTER DEN LINDEN 6, D-10099 BERLIN, GERMANY

AND

MATHEMATISCHES INSTITUT DER HEINRICH–HEINE–UNIVERSITÄT DÜSSELDORF
UNIVERSITÄTSTRASSE 1, D-40225 DÜSSELDORF, GERMANY.

Closed Weingarten Hypersurfaces in Space Forms

Claus Gerhardt

Dedicated to Stefan Hildebrandt on the occasion of his sixtieth birthday

0. Introduction.

In a complete $(n+1)$-dimensional manifold N we want to find closed hypersurfaces M of *prescribed curvature*, so-called *Weingarten* hypersurfaces. To be more precise, let Ω be a connected open subset of N, $f \in C^{2,\alpha}(\overline{\Omega})$, F a smooth, symmetric function defined in the positive cone $\Gamma_+ \subset \mathbf{R}^n$, then we look for a convex hypersurface $M \subset \Omega$ such that

$$F|_M = f(x) \quad \forall x \in M, \tag{0.1}$$

where $F|_M$ means that F is evaluated at the vector $(\kappa_i(x))$ the components of which are the principal curvatures of M.

This is in general a fully nonlinear partial differential equation problem, which is elliptic if we assume F to satisfy

$$\frac{\partial F}{\partial \kappa_i} > 0 \quad \text{in} \quad \Gamma_+. \tag{0.2}$$

Classical examples of curvature functions F are the elementary symmetric polynomials of order k, H_k, defined by

$$H_k = \sum_{i_1<\ldots<i_k} \kappa_{i_1} \ldots \kappa_{i_k}, \quad 1 \le k \le n. \tag{0.3}$$

H_1 is the mean curvature H, H_2 is the scalar curvature—for hypersurfaces in Euclidean space—, and H_n is the Gaussian curvature K.

For technical reasons it is convenient to consider the homogeneous polynomials of degree 1

$$\sigma_k = H_k^{1/k} \tag{0.4}$$

instead of H_k. Then, the σ_k's are not only monotone increasing but also *concave*. Their *inverses* $\widetilde{\sigma}_k$, defined through

$$\widetilde{\sigma}(\kappa_i) = \frac{1}{\sigma_k(\kappa_i^{-1})} \tag{0.5}$$

share these properties; a proof of this non-trivial result can be found in [12]. $\widetilde{\sigma}_1$ is the so-called *harmonic curvature* G, and, evidently, we have $\widetilde{\sigma}_n = \sigma_n$.

To describe the general curvature functions we have in mind, let us define

Definition 0.1. Let $F \in C^0(\overline{\Gamma}_+) \cap C^{2,\alpha}(\Gamma_+)$ be a symmetric function, (positively) homogeneous of degree 1 satisfying

$$F_i = \frac{\partial F}{\partial \kappa_i} > 0 \quad \text{on} \quad \Gamma_+ \tag{0.6}$$

and

$$F \quad \textit{is concave.} \tag{0.7}$$

Then we say

(i) F is of class $(\widetilde{K})$, if

$$F|_{\partial\Gamma_+} = 0; \tag{0.8}$$

(ii) F is of class $(\widetilde{H})$, if

$$\textit{its inverse} \quad \widetilde{F} \quad \textit{is also concave,} \tag{0.9}$$

and

$$F \in C^{2,\alpha}(\overline{\Lambda}_{\varepsilon,c}) \quad \text{and} \quad 0 < F_i \quad \text{in} \quad \overline{\Lambda}_{\varepsilon,c}, \tag{0.10}$$

where $\Lambda_{\varepsilon,c} \subset \Gamma_+$ is defined through

$$\Lambda_{\varepsilon,c} = \{(\kappa_i) \, : \, 0 < \varepsilon \le F, \; 0 < \kappa_i \le c\}. \tag{0.11}$$

(iii) F is of class (H), if

$$\widetilde{F} \quad \textit{is concave.} \tag{0.12}$$

and

$$F \in C^{2,\alpha}(\overline{\Gamma}_+) \quad \text{and} \quad 0 < F_i \quad \text{in} \quad \overline{\Gamma}_+. \tag{0.13}$$

Remark 0.2. Since F_i are homogeneous of degree 0, the condition (0.13) implies that the F_i are also uniformly bounded in $\overline{\Gamma}_+$.

Remark 0.3. Here are some classical curvature functions which satisfy the above definitions.

(i) The $\widetilde{\sigma}_k$'s are of class $(\widetilde{K})$, and also the inverse of the length of the second fundamental form

$$F(\kappa_i) = \frac{1}{\left(\sum_i k_i^{-2}\right)^{1/2}} \tag{0.14}$$

(ii) The σ_k's (and their inverses) are of class $(\widetilde{H})$.

(iii) The mean curvature is of class (H).

Our main assumption in the existence proof is a barrier assumption.

Definition 0.4. Let M_1, M_2 be strictly convex, closed hypersurfaces in N, homeomorphic to S^n and of class $C^{4,\alpha}$ which bound a connected open subset Ω, such that the mean curvature vector of M_1 points outside of Ω and the mean curvature vector of M_2 points inside of Ω. M_1, M_2 are barriers for (F, f) if

$$F|_{M_1} \le f \tag{0.15}$$

and

$$F|_{M_2} \ge f. \tag{0.16}$$

Remark 0.5. In view of the Harnack inequality we deduce from the properties of the barriers that they do not touch, unless both coincide and are solutions of our problem. In this case Ω would be empty.

Then we can prove

Theorem 0.6. *Let N be a space form with curvature $K_N = 0$, let F be of class $(\widetilde{K})$, $0 < f \in C^{2,\alpha}(\overline{\Omega})$ and assume that M_1, M_2 are barriers for (F, f), then the problem*

$$F|_M = f \tag{0.17}$$

has a strictly convex solution $M \subset \overline{\Omega}$ of class $C^{4,\alpha}$.

Theorem 0.7. *Let N be a space form with curvature $K_N = 0$ and $F \in (\widetilde{H})$; let $0 < f \in C^{2,\alpha}(\overline{\Omega})$ be such that $\log f$ is concave and assume that M_1, M_2 are barriers for (F, f), then the problem*

$$F|_M = f \tag{0.18}$$

has a convex solution $M \subset \overline{\Omega}$ of class $C^{4,\alpha}$.

Theorem 0.8. *Let N be a space form with curvature K_N and $F \in (H)$; let $f \in C^{2,\alpha}(\overline{\Omega})$ satisfy*

$$-K_N f \overline{g}_{\alpha\beta} + f_{\alpha\beta} \leq 0 \quad \text{in} \quad \Omega \tag{0.19}$$

and assume that M_1, M_2 are barriers for (F, f), then the problem

$$F|_M = f \tag{0.20}$$

has a convex solution $M \subset \overline{\Omega}$ of class $C^{4,\alpha}$ if $K_N \leq 0$, or—in the case $K_N > 0$—if in addition f is strictly positive in $\overline{\Omega}$.

Remark 0.9. In the first part of Theorem 0.8 ($K_N \leq 0$) f is not supposed to be strictly positive in $\overline{\Omega}$. Though, in view of the barrier condition, f has to be positive in a neighbourhood of M_1. The solution M will be contained in the support of $\max(f, 0)$ and also the assumption (0.19) should only be valid there.

In a separate paper we considered closed Weingarten hypersurfaces in arbitrary Riemannian manifolds with non-positive sectional curvature, cf. [9]. In that paper we have also proved that we can isometrically lift the geometric setting Ω, M_1, M_2 and f to the universal cover $\widetilde{N}$ even in the case of a space form N with $K_N > 0$. Thus, we may—and shall—assume in the following that N is simply connected.

The existence of closed Weingarten hypersurfaces in $\mathbf{R}^{n+1}$ has been studied extensively in previous papers: the case $F = H$ by Bakelman and Kantor [2], Treibergs and Wei [14], the case $F = K$ by Oliker [13], Delanoë [5], and for general curvature functions by Caffarelli, Nirenberg and Spruck [4]. In all papers—except in [5]—the authors imposed a sign condition for the radial derivative of the right-hand side to prove the existence. This condition was necessary for two reasons, first to derive the a priori estimates for the C^1-norm and secondly to apply the inverse function theorem, i.e. the kernel of the linearized operator had to be trivial.

Without this condition the kernel is no longer trivial and the inverse function theorem of Leray–Schauder type arguments fail.

We therefore use the evolution method to approximate stationary solutions. But there is still the difficulty of obtaining the C^1-estimates: either one has to impose some artificial condition on the right-hand side, i.e. the condition depends on the choice of a special coordinate system, or one has to stay in the class of convex hypersurfaces where the C^1-estimates are a trivial consequence of the convexity, but then the preservation of the convexity has to be proved and this can only be achieved for special curvature functions like the Gaussian curvature, or by assuming f to satisfy the condition (0.19).

The paper is organized as follows: In Section 1 we consider general curvature functions and state some basic properties.

In Section 2 we formulate the evolution problem and prove short-time existence.

In Section 3 we derive the evolution equation for some geometric quantities like the metric and the second fundamental form.

In Section 4 we prove that the flow stays in $\overline{\Omega}$.

In Section 5 we state the parabolic equations satisfied by h_{ij} resp. $v = \sqrt{1+|Du|^2}$.

In Section 6 the C^2-estimates are derived, while in Section 7 the convergence to a smooth stationary solution is proved.

1. Curvature Functions.

Let $F \in C^{2,\alpha}(\Gamma_+) \cap C^0(\overline{\Gamma}_+)$ be a symmetric function satisfying the conditions (0.6) and (0.7); then, F can also be viewed as a function defined on the space of symmetric, positive definite matrices $\mathcal{S}_+$, or to be more precise, at least in this section, let $(h_{ij}) \in \mathcal{S}_+$ with eigenvalues κ_i, $1 \le i \le n$, then define $\widehat{F}$ on $\mathcal{S}_+$ by

$$\widehat{F}(h_{ij}) = F(\kappa_i). \tag{1.1}$$

It is well known, see e.g. [3], that $\widehat{F}$ is as smooth as F and that $\widehat{F}^{ij} = \dfrac{\partial F}{\partial h_{ij}}$ satisfies

$$\widehat{F}^{ij}\xi_i\xi_j = \frac{\partial F}{\partial \kappa_i}|\xi_i|^2, \tag{1.2}$$

where we use the summation convention throughout this paper unless otherwise stated.

Moreover, if F is concave then $\widehat{F}$ is also concave, i.e.

$$\widehat{F}^{ij,kl}\eta_{ij}\eta_{kl} \le 0 \tag{1.3}$$

for any symmetric (η_{ij}), where

$$\widehat{F}^{ij,kl} = \frac{\partial^2}{\partial h_{ij}\partial h_{kl}}\widehat{F}. \tag{1.4}$$

An even sharper estimate is valid, namely,

Lemma 1.1. *Let F, $\widehat{F}$ be defined as above, then*

$$\widehat{F}^{ij,kl}\eta_{ij}\eta_{kl} = \frac{\partial^2 F}{\partial\kappa_i\partial\kappa_j}\eta_{ii}\eta_{jj} + \sum_{i\neq j}\frac{F_i - F_j}{\kappa_i - \kappa_j}(\eta_{ij})^2, \tag{1.5}$$

for any $(\eta_{ij}) \in \mathcal{S}$, where $\mathcal{S}$ is the space of all symmetric matrices and where $F_i = \frac{\partial F}{\partial\kappa_i}$. The second term on the right-hand side of (1.5) *is non-positive and has to be interpreted as a limit if $\kappa_i = \kappa_j$.*

Proof. In [7, Lemma 2] it is shown that

$$\left(\frac{\partial F}{\partial\kappa_i} - \frac{\partial F}{\partial\kappa_j}\right)(\kappa_i - \kappa_j) \le 0 \tag{1.6}$$

if F is concave, hence the second term of the right-hand side in (1.5) is non-positive.

A proof of inequality (1.5) can be found in [9, Lemma 1.1].

We also want to mention that F need not to be defined on the positive cone, any open, convex cone will do.

For the rest of the paper we shall no longer distinguish between F and $\widehat{F}$; instead we shall consider F to be defined both on $\mathcal{S}_+$ and Γ_+.

For $(h_{ij}) \in \mathcal{S}_+$ let $(\tilde{h}^{ij}) = (h_{ij})^{-1}$, then we have

Lemma 1.2. *Let F be a curvature function on Γ_+ and $\tilde{F}$ be its inverse, and assume that both F and $\tilde{F}$ are concave, then*

$$F^{ij,kl}\eta_{ij}\eta_{kl} + 2F^{ik}\tilde{h}^{jl}\eta_{ij}\eta_{kl} \ge 2F^{-1}(F^{ij}\eta_{ij})^2 \tag{1.7}$$

for all $(\eta_{ij}) \in \mathcal{S}$.

A proof of the lemma is given in [15, p. 112].

The preceding considerations are also applicable if the κ_i are the principal curvatures of a hypersurface M with metric (g_{ij}). F can then be looked at as being defined on the space of all symmetric tensors (h_{ij}) with eigenvalues κ_i with respect to the metric.

$$F^{ij} = \frac{\partial F}{\partial h_{ij}} \tag{1.8}$$

is then a contravariant tensor of second order. Sometimes, it will be convenient to circumvent the dependence on the metric by considering F to depend on the mixed tensor

$$h^i_j = g^{ik}h_{kj}. \tag{1.9}$$

Then

$$F_i^j = \frac{\partial F}{\partial h_j^i} \tag{1.10}$$

is also a mixed tensor with contravariant index j and covariant index i.

2. The evolution problem.

Let N be a complete $(n+1)$-dimensional Riemannian manifold and M a closed hypersurface. Geometric quantities in N will be denoted by $(\bar{g}_{\alpha\beta})$, $(\bar{R}_{\alpha\beta\gamma\delta})$, etc., and those in M by (g_{ij}), (R_{ijkl}), etc. Greek indices range from 0 to n and Latin from 1 to n; the summation convention is always used. Generic coordinate systems in N resp. M will be denoted by (x^α) resp. (ξ^i). Covariant differentiation will simply be indicated by indices, only in case of possible ambiguity they will be preceded by a semicolon, i.e. for a function u on N, (u_α) will be the gradient and $(u_{\alpha\beta})$ the Hessian, but, e.g. the covariant derivative of the curvature tensor will be abbreviated by $\bar{R}_{\alpha\beta\gamma\delta;\varepsilon}$. We also point out that

$$\bar{R}_{\alpha\beta\gamma\delta;i} = \bar{R}_{\alpha\beta\gamma\delta;\varepsilon} x_i^\varepsilon \tag{2.1}$$

with obvious generalizations to other quantities.

If N is a space of constant curvature, then

$$\bar{R}_{\alpha\beta\gamma\delta} = K_N(\bar{g}_{\alpha\gamma}\bar{g}_{\beta\delta} - \bar{g}_{\alpha\delta}\bar{g}_{\beta\gamma}). \tag{2.2}$$

In local coordinates x^α and ξ^i the geometric quantities of the hypersurface M are connected through the following equations

$$x_{ij}^\alpha = -h_{ij}\nu^\alpha \tag{2.3}$$

the so-called *Gauß formula*. Here, and also in the sequel, a covariant derivative is always a *full* tensor, i.e.

$$x_{ij}^\alpha = x_{,ij}^\alpha - \Gamma_{ij}^k x_k^\alpha + \bar{\Gamma}_{\beta\gamma}^\alpha x_i^\beta x_j^\gamma. \tag{2.4}$$

The comma indicates ordinary partial derivatives.

In this implicit definition (2.3) the *second fundamental form* (h_{ij}) is taken with respect to $-\nu$.

The second equation is the *Weingarten equation*

$$\nu_i^\alpha = h_i^k x_k^\alpha, \tag{2.5}$$

where we remember that ν_i^α is full tensor.

Finally, we have the *Codazzi equation*

$$h_{ij;k} - h_{ik;j} = \overline{R}_{\alpha\beta\gamma\delta}\nu^\alpha x_i^\beta x_j^\gamma x_k^\delta = 0, \tag{2.6}$$

if N is a space of constant curvature, and the *Gauß equation*

$$R_{ijkl} = h_{ik}h_{jl} - h_{il}h_{jk} + \overline{R}_{\alpha\beta\gamma\delta}x_i^\alpha x_j^\beta x_k^\gamma x_l^\delta. \tag{2.7}$$

We want to prove that the equation

$$F = f \tag{2.8}$$

has a solution. For technical reasons it is convenient to solve instead of (2.8) the equivalent equation

$$\Phi(F) = \Phi(f) \tag{2.9}$$

where Φ is real function defined on $\mathbf{R}_+$ such that

$$\dot{\Phi} > 0 \quad \text{and} \quad \ddot{\Phi} \le 0. \tag{2.10}$$

For notational reasons let us abbreviate

$$\tilde{f} = \Phi(f). \tag{2.11}$$

To solve (2.9) we look at the evolution problem

$$\begin{aligned} \dot{x} &= -(\Phi - \tilde{f})\nu \\ x(0) &= x_0 \end{aligned} \tag{2.12}$$

where x_0 is an embedding of an initial strictly convex hypersurface M_0 diffeomorphic to S^n, $\Phi = \Phi(F)$, and F is evaluated for the principal curvatures of the flow hypersurfaces $M(t)$, or, equivalently, we may assume that F depends on the second fundamental form (h_{ij}) and the metric (g_{ij}) of $M(t)$; $x(t)$ is the embedding for $M(t)$.

This is a parabolic problem, so short-time existence is guaranteed—an exact proof is given below—, and under suitable assumptions we shall be able to prove that the solution exists for all time and that the velocity tends to zero if t goes to infinity.

Consider now a tubular neighbourhood $\mathcal{U}$ of the initial hypersurface M_0, then we can introduce so-called *normal Gaussian coordinates* x^α, such that the metric in $\mathcal{U}$ has the form

$$d\bar{s}^2 = dr^2 + \bar{g}_{ij}\, dx^i dx^j \tag{2.13}$$

where $r = x^0$, $\bar{g}_{ij} = \bar{g}_{ij}(r, x)$; here we use slightly ambiguous notation.

A point $p \in \mathcal{U}$ can be represented by its signed distance from M_0 and its base point $x \in M_0$, thus $p = p(r, x)$.

Let $M \subset \mathcal{U}$ be a hypersurface which is a graph over M_0, i.e.

$$M = \{(r, x) \,:\, r = u(x), x \in M_0\}. \tag{2.14}$$

The induced metric of M, g_{ij}, can then be expressed as

$$g_{ij} = \bar{g}_{ij} + u_i u_j \tag{2.15}$$

with inverse

$$g^{ij} = \bar{g}^{ij} - \frac{u^i}{v}\frac{u^j}{v}, \tag{2.16}$$

where $(\bar{g}^{ij}) = (\bar{g}_{ij})^{-1}$ and

$$\begin{aligned} u^i &= \bar{g}^{ij} u_j \\ v^2 &= 1 + \bar{g}^{ij} u_i u_j \end{aligned} \tag{2.17}$$

The normal vector ν of M then takes the form

$$(\nu^\alpha) = v^{-1}(1, -u^i) \tag{2.18}$$

if x^0 is chosen appropriately.

From the Gauß formula we immediately deduce that the second fundamental form of M is given by

$$v^{-1} h_{ij} = -u_{ij} + \bar{h}_{ij}, \tag{2.19}$$

where

$$\bar{h}_{ij} = \frac{1}{2}\dot{\bar{g}}_{ij} = \frac{1}{2}\frac{\partial \bar{g}_{ij}}{\partial r} \tag{2.20}$$

is the second fundamental form of the level surfaces $\{r = \text{const}\}$, and where the second covariant derivatives of u are defined with respect to the induced metric.

At least for small t the hypersurfaces $M(t)$ are graphs over M_0 and the embedding vector looks like

$$\begin{aligned} x^0(t) &= u(t, x^i(t)) \\ x^i(t) &= x^i(t, \xi^i) \end{aligned} \tag{2.21}$$

where the ξ^i are local coordinates for $M(t)$ independent of t.

Furthermore,

$$\dot{x}^0 = \dot{u} = \frac{\partial u}{\partial t} + \dot{x}^i u_i \tag{2.22}$$

and from (2.12) we conclude

$$\begin{aligned} \dot{x}^0 &= -(\Phi - \tilde{f})v^{-1} \\ \dot{x}^i &= v^{-1}u^i(\Phi - \tilde{f}) \end{aligned} \tag{2.23}$$

hence, we obtain

$$\frac{\partial u}{\partial t} = -(\Phi - \tilde{f})v. \tag{2.24}$$

This is a scalar equation, which can be solved on a cylinder $[0, \varepsilon] \times M_0$ for small ε, if the principal curvatures of the initial hypersurface M_0 are strictly positive. The equation (2.23) for the embedding vector is then a classical ordinary differential equation of the form

$$\dot{x} = \varphi(t, x). \tag{2.25}$$

We have therefore proved

Theorem 2.1. *The evolution problem* (2.12) *has a solution on a small time interval* $[0, \varepsilon]$.

3. The evolution equations of some geometric quantities.

In this section we want to show how the metric, the second fundamental form, and the normal vector of the hypersurfaces $M(t)$ evolve. All time derivatives are *total* derivatives.

Lemma 3.1 (Evolution of the metric). *The metric g_{ij} of $M(t)$ satisfies the evolution equation*

$$\dot{g}_{ij} = -2(\Phi - \tilde{f})h_{ij}. \tag{3.1}$$

Proof. Let ξ^i be local coordinates for $M(t)$, then

$$g_{ij} = \bar{g}_{\alpha\beta} x_i^\alpha x_j^\beta \tag{3.2}$$

and thus

$$\dot{g}_{ij} = 2\bar{g}_{\alpha\beta} \dot{x}_i^\alpha x_j^\beta. \tag{3.3}$$

On the other hand, differentiating

$$\dot{x}^\alpha = -(\Phi - \tilde{f})\nu^\alpha \tag{3.4}$$

with respect to ξ^i yields

$$\dot{x}_i^\alpha = -(\Phi - \tilde{f})_i \nu^\alpha - (\Phi - \tilde{f})\nu_i^\alpha \tag{3.5}$$

and the desired result follows from the Weingarten equation.

Lemma 3.2 (Evolution of the normal). *The normal vector ν evolves according to*

$$\dot{\nu} = \nabla_M(\Phi - \tilde{f}) = g^{ij}(\Phi - \tilde{f})_i x_j. \tag{3.6}$$

Proof. Since ν is a unit normal vector we have $\dot{\nu} \in T(M)$. Furthermore, differentiating

$$0 = \langle \nu, x_i \rangle \tag{3.7}$$

with respect to t, we deduce

$$\langle \dot{\nu}, x_i \rangle = -\langle \nu, \dot{x}_i \rangle = (\Phi - \tilde{f})_i. \tag{3.8}$$

Lemma 3.3 (Evolution of the second fundamental form). *The second fundamental form evolves according to*

$$\dot{h}_i^j = (\Phi - \tilde{f})_i^j + (\Phi - \tilde{f})h_i^k h_k^j + (\Phi - \tilde{f})\bar{R}_{\alpha\beta\gamma\delta}\nu^\alpha x_i^\beta \nu^\gamma x_k^\delta g^{kj} \tag{3.9}$$

and

$$\dot{h}_{ij} = (\Phi - \tilde{f})_{ij} - (\Phi - \tilde{f})h_i^k h_{kj} + (\Phi - \tilde{f})\bar{R}_{\alpha\beta\gamma\delta}\nu^\alpha x_i^\beta \nu^\gamma x_j^\delta. \tag{3.10}$$

Proof. We use the *Ricci identities* to interchange the covariant derivatives of ν with respect to t and ξ^i

$$\begin{aligned}\frac{d}{dt}(\nu_i^\alpha) &= (\dot\nu^\alpha)_i - \overline{R}^\alpha{}_{\beta\gamma\delta}\nu^\beta x_i^\gamma \dot x^\delta \\ &= g^{kl}(\Phi-\tilde f)_{ki}x_l^\alpha + g^{kl}(\Phi-\tilde f)_k x_{li}^\alpha - \overline{R}^\alpha{}_{\beta\gamma\delta}\nu^\beta x_i^\gamma \dot x^\delta\end{aligned} \tag{3.11}$$

For the second equality we used (3.6).

On the other hand, in view of the Weingarten equation

$$\frac{d}{dt}(\nu_i^\alpha) = \frac{d}{dt}\left(h_i^k x_k^\alpha\right) = \dot h_i^k x_k^\alpha + h_i^k \dot x_k^\alpha. \tag{3.12}$$

Multiplying the resulting equation with $\bar g_{\alpha\beta}x_j^\beta$ we conclude

$$\dot h_i^k g_{kj} - (\Phi-\tilde f)h_i^k h_{kj} = (\Phi-\tilde f)_{ij} + (\Phi-\tilde f)\overline{R}_{\alpha\beta\gamma\delta}\nu^\alpha x_i^\beta \nu^\gamma x_j^\delta \tag{3.13}$$

or equivalently (3.9).

To derive (3.10), we differentiate

$$h_{ij} = h_i^k g_{kj} \tag{3.14}$$

with respect to t and use (3.3).

Lemma 3.4 (Evolution of $(\Phi-\tilde f)$). *The term $(\Phi-\tilde f)$ evolves according to the equation*

$$\begin{aligned}(\Phi-\tilde f)' - \dot\Phi F^{ij}(\Phi-\tilde f)_{ij} &= \dot\Phi F^{ij}h_{ik}h_j^k(\Phi-\tilde f) + \tilde f_\alpha \nu^\alpha(\Phi-\tilde f) \\ &\quad + \dot\Phi F^{ij}\overline{R}_{\alpha\beta\gamma\delta}\nu^\alpha x_i^\beta \nu^\gamma x_j^\delta(\Phi-\tilde f)\end{aligned} \tag{3.15}$$

where

$$(\Phi-\tilde f)' = \frac{d}{dt}(\Phi-\tilde f) \tag{3.16}$$

and

$$\dot\Phi = \frac{d}{dr}\Phi(r). \tag{3.17}$$

Proof. When we differentiate F with respect to t it is advisable to consider F as a function of the mixed tensor h_j^i; then we obtain

$$(\Phi-\tilde f)' = \dot\Phi F_j^i \dot h_i^j - \tilde f_\alpha \dot x^\alpha. \tag{3.18}$$

The result now follows from (3.9) and (3.4).

Corollary 3.5. *Let N be a space form then the equation (3.15) takes the form*

$$(3.19)\qquad \begin{aligned}(\Phi-\tilde{f})'-\dot{\Phi}F^{ij}(\Phi-\tilde{f})_{ij} &= \dot{\Phi}F^{ij}h_{ik}h_j^k(\Phi-\tilde{f})+\tilde{f}_\alpha\nu^\alpha(\Phi-\tilde{f})\\ &\quad +K_N\dot{\Phi}F^{ij}g_{ij}(\Phi-\tilde{f})\end{aligned}$$

4. Barriers and a priori estimates in the C^0-norm.

In [9, Section 4] we have shown that, if the sectional curvature of the ambient space N is non-positive or if N is a space form with positive curvature, then, the geometric setting of our problem, i.e. Ω, M_1, M_2 and f can be isometrically lifted to the universal cover. If N is space form with $K_N>0$, then $\overline{\Omega}\subset\tilde{N}$ is contained in an open hemisphere. The barriers M_i are boundaries of convex bodies $\langle M_i\rangle$ and, if we introduce geodesic polar coordinates $(x^\alpha)=(r,x^i)=(r,x)$ around a point in $\langle M_1\rangle$ such that

$$(4.1)\qquad d\bar{s}^2=dr^2+\bar{g}_{ij}\,dx^i dx^j$$

then the second fundamental form $\bar{h}_{ij}$ of a geodesic sphere $\{r=\text{const}\}$ that intersects $\overline{\Omega}$ is uniformly positive definite. The M_i are graphs over a fixed geodesic sphere S_0, $M_i=\operatorname{graph} u_i|_{S_0}$.

Moreover, let $M(t)$ be a solution of the evolution problem (2.12) in a maximal time interval $I=[0,T^*)$ such that the hypersurfaces are strictly convex. Then, each $M(t)$ can be represented as a graph over S_0

$$(4.2)\qquad M(t)=\{(r,x)\,:\,r=u(t,x),x\in S_0\}.$$

In [9, Section 5] we also proved the following lemmata

Lemma 4.1. *Choose as initial hypersurface M_0 either M_1 or M_2, then we have for the embedding vector $x=x(t)$*

$$(4.3)\qquad x(t)\in\overline{\Omega}\quad \forall t\in I.$$

and

Lemma 4.2. *Let $M(t)$ be a solution of the evolution problem (2.12) defined on a maximal interval $[0,T^*)$. As initial hypersurface M_0 we choose either M_1 or M_2; then we obtain*

$$(4.4)\qquad \Phi-\tilde{f}\le 0\quad\forall t$$

if $M_0 = M_1$, *and*

$$\Phi - \tilde{f} \ge 0 \quad \forall t \tag{4.5}$$

if $M_0 = M_2$.

5. The evolution equations for h_{ij} and v.

Let $M(t)$ be a solution of problem (2.12); in [9, Section 7] we derived the following evolution equations for h_{ij} resp. h^i_j

Lemma 5.1. *Let* $M(t)$ *be a solution of the problem* (2.12), *then the second fundamental form satisfies*

$$\begin{aligned}\dot{h}_{ij} - \dot{\Phi}F^{kl}h_{ij;kl} = {} & \dot{\Phi}F^{kl}h_{kr}h^r_l h_{ij} - (\Phi - \tilde{f})h^k_i h_{kj} - \dot{\Phi}F h^k_i h_{kj} \\ & - \tilde{f}_{\alpha\beta}x^\alpha_i x^\beta_j + \tilde{f}_\alpha \nu^\alpha h_{ij} + \ddot{\Phi}F_i F_j \\ & + \dot{\Phi}F^{kl,rs}h_{kl;i}h_{rs;j} + (\Phi - \tilde{f})\overline{R}_{\alpha\beta\gamma\delta}\nu^\alpha x^\beta_i \nu^\gamma x^\delta_j \\ & + 2\dot{\Phi}F^{kl}\overline{R}_{\alpha\beta\gamma\delta}x^\alpha_r x^\beta_i x^\gamma_k x^\delta_j h^r_l \\ & - \dot{\Phi}F^{kl}\overline{R}_{\alpha\beta\gamma\delta}x^\alpha_r x^\beta_k x^\gamma_i x^\delta_l h^r_j \\ & - \dot{\Phi}F^{kl}\overline{R}_{\alpha\beta\gamma\delta}x^\alpha_r x^\beta_k x^\gamma_j x^\delta_l h^r_i \\ & + \dot{\Phi}F^{kl}\overline{R}_{\alpha\beta\gamma\delta}\nu^\alpha x^\beta_k \nu^\gamma x^\delta_l h_{ij} \\ & - \dot{\Phi}F\overline{R}_{\alpha\beta\gamma\delta}\nu^\alpha x^\beta_i \nu^\gamma x^\delta_j \\ & + \dot{\Phi}F^{kl}\overline{R}_{\alpha\beta\gamma\delta;\epsilon}\{\nu^\alpha x^\beta_k x^\gamma_l x^\delta_i x^\epsilon_j + \nu^\alpha x^\beta_i x^\gamma_k x^\delta_j x^\epsilon_l\}\end{aligned} \tag{5.1}$$

and

Lemma 5.2. *The evolution equation for* h^i_i *(no summation over* i*) has the form*

$$\begin{aligned}\dot{h}^i_i - \dot{\Phi}F^{kl}h^i_{i;kl} = {} & \dot{\Phi}F^{kl}h_{kr}h^r_l h^i_i + (\Phi - \tilde{f})h^k_i h^i_k - \dot{\Phi}F h^k_i h^i_k \\ & - \tilde{f}_{\alpha\beta}x^\alpha_i x^\beta_k g^{ki} + \tilde{f}_\alpha \nu^\alpha h^i_i + \ddot{\Phi}F_i F^i \\ & + \dot{\Phi}F^{kl,rs}h_{kl;i}h_{rs;m}g^{mi} \\ & + (\Phi - \tilde{f})\overline{R}_{\alpha\beta\gamma\delta}\nu^\alpha x^\beta_i \nu^\gamma x^\delta_m g^{mi} \\ & + 2\dot{\Phi}F^{kl}\overline{R}_{\alpha\beta\gamma\delta}x^\alpha_r x^\beta_i x^\gamma_k x^\delta_m g^{mi} h^r_l \\ & - 2\dot{\Phi}F^{kl}\overline{R}_{\alpha\beta\gamma\delta}x^\alpha_r x^\beta_k x^\gamma_i x^\delta_l h^{ri} \\ & + \dot{\Phi}F^{kl}\overline{R}_{\alpha\beta\gamma\delta}\nu^\alpha x^\beta_k \nu^\gamma x^\delta_l h^i_i \\ & - \dot{\Phi}F\overline{R}_{\alpha\beta\gamma\delta}\nu^\alpha x^\beta_i \nu^\gamma x^\delta_m g^{mi} \\ & + \dot{\Phi}F^{kl}\overline{R}_{\alpha\beta\gamma\delta;\epsilon}\{\nu^\alpha x^\beta_k x^\gamma_l x^\delta_i x^\epsilon_m + \nu^\alpha x^\beta_i x^\gamma_k x^\delta_m x^\epsilon_l\}g^{mi}\end{aligned} \tag{5.2}$$

In case N is a space form with curvature K_N, the preceding evolution equations simplify to

Corollary 5.3. *Let N be a space form with curvature K_N, then equation (5.1) takes the form*

$$
\begin{aligned}
\dot{h}_{ij} - \dot{\Phi}F^{kl}h_{ij;kl} = {} & \dot{\Phi}F^{kl}h_{kr}h_l^r h_{ij} - (\Phi - \tilde{f})h_i^k h_{kj} - \dot{\Phi}F h_i^k h_{kj} \\
& -\tilde{f}_{\alpha\beta}x_i^\alpha x_j^\beta + \tilde{f}_\alpha \nu^\alpha h_{ij} + \ddot{\Phi}F_i F_j \\
& +\dot{\Phi}F^{kl,rs}h_{kl;i}h_{rs;j} \\
& +K_N\{(\Phi - \tilde{f}) + \dot{\Phi}F\}g_{ij} - K_N\dot{\Phi}F^{kl}g_{kl}h_{ij}
\end{aligned}
\tag{5.3}
$$

and

Corollary 5.4. *Let N be a space form with curvature K_N, then equation (5.2) takes the form*

$$
\begin{aligned}
\dot{h}_i^j - \dot{\Phi}F^{kl}h_{i;kl}^j = {} & \dot{\Phi}F^{kl}h_{kr}h_l^r h_i^j + (\Phi - \tilde{f})h_i^k h_k^j - \dot{\Phi}F h_i^k h_k^j \\
& -\tilde{f}_{\alpha\beta}x_i^\alpha x_k^\beta g^{kj} + \tilde{f}_\alpha \nu^\alpha h_i^j + \ddot{\Phi}F_i F^j \\
& +\dot{\Phi}F^{kl,rs}h_{kl;i}h_{rs;m}g^{mj} \\
& +K_N\{(\Phi - \tilde{f}) + \dot{\Phi}F\}\delta_i^j - K_N\dot{\Phi}F^{kl}g_{kl}h_i^j
\end{aligned}
\tag{5.4}
$$

The proof is straightforward, if one observes that F^{ij} and h_{ij} can be diagonalized simultaneously, cf. [9, equ. (1.12)].

Suppose now, that we have introduced geodesic polar coordinates $(x^\alpha) = (r, x^i)$ such that the hypersurfaces $M(t)$ are graphs over a geodesic sphere S_0. From the relation (2.18) we conclude

$$
v = \sqrt{1 + |Du|^2} = (r_\alpha \nu^\alpha)^{-1}. \tag{5.5}
$$

We know, that as long as the hypersurfaces are convex, the quantity v is uniformly bounded, or more precisely, cf. [9, Lemma 6.1]

Lemma 5.5. *Let $M = \operatorname{graph} u|_{S_0}$ be a closed convex hypersurface represented in normal Gaussian coordinates then the quantity $v = \sqrt{1 + |Du|^2}$ can be estimated by*

$$
v \le c(|u|, S_0, \bar{g}_{ij}). \tag{5.6}
$$

Furthermore, the function u and the quantity v satisfy the following evolution equations

Lemma 5.6. *Consider the flow in a normal Gaussian coordinate system where the $M(t)$ can be written as graphs of a function $u(t)$. Then u resp. $\tilde v$ satisfy the evolution equations*

$$\dot u - \dot\Phi F^{ij} u_{ij} = -(\Phi - \tilde f)v^{-1} + \dot\Phi F v^{-1} - \dot\Phi F^{ij}\bar h_{ij} \tag{5.7}$$

resp.

$$\begin{aligned} \dot{\tilde v} - \dot\Phi F^{ij}\tilde v_{ij} &= -\dot\Phi F^{ij} h_{ik} h_j^k \tilde v - 2\tilde v^{-1}\dot\Phi F^{ij}\tilde v_i \tilde v_j \\ &\quad + r_{\alpha\beta}\nu^\alpha\nu^\beta\big[(\Phi - \tilde f) - \dot\Phi F\big]\tilde v^2 \\ &\quad + \dot\Phi F^{ij}\bar R_{\alpha\beta\gamma\delta}\nu^\alpha x_i^\beta x_j^\gamma x_k^\delta r_\varepsilon x_m^\varepsilon g^{mk}\tilde v^2 \\ &\quad + 2\dot\Phi F^{ij} r_{\alpha\beta} h_i^k x_k^\alpha x_j^\beta \tilde v^2 \\ &\quad + \dot\Phi F^{ij} r_{\alpha\beta\gamma}\nu^\alpha x_i^\beta x_j^\gamma \tilde v^2 + \tilde f_\alpha x_m^\alpha g^{mk} r_\beta x_k^\beta \tilde v^2 \end{aligned} \tag{5.8}$$

cf. [9, equ. (8.2) resp. Lemma 7.3].

In a simply connected space form we can deduce a considerable simpler and more aesthetic form of (5.8).

First, we observe that by symmetry

$$\bar g_{ij} = h(r)\sigma_{ij}, \tag{5.9}$$

where σ_{ij} is the metric of a geodesic sphere of radius 1. Then, we fix a point in N and choose the coordinates (x^i) such that in that point

$$\bar g_{ij,k} = 0. \tag{5.10}$$

Let us calculate the corresponding Christoffel symbols in N. We have

$$\bar\Gamma^0_{ij} = -\frac{1}{2}\dot{\bar g}_{ij} = -\bar h_{ij}, \tag{5.11}$$

$$\bar\Gamma^0_{00} = \bar\Gamma^0_{0i} = \bar\Gamma^i_{jk} = 0, \tag{5.12}$$

and

$$\bar\Gamma^i_{0j} = \bar h^i_j, \tag{5.13}$$

from which we conclude

Lemma 5.7. *In the above coordinate system the covariant derivatives of r can be expressed as follows*

$$\begin{cases} r_{0\alpha} = r_{\alpha 0} = 0 \\ r_{ij} = \bar h_{ij} \end{cases} \tag{5.14}$$

$$\begin{cases} r_{0ij} = -\frac{\bar{H}^2}{n^2}\bar{g}_{ij} \\ r_{ij0} = \frac{\dot{\bar{H}}}{n}\bar{g}_{ij} \\ r_{i0j} = -\frac{\bar{H}^2}{n^2}\bar{g}_{ij} \end{cases} \tag{5.15}$$

and

$$r_{00j} = r_{0i0} = r_{ijk} = 0. \tag{5.16}$$

Proof. To prove (5.14), we use $|Dr| = 1$ to obtain

$$0 = r_{\alpha\beta}r^{\beta} = r_{\alpha 0} = r_{0\alpha} \tag{5.17}$$

and

$$r_{ij} = -\bar{\Gamma}^{\alpha}_{ij}r_{\alpha} = -\bar{\Gamma}^{0}_{ij} = \bar{h}_{ij}. \tag{5.18}$$

The covariant derivatives of the third order are defined by

$$r_{\alpha\beta\gamma} = r_{\alpha\beta,\gamma} - \bar{\Gamma}^{m}_{\alpha\gamma}r_{\beta m} - \bar{\Gamma}^{m}_{\beta\gamma}r_{m\alpha}, \tag{5.19}$$

where we already used (5.17). The relation (5.16) now follows immediately and also

$$r_{0ij} = r_{i0j} = \bar{h}^{m}_{i}\bar{h}_{mj} \tag{5.20}$$

and

$$r_{ij0} = \dot{\bar{h}}_{ij} - 2\bar{h}^{m}_{i}\bar{h}_{mj}. \tag{5.21}$$

To complete the proof, we observe that the geodesic spheres are totally umbilical, i.e.

$$\bar{h}_{ij} = \frac{\bar{H}}{n}\bar{g}_{ij} \tag{5.22}$$

and hence

$$\dot{\bar{h}}_{ij} = \frac{\dot{\bar{H}}}{n}\bar{g}_{ij} + 2\frac{\bar{H}^2}{n^2}\bar{g}_{ij}. \tag{5.23}$$

To derive a simpler version of equation (5.8), let $\eta = \eta(r)$ be a positive solution of

$$\dot\eta = -\frac{\bar H}{n}\eta, \quad r > 0, \tag{5.24}$$

wherever it is defined and set

$$\chi = v\eta(u). \tag{5.25}$$

Then, we can prove

Lemma 5.8. *The function χ satisfies the evolution equation*

$$\begin{aligned} \dot\chi - \dot\Phi F^{ij}\chi_{ij} &= -\dot\Phi F^{ij}h_{ik}h_j^k\chi - 2\chi^{-1}\dot\Phi F^{ij}\chi_i\chi_j \\ &\quad + \{\dot\Phi F + (\Phi - \tilde f)\}\frac{\bar H}{n}v\chi + \tilde f_\alpha x_k^\alpha g^{ik}u_i v\chi \end{aligned} \tag{5.26}$$

Proof. Using the same notation as before, we obtain

$$\begin{aligned} \dot\chi - \dot\Phi F^{ij}\chi_{ij} &= \{\dot v - \dot\Phi F^{ij}v_{ij}\}\eta + \{\dot u - \dot\Phi F^{ij}u_{ij}\}v\dot\eta \\ &\quad -2\dot\eta\dot\Phi F^{ij}v_i u_j - v\ddot\eta\dot\Phi F^{ij}u_i u_j \end{aligned} \tag{5.27}$$

We then rewrite the equation (5.8) using the expressions in (5.14) to (5.16) to deduce

$$\begin{aligned} \dot v - \dot\Phi F^{ij}v_{ij} &= -\dot\Phi F^{ij}h_{ik}h_j^k v - 2v^{-1}\dot\Phi F^{ij}v_i v_j \\ &\quad +2\dot\Phi F^{ij}v_i u_j\frac{\bar H}{n} \\ &\quad -\dot\Phi F^{ij}g_{ij}\frac{\bar H^2}{n^2}v - \dot\Phi F^{ij}u_i u_j\frac{\dot{\bar H}}{n}v \\ &\quad +\frac{\bar H}{n}|Du|^2\left[(\Phi - \tilde f) - \dot\Phi F\right] \\ &\quad +2\dot\Phi F\frac{\bar H}{n}v^2 + \tilde f_\alpha x_k^\alpha u^k v^2 \end{aligned} \tag{5.28}$$

where we also took into account that

$$\begin{aligned} v_i &= -v^2 h_i^k u_k + v^3\bar h_{ik}u^k \\ &= -v^2 h_i^k u_k + v\frac{\bar H}{n}u_i \end{aligned} \tag{5.29}$$

Inserting (5.28) and (5.7) in (5.27) and observing that in view of (5.24)

$$\ddot\eta = -\frac{\dot{\bar H}}{n}\eta - \frac{\bar H}{n}\dot\eta, \tag{5.30}$$

the equation (5.26) can be easily deduced.

As we have already remarked before, the mean curvature $\overline{H}$ of the geodesic spheres in question is uniformly strictly positive.

6. A priori estimates in the C^2-norm.

Let $M(t)$ be a solution of the evolution problem (2.12) with initial hypersurface $M_0 = M_1$ or $M_0 = M_2$ defined in a maximal time interval $I = [0, T^*)$. We assume $M(t)$ to be represented as the graph of a function u in geodesic polar coordinates. We know that during the evolution the flow stays in the compact set $\overline{\Omega}$ and that the hypersurfaces are strictly convex—this is contained in the definition of the maximal time-interval—, and, hence, Du is uniformly bounded.

We want to show that the second derivatives of u are uniformly bounded or equivalently that the principal curvatures of the flow hypersurfaces are uniformly bounded and positive.

1. The Case of Theorem 0.6.

We first prove

Lemma 6.1. *Let $F \in (\widetilde{K})$, $M_0 = M_1$, $\Phi(t) = \log t$ and $K_N = 0$, then the principal curvatures of the evolution hypersurfaces are uniformly bounded from above.*

Proof. First, we observe, that

$$\Phi \le \tilde{f} \quad \text{or} \quad F \le f \tag{6.1}$$

in view of the results in Lemma 4.2.

Next, let φ be defined by

$$\varphi = \sup\{h_{ij}\eta^i\eta^j \; : \; \|\eta\| = 1\} \tag{6.2}$$

and w by

$$w = \log\varphi + \log\chi. \tag{6.3}$$

We claim that w is bounded. Let $0 < T < T^*$, and $x_0 = x(t_0)$, $0 < t_0 \le T$, be a point in $M(t_0)$ such that

$$\sup_{M_0} w < \sup\{\sup_{M(t)} w \; : \; 0 < t \le T\} = w(x_0). \tag{6.4}$$

We then can introduce a Riemannian normal coordinate system ξ^i at $x_0 \in M(t_0)$ such that at $x_0 = x(t_0, \xi_0)$ we have

$$g_{ij} = \delta_{ij} \quad \text{and} \quad \varphi = h^n_n. \tag{6.5}$$

Let $\eta = (\eta^i)$ be the contravariant vector defined by

$$\eta = (0, \dots, 0, 1) \tag{6.6}$$

and set

$$\tilde{\varphi} = \frac{h_{ij}\eta^i\eta^j}{g_{ij}\eta^i\eta^j}. \tag{6.7}$$

$\tilde{\varphi}$ is well defined in a neighbourhood of (t_0, ξ_0).

Now, define $\tilde{w}$ by replacing φ by $\tilde{\varphi}$ in (6.3); then $\tilde{w}$ assumes its maximum at (t_0, ξ_0). Moreover, at (t_0, ξ_0) we have

$$\dot{\tilde{\varphi}} = \dot{h}^n_n \tag{6.8}$$

and the spacial derivatives do also coincide; in short, $\tilde{\varphi}$ satisfies at (t_0, ξ_0) the same differential equation (5.2) as h^n_n. For the sake of greater clarity, let us therefore treat h^n_n like a scalar and pretend that w is defined by

$$w = \log h^n_n + \log \chi. \tag{6.9}$$

At (t_0, ξ_0) we have $\dot{w} \geq 0$, and, in view of the maximum principle, we deduce from (5.4) and (5.26)

$$0 \leq -h^n_n + c, \tag{6.10}$$

where we have estimated bounded terms by a constant c.

Thus, the principal curvatures are bounded from above.

We further claim that the principal curvatures are uniformly strictly positive, or equivalently—because of the condition (0.8)—

Lemma 6.2. *Under the assumptions of the preceding lemma, we have*

$$0 < \varepsilon_0 \leq F \quad \forall t \tag{6.11}$$

with a given ε_0.

Proof. Consider the function

$$w = \log(-(\Phi - \tilde{f})) + \log \chi. \tag{6.12}$$

Let $0 < T < T^*$ and suppose

$$\sup_{M_0} w < \sup\{\sup_{M(t)} w : 0 \le t \le T\}. \tag{6.13}$$

Then, there is $x_0 = x(t_0)$, $0 < t_0 \le T$, such that

$$w(x_0) = \sup\{\sup_{M(t)} w : 0 \le t \le T\}. \tag{6.14}$$

From (3.19), (5.26) and the maximum principle we then infer

$$0 \le (\Phi - \tilde{f})\frac{\overline{H}}{n} v + c, \tag{6.15}$$

i.e. w is a priori bounded.

2. The Case of Theorem 0.7.

First, we obtain by the same arguments as before

Lemma 6.3. *Let $K_N = 0$, $F \in (\tilde{H})$, $M_0 = M_1$, $\Phi(t) = \log t$, and $0 < f \in C^{2,\alpha}(\overline{\Omega})$, then the principal curvatures of the evolution hypersurfaces are bounded from above as long as they are non-negative.*

Lemma 6.4. *Let $K_N = 0$, $F \in (\tilde{H})$, $M_0 = M_1$, $\Phi(t) = \log t$, and $0 < f \in C^{2,\alpha}(\overline{\Omega})$, then there exists ε_0 such that*

$$0 < \varepsilon_0 \le F \quad \forall t \tag{6.16}$$

as long as the evolution hypersurfaces are convex.

It remains to prove that the principal curvatures stay positive during the evolution. For this achievement we need to know the evolution equation for the inverse of the second fundamental form.

Lemma 6.5. *Let $(\tilde{h}^{ij}) = (h_{ij})^{-1}$ in contravariant form, then the mixed tensor $(\tilde{h}^i_j)$ satisfies the evolution equation (no summation over i)*

$$\begin{aligned}
\dot{\tilde{h}}^i_i - \dot{\Phi} F^{kl} \tilde{h}^i_{i;kl} = {} & -\dot{\Phi} F^{kl} h_{kr} h^r_l \tilde{h}^i_i + \{\dot{\Phi} F - (\Phi - \tilde{f})\}\delta^i_i \\
& - K_N \{\dot{\Phi} F + (\Phi - \tilde{f})\} \tilde{h}_{ki} \tilde{h}^{ki} + K_N \dot{\Phi} F^{kl} g_{kl} \tilde{h}^i_i \\
& + \tilde{f}_{\alpha\beta} x^\alpha_k x^\beta_l \tilde{h}^{ki} \tilde{h}^l_i - \tilde{f}_\alpha \nu^\alpha \tilde{h}^i_i \\
& - \{\dot{\Phi} F^{rs,kl} h_{rs;p} h_{kl;q} + 2\dot{\Phi} F^{rl} \tilde{h}^{ks} h_{rs;p} h_{kl;q} \\
& + \ddot{\Phi} F_p F_q\} \tilde{h}^{pi} \tilde{h}^q_i
\end{aligned} \tag{6.17}$$

Proof. We write

$$\tilde{h}_i^i = g_{ij}\tilde{h}^{ij} \tag{6.18}$$

and use the rule for differentiation of the inverse of a second order tensor to obtain the desired result in view of Corollary 5.3 and the evolution equation of the metric, cf. equation (3.1).

We can then prove

Lemma 6.6. *Let $K_N = 0$, $F \in (\tilde{H})$, $M_0 = M_1$, $\Phi(t) = \log t$, and $0 < f \in C^{2,\alpha}(\bar{\Omega})$ be such that $\log f$ is concave, then there exists a constant λ such that the principal curvatures κ_i of the evolution hypersurfaces $M(t)$ are bounded below by*

$$0 < e^{-\lambda t} \le \kappa_i. \tag{6.19}$$

Proof. Since M_0 is strictly convex the inverse $\tilde{h}^{ij}$ is well-defined during the evolution. We shall show that the eigenvalues of $\tilde{h}^{ij}$ (with respect to g_{ij}) grow at most exponential in t.

Define

$$\varphi = \sup\{\tilde{h}_{ij}\eta^i\eta^j \,:\, \|\eta\| = 1\} \tag{6.20}$$

and w by

$$w = \varphi e^{-\lambda t}, \quad \lambda > 0. \tag{6.21}$$

We claim that w is bounded. Let $0 < T < T^*$, and $x_0 = x(t_0)$, $0 < t_0 \le T$, be a point in $M(t_0)$ such that

$$\sup_{M_0} w < \sup\{\sup_{M(t)} w \,:\, 0 < t \le T\} = w(x_0). \tag{6.22}$$

Arguing as in the proof of Lemma 6.1, we introduce Riemannian normal coordinates ξ^i in $x_0 \in M(t_0)$ such that

$$\varphi(x_0) = \tilde{h}_n^n \tag{6.23}$$

and we may pretend as before that w is defined by

$$w = \tilde{h}_n^n e^{-\lambda t}. \tag{6.24}$$

Applying the maximum principle we deduce from (6.17)

$$0 \le -\lambda w + c + cw, \tag{6.25}$$

where we used that $\tilde{f}$ is concave, the estimates in Lemma 6.3 and 6.4, and also the inequality (1.7) to estimate the terms involving the derivatives of the second fundamental form; we should also point out that, because of the homogeneity of F,

$$F_i = F^{kl} h_{kl;i}. \tag{6.26}$$

Thus, the lemma is proved if λ is chosen large enough.

3. The Case of Theorem 0.8.

We first consider the case $K_N \le 0$.

Lemma 6.7. *Let $F \in (H)$, $K_N \le 0$, $\Phi(t) = t$, $M_0 = M_1$, and suppose that $f \in C^{2,\alpha}(\overline{\Omega})$ satisfies (0.19). Let $M(t)$ be strictly convex solutions of the evolution problem in a maximal time-interval $[0, T^*)$, then there are constants λ and c such that the principal curvatures can be estimated by*

$$e^{-\lambda t} \le \kappa_i \le c. \tag{6.27}$$

Proof. First, we observe that in view of Lemma 4.2

$$F \le f \tag{6.28}$$

and hence

$$0 < \kappa_i \le c, \tag{6.29}$$

because

$$F = F^{ij} h_{ij} \tag{6.30}$$

and F^{ij} is by assumption uniformly positive definite in $\overline{\Gamma}_+$.

Thus, it remains to prove the lower estimate in (6.27). The proof is identical to that of Lemma 6.6 with the only exception, that, when we apply the maximum principle, we have to use the assumption (0.19) in order to neglect the quadratic terms in w.

Consider now the second part of Theorem 0.8, $K_N > 0$.

Lemma 6.8. *Let $F \in (H)$, $K_N > 0$, $\Phi(t) = \log t$, $M_0 = M_2$, and suppose that $0 < f \in C^{2,\alpha}(\overline{\Omega})$. Let $M(t)$ be strictly convex solutions in a maximal time-interval $[0, T^*)$, then there are constants ε_0 and c such that*

$$0 < \varepsilon_0 \le F \le c \quad \forall t. \tag{6.31}$$

Proof. First, we obtain from Lemma 4.2

$$\Phi \ge \tilde{f} \quad \forall t, \tag{6.32}$$

hence, the lower estimate in (6.31). To prove the upper estimate, we define

$$w = \log(\Phi - \tilde{f}) + \log \chi + \lambda u, \tag{6.33}$$

where λ is supposed to be large and χ is the function in Lemma 5.8. We claim that w is bounded from above. Let $0 < T < T^*$, and $x_0 = x(t_0)$, $0 < t_0 \le T$, be a point in $M(t_0)$ such that

$$\sup_{M_0} w < \sup\{\sup_{M(t)} w \,:\, 0 < t \le T\} = w(x_0). \tag{6.34}$$

Combining the equations (3.19), (5.26) and (5.7) we conclude from the maximum principle

$$\begin{aligned} 0 \le \dot{w} - \dot{\Phi} F^{ij} w_{ij} \le\; & \dot{\Phi} F^{ij} \log(\Phi - \tilde{f})_i \log(\Phi - \tilde{f})_j \\ & - \dot{\Phi} F^{ij} \log \chi_i \log \chi_j \\ & + (\Phi - \tilde{f}) \frac{\bar{H}}{n} v + K_N \dot{\Phi} F^{ij} g_{ij} \\ & - \lambda (\Phi - \tilde{f}) v^{-1} - \lambda \dot{\Phi} F^{ij} \bar{h}_{ij} + c\lambda \end{aligned} \tag{6.35}$$

Let us first consider the terms involving the derivatives; since $Dw = 0$ they are equal to

$$2\lambda \dot{\Phi} F^{ij} \chi_i u_j \chi^{-1} + \lambda^2 \dot{\Phi} F^{ij} u_i u_j. \tag{6.36}$$

The first term is non-positive, since

$$\begin{aligned} F^{ij} \chi_i u_j &= F^{ij} v_i u_j \eta + \dot{\eta} F^{ij} u_i u_j v \\ &= -F^{ij} h_i^k u_k u_j \eta v^2 + \frac{\bar{H}}{n} F^{ij} u_i u_j v \eta + \dot{\eta} F^{ij} u_i u_j v \\ &= -F^{ij} h_i^k u_k u_j \eta v^2 \le 0 \end{aligned} \tag{6.37}$$

where we used (5.29) and (5.24).

Thus the right-hand side of inequality (6.35) can be estimated from above by

$$(\Phi - \tilde{f})\frac{\overline{H}}{n}v - \lambda(\Phi - \tilde{f})v^{-1} + c(1 + \lambda^2) \tag{6.38}$$

which yields the desired estimate if λ is chosen large enough. Here, we also used the assumption that F^{ij} is uniformly bounded in $\overline{\Gamma}_+$.

Next, let us prove the a priori estimates for the principal curvatures.

Lemma 6.9. *Suppose that the assumptions of the preceding lemma are valid and that in addition f satisfies (0.19), then the principal curvatures of the evolution hypersurfaces can be estimated by*

$$e^{-\lambda t} \le \kappa_i \le c \quad \forall t \tag{6.39}$$

for suitable constants λ and c.

The proof is identical to that of Lemma 6.7 since we know already upper and lower bounds for F.

7. Convergence to a stationary solution.

We are ready to prove the Theorems. Let $M(t)$ be a flow with initial hypersurface $M_0 = M_1$ or $M_0 = M_2$. Let us look at the scalar version of the flow, cf. (2.24),

$$\frac{\partial u}{\partial t} = -(\Phi - \tilde{f})v. \tag{7.1}$$

This is a scalar parabolic differential equation defined on the cylinder

$$Q_{T^*} = [0, T^*) \times S_0 \tag{7.2}$$

with initial value $u_0 \in C^{4,\alpha}(S_0)$, where $u_0 = u_i$, $i \in \{1, 2\}$. S_0 is a geodesic sphere equipped with the induced metric. In view of the a priori estimates we have proved in the preceding sections, we know that

$$|u|_{2,0,S_0} \le c \tag{7.3}$$

and

$$F \quad \textit{is uniformly elliptic in} \quad u \tag{7.4}$$

independent of t. Furthermore, F is concave and thus, we can apply the regularity results in Krylov [11, Chapter 5.5] to conclude that uniform $C^{2,\alpha}$-estimates are valid, leading further to uniform $C^{4,\alpha}$-estimates in view of the regularity results for linear operators.

Therefore, the maximal time interval is unbounded, i.e. $T^* = \infty$.

Now, integrate (7.1) and observe that the right-hand side has a sign to obtain

$$|u(t,x) - u(0,x)| = \int_0^t |\Phi - \tilde{f}|v \geq \int_0^t |\Phi - \tilde{f}|, \tag{7.5}$$

i.e.

$$\int_0^\infty |\Phi - \tilde{f}| < \infty \quad \forall x \in S_0. \tag{7.6}$$

Thus, for any $x \in S_0$ there is a sequence $t_k \to \infty$ such that $(\Phi - \tilde{f}) \to 0$. On the other hand, $u(\cdot, x)$ is monotone and therefore

$$\lim_{t\to\infty} u(t,x) = \tilde{u}(x) \tag{7.7}$$

exists and is of class $C^{4,\alpha}(S_0)$ in view of the a priori estimates. We finally deduce that $\tilde{u}$ is a stationary solution of our problem and that

$$\lim_{t\to\infty} (\Phi - \tilde{f}) = 0. \tag{7.8}$$

References.

[1] B. Andrews, *Contraction of convex hypersurfaces in Euclidean space*, preprint.

[2] I. Bakelman & B. Kantor, *Existence of spherically homeomorphic hypersurfaces in Euclidean space with prescribed mean curvature*, Geometry and Topology, Leningrad, **1** (1974) 3–10.

[3] L. Caffarelli, L. Nirenberg & J. Spruck, *The Dirichlet problem for nonlinear second order elliptic equations, III: Functions of the eigenvalues of the Hessian*, Acta Math. **155** (1985) 261–301.

[4] L. Caffarelli, L. Nirenberg & J. Spruck, *The Dirichlet problem for nonlinear second order elliptic equations, IV: Starshaped compact Weingarten hypersurfaces*, In: Current topics in partial differential equations, Y. Ohya, K. Kasahara & N. Shimakura (eds.), Kinokunize, Tokyo, 1985.

[5] P. Delanoë, *Plongements radiaux* $S^n \to \mathbf{R}^{n+1}$ *à courbure de Gauss positive prescrite*, Ann. Sci. École Norm. Sup. (4) **18** (1985), no 4, 635–649.

[6] M. P. Do Carmo & F. W. Warner, *Rigidity and convexity of hypersurfaces in spheres*, J. Diff. Geom. **4** (1970) 133–144.

[7] K. Ecker & G. Huisken, *Immersed hypersurfaces with constant Weingarten curvature*, Math. Ann. **283** (1989) 329–332.

[8] C. Gerhardt, *Flow of nonconvex hypersurfaces into spheres*, J. Diff. Geom. **32** (1990) 299–314.

[9] C. Gerhardt, *Closed Weingarten hypersurfaces in Riemannian manifolds*, preprint.

[10] D. Gromoll, W. Klingenberg & W. Meyer, *Riemannsche Geometrie im Großen*, Lecture Notes in Mathematics, Vol. 55, Springer, Berlin–Heidelberg–New York, 1975.

[11] N. V. Krylov, *Nonlinear elliptic and parabolic equations of second order*, Reidel, Dordrecht, 1987.

[12] D. S. Mitrinovic, *Analytic inequalities*, Springer, Berlin–Heidelberg–New York, 1970.

[13] V. I. Oliker, *Hypersurfaces in* $\mathbf{R}^{n+1}$ *with prescribed Gaussian curvature and related equations of Monge–Ampère type*, Commun. Partial Diff. Equations **9** (1984) 807–838.

[14] A. E. Treibergs & S. W. Wei, *Embedded hypersurfaces with prescribed mean curvature*, J. Diff. Geom. **18** (1983) 513–521.

[15] J. I. E. Urbas, *On the expansion of convex hypersurfaces by symmetric functions of their principal radii of curvature*, J. Diff. Geom. **33** (1991) 91–125.

Received July 4, 1994.

Ruprecht–Karls–Universität Heidelberg
Institut für Angewandte Mathematik
Im Neuenheimer Feld 294
D–69120 Heidelberg
E-mail address: gerhardt@math.uni-heidelberg.de

Asymptotic limits of a Ginzburg-Landau type functional

MIN-CHUN HONG, JÜRGEN JOST, AND MICHAEL STRUWE

Dedicated to Stefan Hildebrandt on the occasion of his 60th birthday

We consider a Ginzburg-Landau type functional involving a section of a line bundle over a Riemann surface and a connection on this bundle. We select a scaling parameter ε in such a way that self-duality is preserved and that the infimum of the functional stays bounded as ε tends to 0, and we perform a corresponding limit analysis.

1. Introduction.

Let Ω be an open domain on some Riemann surface Σ, with – possibly empty – smooth boundary $\partial\Omega$. Let L be a complex line bundle over $\bar{\Omega}$, equipped with a Hermitian metric $\langle\cdot,\cdot\rangle$. For a section u of L, we write

$$|u(x)| = \langle u(x), u(x)\rangle^{\frac{1}{2}}.$$

Ginzburg-Landau functionals are defined for a section u of L and a unitary connection A on L. That A is unitary means that

$$d < u, v >= \langle \nabla_A u, v\rangle + \langle u, \nabla_A v\rangle \tag{1.1}$$

for sections u, v of L, where d is the exterior derivative and ∇_A is the covariant derivative defined by A. We employ the physical convention

$$\nabla_A u := (d - iA)u,$$

i.e. we let A be real valued. In local coordinates (x^1, x^2) on Σ, we write

$$\nabla_A^k := \nabla_A\left(\frac{\partial}{\partial x^k}\right) =: \partial_k - iA^k \quad (k = 1, 2).$$

The curvature of A is

$$F := dA,$$

i.e.

$$F^{kj} = \partial_k A^j - \partial_j A^k = i\left(\nabla_A^k \nabla_A^j - \nabla_A^j \nabla_A^k\right). \tag{1.2}$$

Again, this is the physical convention. The prototype of a Ginzburg-Landau functional is the Yang-Mill-Higgs functional

$$E(u, A, \Omega) = \int_\Omega \left\{ |dA|^2 + |\nabla_A u|^2 + \frac{1}{4}\left(1 - |u|^2\right)^2 \right\} \mathrm{dvol}(x), \tag{1.3}$$

where dvol is the volume form of some fixed Kähler metric on Σ. The Euler-Lagrange equations for E are

$$\Delta_A u = -\frac{1}{2}\left(1 - |u|^2\right) u \tag{1.4}$$

$$\partial_k F^{kj} = -\mathrm{Im}\left\langle \left(\partial_j - iA^j\right) u, u \right\rangle, \tag{1.5}$$

where

$$\Delta_A = \nabla_A^k \nabla_A^k \tag{1.6}$$

is the Laplacian defined by A, with the analysts' sign convention, and where we employ the usual summation convention.

In more invariant form, (1.5) may be rewritten as

$$- * dh = Re\,\langle iu, \nabla_A u \rangle \tag{1.7}$$

with

$$h := *dA,$$

$*$ being the usual star operator defined by the conformal structure. An important feature of E is its gauge invariance, i.e. its invariance under the substitution

$$(u, A) \to (u \exp(i\psi), A + d\psi) \tag{1.8}$$

for a real valued function ψ.

Another important feature of E is the self-duality. Namely, decomposing ∇_A into its $(1,0)$ and $(0,1)$ parts,

$$\nabla_A = \partial_A + \bar{\partial}_A,$$

in case $\Omega = \mathbb{R}^2$, and if $|u(x)| \to 1$, $\nabla_A u(x) \to 0$ sufficiently fast as $|x| \to \infty$, then E can be rewritten as

$$E(u, A) = \int_{\mathbb{R}^2} \left\{ 2 \left| \bar{\partial}_A u \right|^2 + \left| *F - \frac{1}{2}(1 - |u|^2) \right|^2 \right\} dx + 2\pi d, \tag{1.9}$$

for some integer d, the so-called vortex number; see [18], p. 54. Thus, we see that the infimum of E, namely $2\pi d$, is attained iff the vortex equations

$$\bar{\partial}_A u = 0 \tag{1.10}$$

and

$$*F = \frac{1}{2}(1 - |u|^2) \tag{1.11}$$

are satisfied. Of course, since E is nonnegative, this is possible only if $d \geq 0$. (If $d < 0$, one should consider antiholomorphic sections instead of holomorphic ones.) Taubes ([28]) showed that for any collection of d points $x_j \in \mathbb{R}^2$, possibly with multiplicities, there exists a solution, unique up to gauge equivalence, of the vortex equations with

$$u(x_j) = 0 \quad j = 1, \dots, d. \tag{1.12}$$

Likewise, if Ω is a compact Riemann surface Σ, E can be rewritten as

$$E(u, A, \Sigma) = \int_{\Sigma} \left\{ 2 \left| \bar{\partial}_A u \right|^2 + \left| \Lambda F - \frac{1}{2}(1 - |u|^2) \right|^2 \right\} \mathrm{dvol}(x) + 2\pi \deg L, \tag{1.13}$$

where $\deg L$ is of course the degree of L and Λ denotes contraction with the Kähler form of Σ. Thus, again the infimum $2\pi \deg L$ of E is realized by the solutions of the vortex equations

$$\bar{\partial}_A u = 0 \tag{1.14}$$

$$\Lambda F = \frac{1}{2}(1 - |u|^2). \tag{1.15}$$

Integrating (1.15) over Σ, one sees that a solution can only exist if

$$2\pi \deg L < \frac{1}{2} \mathrm{Vol} \Sigma. \tag{1.16}$$

This obstruction, however, is easily circumvented by replacing the term $(1-|u|^2)$ by $(\tau-|u|^2)$, for $\tau\in\mathbb{R}$ satisfying

$$2\pi\deg L<\frac{\tau}{2}\mathrm{Vol}\Sigma. \tag{1.17}$$

The resulting equations have been solved and studied by Bradlow and García-Prada, see e.g. [5], [6], [12], [13], [14].

The Ginzburg-Landau functional originated in the theory of superconductivity, where Ω represents the cross section of a wire and $u(x)$ is a complex order parameter. $|u(x)|=1$ corresponds to a superconducting phase, and therefore, one wishes to constrain u to have absolute value 1. Of course, there are topological obstructions for that; namely if d in (1.9) or $d:=\deg L$ in (1.13) is not 0, then any section u will have at least $|d|$ zeroes (counted with multiplicity), the so-called vortices. For this reason, the family of functionals

$$\tilde{E}_\varepsilon(u,A,\Omega):=\int_\Omega\left\{|dA|^2+|\nabla_A u|^2+\frac{1}{4\varepsilon^2}\left(1-|u|^2\right)^2\right\}\mathrm{dvol}(x), \tag{1.18}$$

depending on a real parameter $\varepsilon>0$, has been studied, and the limiting behavior as $\varepsilon\to 0$ has been investigated. The first mathematical treatment is due to Berger-Chen [1]. They performed the limit analysis in the class of rotationally symmetric solutions $u_\varepsilon, A_\varepsilon$ on $\mathbb{R}^2$. Their results are quite explicit, and as $\varepsilon\to 0$, they found a "nonlinear desingularization"; namely, $h_\varepsilon:=*dA_\varepsilon$ tends to a limit h that satisfies the London equation

$$\Delta h-h=-2\pi d\delta(x)\text{ und }h(x)\to 0\quad\text{for }|x|\to\infty,$$

where $\delta(x)$ is the Dirac delta function, and d is the vortex number. Of course, since only rotationally symmetric solutions are considered, in the limit, one obtains a singularity of multiplicity d at the origin.

In the nonsymmetric case, Chen [7] also obtained an existence result for solutions on bounded domains $\Omega\subset\mathbb{R}^2$ with prescribed d for any fixed positive ε.

More recently, Bethuel-Brézis-Hélein ([2], [3]) simplified the functional $\tilde{E}_\varepsilon$ by dropping the term $|dA|^2$ and succeeded in performing the limit analysis without the symmetry assumption. They found, in particular, that the singularities of minimizers $u_\varepsilon, A_\varepsilon$ decouple in the limit, i.e. that all singularities of the limit have multiplicity ± 1. Thus, in particular, for $|d|\neq 1$, the rotationally symmetric solutions of Berger-Chen cannot be minimizing (for the modified functional) for sufficiently small ε. The results in [2], [3] were obtained only for star-shaped domains in $\mathbb{R}^2$. This restriction was removed, and some of the arguments were considerably simplified by Struwe ([25], [26]).

Related results were obtained by Hardt-Lin [16], Chen-Lin [9], Lin [19], and, more recently, by del Pino-Felmer [11].

Bethuel-Riviére [4], and more recently Orlandi [22] and J. Qing [23], studied the original functionals $\tilde{E}_\varepsilon$ and obtained results analogous to those of Bethuel-Brézis-Hélein and Struwe.

A characteristic feature of the functionals $\tilde{E}_\varepsilon$ is that self-duality is lost for $\varepsilon \neq 1$, and that

$$\liminf_{\varepsilon\to 0}\left(\inf \tilde{E}_\varepsilon\right) \tag{1.19}$$

is infinite under fixed nontrivial natural boundary conditions. In order to restore self-duality (and to thus make the results better applicable in Riemann surface theory) and to get finite limits for the infima of the functionals, we consider here instead the functionals

$$E_\varepsilon(u, A, \Omega) := \int_\Omega \left\{\varepsilon^2 |dA|^2 + |\nabla_A u|^2 + \frac{1}{4\varepsilon^2}\left(1 - |u|^2\right)^2\right\} \mathrm{dvol}(x). \tag{1.20}$$

The Euler-Lagrange equations for (1.20) are

$$\Delta_A u = -\frac{1}{2\varepsilon^2}\left(1 - |u|^2\right) u \tag{1.21}$$

$$\varepsilon^2 \partial_k F^{kj} = -\mathrm{Im}\left\langle\left(\partial_j - iA_j\right) u, u\right\rangle, \tag{1.22}$$

or, with $h := *dA$, equivalently

$$-\varepsilon^2 * dh = \mathrm{Re}\left\langle iu, \nabla_A u\right\rangle. \tag{1.23}$$

Again, the latter are satisfied by solutions of the vortex equations on $\mathbb{R}^2$,

$$\overline{\partial}_A u = 0 \tag{1.24}$$

$$-i\varepsilon^2 * F = \frac{1}{2}\left(1 - |u|^2\right). \tag{1.25}$$

On a general Riemann surface Σ, the second equation becomes

$$\varepsilon^2 \Lambda F = \frac{1}{2}\left(1 - |u|^2\right). \tag{1.26}$$

The compatibility condition (1.16) now becomes

$$2\pi\varepsilon^2 \deg L < \frac{1}{2}\mathrm{Vol}(\Sigma), \tag{1.27}$$

and this is obviously satisfied for sufficiently small $\varepsilon > 0$.

In fact, if $\Omega \subset \mathbb{R}^2$ is a flat domain, E_ε can be obtained from $E = E_1$ by a simple rescaling of the domain. Namely,

$$E_\varepsilon(u_\varepsilon, A_\varepsilon, \Omega) = E(v_\varepsilon, B_\varepsilon, \Omega_\varepsilon), \tag{1.28}$$

if the various quantities are related via

$$\begin{aligned} v_\varepsilon(x) &= u_\varepsilon(x_0 + \varepsilon x), \\ B_\varepsilon(x) &= A(x_0 + \varepsilon x) = \varepsilon A_1(x_0 + \varepsilon x)dx_1 + \varepsilon A_2(x_0 + \varepsilon x)dx_2, \end{aligned}$$

and $\Omega = \{x_0 + \varepsilon x : x \in \Omega_\varepsilon\}$ for some fixed $x_0 \in \Omega$; that is,

$$\Omega_\varepsilon = \left\{\frac{1}{\varepsilon}(y - x_0) : y \in \Omega\right\}.$$

Of course, solutions of the corresponding equations are likewise related by rescaling. This rescaling with parameter ε will therefore be often applied in the present paper. In particular, the limiting analysis as $\varepsilon \to 0$ becomes equivalent to an analysis of solutions of (1.4), (1.5) near vortices, i.e. where $u(x) = 0$. Since these equations admit solutions having vortices of arbitrary degree, in contrast to [3], no decoupling of the singularities will occur.

If Ω happens to have a nonempty boundary, one wishes to impose boundary conditions on u and A. Because of the gauge invariance (1.8), it is not meaningful to impose Dirichlet boundary conditions for both u and A. Instead, natural boundary conditions are

$$|u| = 1 \quad \text{on } \partial\Omega \tag{1.29}$$

and

$$\langle iu, \nabla_A(\tau)u\rangle = g \quad \text{on } \partial\Omega \tag{1.30}$$

where τ denotes a unit tangent vector field along $\partial\Omega$, and g is given.

In view of the scaling properties and of (1.23), one might also replace (1.30) by the condition

$$\langle iu, \nabla_A(\tau)u\rangle = \varepsilon g \quad \text{on } \partial\Omega. \tag{1.31}$$

This would lead to a nicer behavior of the curvature dA for minimizers of E_ε in the limit $\varepsilon \to 0$.

If Ω is diffeomorphic to the disc, we may consider a section u as a complex valued function with prescribed degree $\deg(u, \partial\Omega) = d$. In the general case, the topology of the bundle L also imposes a global condition on sections.

In this paper we will focus on the latter topological conditions and we consider the asymptotic behavior as $\varepsilon \to 0$ of sections of a line bundle L over a compact Riemann surface Σ without boundary which minimize the scaled, self-dual Ginzburg-Landau energy E_ε.

In this self-dual case, the minimizers of E_ε are precisely the solutions of the first-order vortex equations (1.24) and (1.25), respectively (1.26). For these solutions, we have the following result.

Theorem 1.1. *Let $(u_\varepsilon, A_\varepsilon)$ be solutions of (1.24), (1.26) on some compact Riemann surface Σ, with fixed $d = \deg L \geq 0$. Then for some sequence $\varepsilon_n \to 0$, there exist points x_j, $j = 1, \dots, l \leq d$, such that*

$$|u_\varepsilon| \to 1, \ \nabla_{A_\varepsilon} u_\varepsilon \to 0, \ dA_\varepsilon \to 0$$

uniformly on compact subsets of $\Sigma \setminus \{x_1, \dots, x_l\}$. Moreover, for $h_\varepsilon := *dA_\varepsilon$, we have

$$h_\varepsilon \to 2\pi \sum_{j=1}^{l} \delta(x_j)$$

in the sense of measures, where the delta functions have to be counted with multiplicity.

An analogous result holds on $\mathbb{R}^2$; however, one has to deal with dichotomy, that is, vortices moving off to infinity.

Theorem 1.1 yields a method for degenerating a line bundle L on Σ of degree d into a flat line bundle with $|d|$ singular points (counted with multiplicity) and a covariantly constant section.

The proof of Theorem 1.1 gives similar but slightly weaker results on the asymtotic behaviour in the interior of minimizers of E_ε in the general case.

If u were real instead of complex valued (and A trivial), one would obtain a functional of the type studied by Modica and Mortola ([21], [20]) and many other authors. Such functionals are of great physical importance and mathematical interest. For real valued u, in the limit $\varepsilon \to 0$, a phase transition takes place along a real hypersurface, whereas in our context of a complex valued u, a different phase is realized only on a set of real codimension 2. One might expect such a behavior also for similar functionals on domains of higher dimension.

Most of the research represented in the present paper was carried out during the summer 1994 when the three authors met in Zürich. Since then, a very interesting independent development has taken place that relates to our work. Namely, the above self-duality equations are the two-dimensional analogue of the Seiberg-Witten equations that Taubes [30], [31] used to relate the Seiberg-Witten and Gromov invariants in four-dimensional geometry through a similar change of scale.

Acknowledgement. The first author gratefully acknowledges the hospitality of the Mathematik Department of the ETH Zürich where he was a Postdoc, and the second one the hospitality of the Forschungsinstitut für Mathematik at the ETH Zürich and, in particular, its director Jürgen Moser. The second author was also generously supported by the Leibniz program and the SFB 237 of the DFG.

2. Preliminary estimates.

Most of the results in this section are well-known from previous work cited above.

Since the space of solutions of our equations is invariant under gauge transformations, one cannot have regularity results without fixing the gauge. The most common gauge is the Coulomb gauge where one requires

$$d^*A = 0. \tag{2.1}$$

This may be supplemented by the boundary condition

$$A(\nu) = 0 \quad \text{on } \partial\Omega, \tag{2.2}$$

where ν denotes a unit normal vector field. Another gauge that is more useful for rotationally symmetric solutions was discovered by Cronström [10]. If we use standard polar coordinates (r, θ) on $\mathbb{R}^2$ and write

$$A = S(r,\theta)d\theta + T(r,\theta)dr, \tag{2.3}$$

then Cronström's gauge condition is simply

$$T \equiv 0, \tag{2.4}$$

that is,

$$A = S(r,\theta)d\theta. \tag{2.5}$$

If such an A is radially symmetric (which holds if in addition (2.1)is satisfied), then

$$A = S(r)d\theta. \tag{2.6}$$

Returning to the Coulomb gauge, we note that (2.1) and (1.7) yield

$$-\Delta A = (d^*d + dd^*)A = \langle iu, \nabla_A u \rangle , \tag{2.7}$$

and (1.4) and (2.7) together constitute an elliptic system; similarly (1.21) and (1.22) together with (2.1). Therefore, whenever we fix the Coulomb gauge, we may apply elliptic regularity theory. In particular, smoothness of solutions is automatic.

Lemma 2.1. *Any solution $(u_\varepsilon, A_\varepsilon)$ of the Ginzburg-Landau equations (1.21), (1.22) satisfies*

$$|u_\varepsilon| \leq 1 \quad in \;\; \Omega. \tag{2.8}$$

Proof. As noted, in the Coulomb gauge, u_ε is smooth. We compute

$$\frac{1}{2}\Delta |u_\varepsilon|^2 = -\frac{1}{2\varepsilon^2} |u_\varepsilon|^2 \left(1 - |u|^2\right) + |\nabla_{A_\varepsilon} u_\varepsilon|^2 , \tag{2.9}$$

and (2.8) follows from the maximum principle. □

For small radii $r > 0$, $x_0 \in \Sigma$ let $B_r(x_0;\Sigma)$ denote the geodesic ball of radius r around x_0 in Σ and let $B_r(x_0) = B_r(x_0;\mathbb{R}^2)$ for brevity. Also denote by $B = B_1(0;\mathbb{R}^2)$ the unit disc.

Lemma 2.2. *For any $E_0 > 0$ there exists a constant $C = C(\Sigma, E_0)$ such that for $0 < \epsilon < 1$ any solution (u_ϵ, A_ϵ) of the Ginzburg-Landau equation (1.21), (1.22) on a ball $B_\epsilon(x_0;\Sigma)$ with*

$$E_\epsilon\big(u_\epsilon, A_\epsilon; B_\epsilon(x_0;\Sigma)\big) \leq E_0$$

satisfies

$$|\nabla |u_\epsilon|| \, (x_0) \leq C\epsilon^{-1}.$$

(We may assume $C \geq 1$.)

Proof. Rescaling $v(x) = u_\epsilon(x_0 + \epsilon x)$, etc., we may assume $\epsilon = 1$, $B_\epsilon(x_0; \Sigma) = B \subset \mathbb{R}^2$, and we denote $u_\epsilon = u$, $A_\epsilon = A$ for convenience. For simplicity, we assume that on B we have the standard Euclidean metric.

Specifying the Coulomb gauge

$$d^* A = 0 \qquad \text{on } B$$

with boundary condition

$$A(\nu) = 0 \qquad \text{on } \partial B,$$

by a result of Uhlenbeck [32], p. 35 f., we can estimate

$$\int_{\partial B} |A|^2 \, do + \int_B |\nabla A|^2 \, dx = \int_B |dA|^2 \, dx \leq E_1(u, A; B) \leq E_0.$$

Moreover, equation (1.22) – or rather (1.23) – becomes

$$-\Delta A = (d^* d + dd^*)A = \mathrm{Im}\langle u, \nabla_A u\rangle \qquad \text{in } B,$$

and we conclude that $A \in H^{2,2}_{loc}(B)$ with

$$\|A\|_{H^{2,2}(B')} \leq C\left(\|A\|_{L^2(B)} + \|\Delta A\|_{L^2(B)}\right) \leq CE_1(u, A; B) \leq CE_0$$

for any domain $B' \subset\subset B$, where $C = C(B')$. In particular, $A \in L^\infty(B') \cap W^{1,p}(B')$ for any $p < \infty$, any $B' \subset\subset B$, with bounds depending only on E_0, B', and p.

Now rewrite equation (1.21) in the form

$$-\Delta u = \frac{1}{2} u(1 - |u|^2) + (\Delta_A u - \Delta u)$$

and observe that

$$|\Delta_A u - \Delta u| \leq C\big(|A|\,|\nabla_A u| + (|\nabla A| + |A|^2)\,|u|\big)$$

by the above is locally bounded in L^2.

Hence, using also Lemma 2.1, in a first step we obtain that $u \in H^{2,2}_{loc}(B) \hookrightarrow W^{1,p}_{loc}(B)$ with uniform local bounds for any $p < \infty$. Thus, also the error term $|\Delta_A u - \Delta u|$ is locally bounded in L^p for any $p < \infty$. In particular, if we fix some $p > 2$ we find that $u \in W^{2,p}_{loc}(B) \hookrightarrow C^1(B)$ together with the (gauge-invariant) bound $|\nabla\,|u||\,(0) \leq C = C(E_0)$. □

Lemma 2.3. *There exists a constant* $\epsilon_1 > 0$ *with the following property: If* $0 < \epsilon \leq \rho < 1$ *and* (u_ϵ, A_ϵ) *solves (1.21), (1.22) on a ball* $B_\rho(x_0) \subset \Sigma$ *with* $E_\epsilon\big(u_\epsilon, A_\epsilon; B_\rho(x_0)\big) < \epsilon_1$, *then* $|u(x_0)| \geq \frac{1}{2}$.

Proof. Suppose $|u(x_0)| \leq \frac{1}{2}$. We may assume $E_\epsilon\big(u_\epsilon, A_\epsilon; B_\rho(x_0)\big) \leq 1$. Then by Lemma 2.2 for all $x \in \Sigma$ such that

$$|x - x_0| \leq C^{-1}\epsilon \leq \rho, \quad C = C(\Sigma),$$

we have $|u(x)|^2 \leq \frac{1}{2}$ and hence

$$E\big(u_\epsilon, A_\epsilon; B_\rho(x_0)\big) \geq \int_{B_{C^{-1}\epsilon}(x_0)} \frac{(1-|u|^2)^2}{\epsilon^2}\,dx \geq \frac{\pi}{4C^2}.$$

Thus, if we let $\epsilon_1 = \frac{\pi}{4C^2} < 1$, the claim follows. □

3. A Bochner type formula and consequences.

The proof of Theorem 1.1 will be a consequence of a Bochner-type formula and an ε_0-regularity estimate for equations (1.4), (1.5). For a Ginzburg-Landau type system without magnetic field in a different context these tools were developed by Chen-Struwe [8]. They were first applied in the present context (still without a magnetic field and in the case $d = 0$) by Chen-Lin [9]. Here we extend this method to problems with magnetic field; moreover, we allow vortex behaviour, that is, $d > 0$. Moreover, we scale $\epsilon = 1$.

For reference, we recall equations (1.4), (1.5), that is

$$\partial_k F^{kj} = -\mathrm{Im}\left\langle \left(\partial_j - iA^j\right) u, u\right\rangle \tag{3.1}$$

$$\Delta_A u = -\frac{1}{2}\left(1 - |u|^2\right) u. \tag{3.2}$$

Our Laplacian on functions is

$$\Delta := \partial_j \partial_j \quad (= -d^* d) \quad (\text{"analysts' Laplacian"})$$

For a section u of L, and function s, we have the product rule

$$\nabla_A(su) = s\nabla_A u + (ds)u.$$

As always, A is a unitary connection on L. In geodesic normal coordinates at the point under consideration and denoting the curvature tensor of the metric of Σ by

$$R_{jk} = R\left(\frac{\partial}{\partial x^j}, \frac{\partial}{\partial x^k}\right),$$

we derive the following identities

$$\frac{1}{2}\Delta\left|\nabla_A u\right|^2 = \left\langle\left(\nabla_A\right)^2 u, \left(\nabla_A\right)^2 u\right\rangle + \mathrm{Re}\langle\Delta_A\nabla_A u, \nabla_A u\rangle, \tag{3.3}$$

$$\begin{aligned} \mathrm{Re}\ \langle\Delta_A\nabla_A u, \nabla_A u\rangle &= \mathrm{Re}\langle\nabla_A^k\nabla_A^k\nabla_A^j u, \nabla_A^j u\rangle \\ &= \mathrm{Re}\langle\nabla_A^k\nabla_A^j\nabla_A^k u, \nabla_A^j u\rangle + \mathrm{Re}\langle\nabla_A^k\left(-iF^{kj}u\right), \nabla_A^j u\rangle \\ &= \mathrm{Re}\langle\nabla_A^j\Delta_A u, \nabla_A^j u\rangle + 2\mathrm{Re}\langle -iF^{kj}\nabla_A^k u, \nabla_A^j u\rangle \\ &\quad +\mathrm{Re}\langle -i\left(\partial_k F^{kj}\right)u, \nabla_A^j u\rangle + \mathrm{Re}\langle R_{kj}\nabla_A^k u, \nabla_A^j u\rangle. \end{aligned} \tag{3.4}$$

Moreover, by (3.1), we have

$$\begin{aligned} \mathrm{Re}\langle -i\left(\partial_k F^{kj}\right)u, \nabla_A^j u\rangle &= -\mathrm{Im}\langle\nabla_A^j u, u\rangle\mathrm{Re}(-i\langle u, \nabla_A^j u\rangle) \\ &= \left(\mathrm{Im}\langle\nabla_A u, u\rangle\right)^2, \end{aligned} \tag{3.5}$$

and by (3.2)

$$\mathrm{Re}\left\langle\nabla_A\Delta_A u, \nabla_A u\right\rangle = -\frac{1}{2}\left(1-|u|^2\right)\left|\nabla_A u\right|^2 + \frac{1}{4}\left|d\,|u|^2\right|^2. \tag{3.6}$$

(Observe that $d\,|u|^2 = 2\mathrm{Re}\,\langle u, \nabla_A u\rangle$). From (3.1)–(3.4) we conclude

$$\frac{1}{2}\Delta\left|\nabla_A u\right|^2 \geq -\frac{1}{2}\left(1-|u|^2\right)\left|\nabla_A u\right|^2 - 2\,|F|\left|\nabla_A u\right|^2 - 2\,|R|\left|\nabla_A u\right|^2. \tag{3.7}$$

Next we consider the term

$$\frac{1}{2}\Delta\,|F|^2 = |\nabla F|^2 + F\Delta F. \tag{3.8}$$

Note that in order to be consistent with the convention $F = dA$ and $|dA|^2 = (\partial_1 A_2 - \partial_2 A_1)^2$, we have $|F|^2 = \left|F^{12}\right|^2$. By (3.3) there holds

$$\begin{aligned} \Delta F^{12} &= \partial_1\partial_1 F^{12} + \partial_2\partial_2 F^{12} = \partial_1\partial_1 F^{12} - \partial_2\partial_2 F^{21} \\ &= -\mathrm{Im}\left\langle\nabla_A^1\nabla_A^2 u, u\right\rangle - \mathrm{Im}\left\langle\nabla_A^2 u, \nabla_A^1 u\right\rangle \\ &\quad +\mathrm{Im}\left\langle\nabla_A^2\nabla_A^1 u, u\right\rangle + \mathrm{Im}\left\langle\nabla_A^1 u, \nabla_A^2 u\right\rangle. \\ &= F^{12}\,|u|^2 + 2\mathrm{Im}\left\langle\nabla_A^1 u, \nabla_A^2 u\right\rangle \end{aligned} \tag{3.9}$$

From (3.9) we derive

$$\begin{aligned} F\Delta F &= F^{12}\Delta F^{12} \\ &= (F^{12})^2 |u|^2 + 2F^{12}\mathrm{Im}\left\langle \nabla_A^1 u, \nabla_A^2 u \right\rangle \\ &= |F|^2 |u|^2 + F^{kj}\mathrm{Im}\langle \nabla_A^k u, \nabla_A^j u \rangle. \end{aligned} \tag{3.10}$$

Note incidentally that the term in (3.2) also equals

$$2\mathrm{Re}\langle -iF^{kj}\nabla_A^k u, \nabla_A^j u\rangle = 2F^{kj}\mathrm{Im}\langle \nabla_A^k u, \nabla_A^j u\rangle.$$

(3.8) and (3.10) imply

$$\frac{1}{2}\Delta |F|^2 \geq |F|^2 |u|^2 - 2|F| |\nabla_A u|^2. \tag{3.11}$$

Combining (3.7) and (3.11), we find

$$\begin{aligned} &\frac{1}{2}\Delta\left(|\nabla_A u|^2 + |F|^2\right) \\ &\geq |F|^2 |u|^2 - 6|F| |\nabla_A u|^2 - \frac{1}{2}\left(1 - |u|^2\right)|\nabla_A u|^2 - 2|R| |\nabla_A u|^2. \end{aligned} \tag{3.12}$$

It may also be useful to note the equation (which also follows from (3.3)–(3.6), (3.8), (3.10))

$$\begin{aligned} &\frac{1}{2}\Delta\left(|\nabla_A u|^2 + |F|^2\right) \\ &= \left|(\nabla_A)^2 u\right|^2 + \langle \nabla_A u, u\rangle^2 + |\nabla F|^2 - 3F^{kj}\mathrm{Im}\left\langle \nabla_A^k u, \nabla_A^j u\right\rangle \\ &\quad + |F|^2 |u|^2 - \frac{1}{2}\left(1 - |u|^2\right)|\nabla_A u|^2 + \mathrm{Re}\left\langle R_{kj}\nabla_A^k u, \nabla_A^j u\right\rangle. \end{aligned} \tag{3.13}$$

Finally, we consider the term

$$\frac{1}{2}\Delta\left(1 - |u|^2\right)^2 = \left|\nabla |u|^2\right|^2 + \left(|u|^2 - 1\right)\Delta |u|^2. \tag{3.14}$$

From (3.2) we derive

$$\Delta |u|^2 = -|u|^2\left(1 - |u|^2\right) + 2|\nabla_A u|^2. \tag{3.15}$$

Equations (3.14), (3.15) yield

$$\frac{1}{2}\Delta\left(1 - |u|^2\right)^2 = \left|\nabla |u|^2\right|^2 + 2\left(|u|^2 - 1\right)|\nabla_A u|^2 + |u|^2\left(1 - |u|^2\right)^2. \tag{3.16}$$

Finally, denoting

$$e(u,A) := |\nabla_A u|^2 + |F|^2 + \frac{1}{4}\left(1-|u|^2\right)^2,$$

from (3.12) and (3.16) we obtain the differential inequality

$$\begin{aligned} \frac{1}{2}\Delta e(u,A) \ \geq \ & |u|^2\left(|F|^2 + \frac{1}{4}\left(1-|u|^2\right)^2\right) && (3.17) \\ & - |\nabla_A u|^2\left(6|F| + \left(1-|u|^2\right) + 2|R|\right). && (3.18) \end{aligned}$$

Suppose $|u| \geq \frac{1}{2}$. Then we can estimate

$$\begin{aligned} 6|\nabla_A u|^2|F| \ &\leq \ |u|^2|F|^2 + 36|\nabla_A u|^4 \\ |\nabla_A u|^2(1-|u|^2) \ &\leq \ \frac{1}{4}|u|^2(1-|u|^2)^2 + 4|\nabla_A u|^4; \end{aligned}$$

that is, we have proved:

Proposition 3.1. *Let (u,A) be a solution of (3.3), (3.4) with $|u| \geq \frac{1}{2}$ on Ω. Then there holds*

$$\Delta e(u,A) \geq -Ce(u,A)^2 - 4|R|\,e(u,A), \qquad (3.19)$$

for some absolute constant C.

Observe that estimate (3.19) remains true for any $\epsilon \in]0,1]$ with the same constant C and the same bound for the curvature of Σ, as is easily seen by scaling.

As a consequence of Proposition 3.1 we derive

Theorem 3.2 (ε_0-regularity estimate). *Let (u,A) be a solution of equations (3.1) and (3.2) on B_{2R} for $\varepsilon = 1$ with $|u| \geq \frac{1}{2}$. There exists $\varepsilon_0 > 0$ and a constant $C_0 > 0$ with the following property: If $E(u,A;B_{2R}) < \varepsilon_0$, then*

$$\sup_{B_{R/2}} e(u,A) \leq C_0 R^{-2} E(u,A;B_R). \qquad (3.20)$$

Proof. As in [17] and [24], we choose $r_0 < R$ such that

$$(R-r_0)^2 \sup_{B_{r_0}} e(u,A) = \max_{0 \leq r \leq R}\left\{(R-r)^2 \sup_{B_r} e(u,A)\right\}$$

and let $x_0 \in \bar{B}_{r_0}$ be determined such that

$$e_0 := (e(u,A))(x_0) = \sup_{B_{r_0}} e(u,A).$$

Now we are going to prove that $e_0 \leq 4((R-r_0)^{-2})$. Assume by contradiction that $\varrho_0 = e_0^{-\frac{1}{2}} \leq \frac{R-r_0}{2}$. We rescale

$$v(x) = u(x_0 + \varrho_0 x),\ B(x) = \varrho_0 A(x_0 + \varrho_0 x),\ \nabla_B v = \varrho_0 \nabla_A u,\ dB = \varrho_0^2 dA.$$

Let

$$e_{\varrho_0}(v,B) = |\nabla_B v|^2 + \varrho_0^{-2}|dB|^2 + \frac{\varrho_0^2}{4}\left(1-|v|^2\right)^2 = \varrho_0^2 e(u,A).$$

Then we have

$$1 = e_{\varrho_0}(v,B)(0)$$

while

$$\sup_{B_1} e_{\varrho_0}(v,B) = \varrho_0^2 \sup_{B_{\varrho_0}(x_0)} e(u,A) \leq \varrho_0^2 \sup_{B_{\frac{R+r_0}{2}}} e(u,A) \leq 4\varrho_0^2 \sup_{B_{r_0}} e(u,A) \leq 4.$$

From Proposition 3.1, we get

$$\Delta e_{\varrho_0} \geq -C e_{\varrho_0}$$

where C is a constant. Then using Moser's sup-estimate, we have

$$\begin{aligned} 1 = (e_{\varrho_0}(v,B))(0) &\leq C\int_{B_1} e_{\varrho_0}(v,B)dx \\ &= C\int_{B_{\varrho_0}} e(u,A)dx \\ &\leq C\int_{B_R} e(u,A)dx < 1, \end{aligned} \tag{3.21}$$

if $\varepsilon_0 = C^{-1}$. This is a contradiction. Hence

$$e_0\left(\frac{R-r_0}{2}\right)^2 \leq 1,$$

and we get

$$\left(\frac{R}{2}\right)^2 \sup_{B_{R/2}} e(u,A) \leq (R-r_0)^2 e_0 \leq 4;$$

that is,

$$\sup_{B_{R/2}} e(v, A) \leq 16R^{-2}.$$

Scaling with R instead of ϱ_0, the desired conclusion then follows from (3.20).
□

We may assume that $\epsilon_0 < \epsilon_1$, where $\epsilon_1 > 0$ is the constant determined in Lemma 2.3. As a corollary we obtain a simple proof for the following result of Taubes [29].

Corollary 3.3 (Gap theorem). *Let (u, A) be a solution of equations (1.4) and (1.5) on $\mathbb{R}^2$. Then, if $E(u, A; \mathbb{R}^2) < \epsilon_0$, it follows that*

$$|u|^2 = 1, \ \nabla_A u = 0 \ und \ dA = 0 \quad on \ \mathbb{R}^2.$$

Proof. By Lemma 2.3 we have $|u| \geq \frac{1}{2}$ everywhere. Thus the claim follows if we let $R \to \infty$ in Theorem 3.2. □

4. Proof of Theorem 1.1.

As in [25], [26], we obtain the concentration points x_j from the following Lemma.

For $0 < \varepsilon < \varepsilon_0$, $\varrho > 0$, and minimizers $(u_\varepsilon, A_\varepsilon)$ of E_ε, consider

$$\Sigma_\varepsilon = \{x \in \Omega : E_\varepsilon(u_\varepsilon, A_\varepsilon; B_\varrho(x)) \geq \varepsilon_0\}$$

and its cover $(B_\varrho(x))_{x \in \Sigma_\varepsilon}$. By Vitali's covering lemma there exists a finite collection of disjoint balls $B_i = B_\varrho(x_i)$, $x_i \in \Sigma_\varepsilon$, $1 \leq i \leq I = I(\varepsilon)$, such that

$$\bigcup_{x \in \Sigma_\varepsilon} B_\varrho(x) \subset \bigcup_i B_{5\varrho}(x_i).$$

Lemma 4.1. *Let ε_0 be given as in Theorem 3.2. There exists a number $J_0 = J_0(L)$ such that for any $\varrho > 0$ any disjoint collections of balls $B_\varrho(x_j)$, $x_j \in \Sigma$, $1 \leq j \leq J$, with $E(u_\varepsilon, A_\varepsilon; B_\varrho(x_j)) \geq \varepsilon_0$ there holds $J \leq J_0$.*

Proof. As the $(u_\varepsilon, A_\varepsilon)$ are solutions to the self-duality equations, there holds

$$E_\varepsilon(u_\varepsilon, A_\varepsilon) \le 2\pi \deg L =: C_1.$$

Hence

$$J \le \sum_i \frac{8E_\varepsilon\left(u_\varepsilon, A_\varepsilon; B_\varrho(x_i)\right)}{\varepsilon_0} \le 8C_1\varepsilon_0^{-1}.$$

□

As a consequence of Lemma 4.1, the number $I(\varepsilon)$ introduced above is bounded independently of ε and ϱ. In particular, we may choose $\varrho = \sqrt{\varepsilon}$. Then we may assume that $I(\varepsilon) = I_0$ is independent of ε and that $x_i = x_i^\varepsilon \to x_i^0$ as $\varepsilon \to 0$ for $1 \le i \le I_0$.

Let $\Sigma_0 = \{x_i^0; 1 \le i \le I_0\}$ be the set of limits of concentration points.

For $0 < \varepsilon < 1$ consider any point x_0 such that

$$\inf_i |x_0 - x_i^\varepsilon| \ge 5\sqrt{\varepsilon}.$$

If follows that $x_0 \notin \Sigma_\varepsilon$ and therefore

$$E_\varepsilon\left(u_\varepsilon, A_\varepsilon; B_{\sqrt{\varepsilon}}(x_0)\right) < \varepsilon_0.$$

Rescale $(u_\varepsilon, A_\varepsilon)$ around x_0. Note that we also have to rescale the Kähler metric and the rescaled metric converges smoothly locally to the standard metric on $\mathbb{R}^2$ as $\varepsilon \to 0$. The rescaled solutions $(v_\varepsilon, B_\varepsilon)$ are defined on the ball $B_{1/\sqrt{\varepsilon}}(0)$. Applying Theorem 3.2 with $R = \frac{1}{2\sqrt{\varepsilon}}$ we obtain

$$\frac{\left(1 - |u_\varepsilon|^2\right)}{4}(x_0) \le \sup_{B_{1/2\sqrt{\varepsilon}}(0)} e(v_\varepsilon, B_\varepsilon) \le C_0\varepsilon_0\varepsilon. \tag{4.1}$$

In particular, for any set $\Sigma' \subset\subset \Sigma \setminus \Sigma_0$ there holds

$$\sup_{\Sigma'} \left(1 - |u_\varepsilon|^2\right) \le C\varepsilon \to 0$$

as $\varepsilon \to 0$, and it follows that for small $\varepsilon > 0$ the total degree d of L, endowed with the connection A_ε, equals the sum of the local degrees d_i of (L, A_ε), restricted to a neighborhood of x_i^0. Finally, on a blown-up neighborhood $B_\varrho(x_i^0)$ of a point x_i^0, the rescaled equations (1.24), (1.26) approximate the vortex equations (1.10), (1.11) on $\mathbb{R}^2$. We thus expect the energy of $(u_\varepsilon, A_\varepsilon)$, restricted to any such ball $B_\varrho(x_i^0)$, to be bounded from below by the energy of a vortex configuartion on $\mathbb{R}^2$ with degree d_i.

In fact, we have

Lemma 4.2. *For any index i there holds*

$$\liminf_{\varepsilon\to 0} E_\varepsilon\big(u_\varepsilon, A_\varepsilon; B_{\sqrt{\varepsilon}}(x_i^0)\big) \geq 2\pi\, |d_i|\,.$$

Proof. Again denote by $(v_\varepsilon, B_\varepsilon)$ the rescaled solution, defined on a ball $B_{1/\sqrt{\varepsilon}}(0)$. Note that the difference from the scaled metric to the Euclidean metric vanishes as $\varepsilon \to 0$, uniformly on $B_{1/\sqrt{\varepsilon}}(0)$.

Hence we have

$$E_\varepsilon\big(u_\varepsilon, A_\varepsilon; B_{\sqrt{\varepsilon}}(x_i^0;\Sigma)\big) \geq E_1\big(v_\varepsilon, B_\varepsilon; B_{1/\sqrt{\varepsilon}}(0;\mathbb{R}^2)\big) + o(1)$$

where $o(1) \to 0$ as $\varepsilon \to 0$ and where we emphasize the use of the Kähler metric on Σ and the Euclidean metric on $\mathbb{R}^2$, respectively.

Suppose that $d_i \geq 0$. Then, via an integration by parts as in [18], p. 54, we obtain

$$\begin{aligned} E_1\big(v_\varepsilon, B_\varepsilon; B_{1/\sqrt{\varepsilon}}(0)\big) &= \int_{B_{1/\sqrt{\varepsilon}}(0)} \Big\{ \big|\bar\partial_{B_\varepsilon} v_\varepsilon\big|^2 + \Big|F_\varepsilon + \frac{1}{2}\big(|v_\varepsilon|^2 - 1\big)\Big|^2 \Big\}\, dx \\ &\quad + \int_{\partial B_{1/\sqrt{\varepsilon}}(0)} \mathrm{Im}\big\langle v_\varepsilon, \tau\cdot\nabla_{B_\varepsilon} v_\varepsilon\big\rangle\, do + \int_{B_{1/\sqrt{\varepsilon}}(0)} F_\varepsilon\, dx, \end{aligned}$$

where $F_\varepsilon = *dB_\varepsilon$ is the curvature of B_ε and τ is a (positively oriented) unit tangent vector field along $\partial B_{1/\sqrt{\varepsilon}}(0)$. Note that by (4.1) we can write

$$v_\varepsilon = \varrho_\varepsilon e^{i\psi_\varepsilon} \qquad \text{on } \partial B_{1/\sqrt{\varepsilon}}(0)$$

with $\varrho_\varepsilon^2 \geq 1 - C\varepsilon$. Moreover,

$$\mathrm{Im}\big\langle v_\varepsilon, \tau\cdot\nabla_{B_\varepsilon} v_\varepsilon\big\rangle = \varrho_\varepsilon^2\big(d\psi - B_\varepsilon(\tau)\big).$$

Thus, and using (4.1), we deduce that

$$\int_{\partial B_{1/\sqrt{\varepsilon}}(0)} \mathrm{Im}\big\langle v_\varepsilon, \tau\cdot\nabla_{B_\varepsilon} v_\varepsilon\big\rangle\, do = \int_{\partial B_{1/\sqrt{\varepsilon}}(0)} \big(d\psi - B_\varepsilon(\tau)\big)\, do - \mu(\varepsilon),$$

where

$$\begin{aligned} \mu(\varepsilon) &= \int_{\partial B_{1/\sqrt{\varepsilon}}(0)} (1-\varrho_\varepsilon^2)\frac{\mathrm{Im}\big\langle v_\varepsilon, \tau\cdot\nabla_{B_\varepsilon} v_\varepsilon\big\rangle}{\varrho_\varepsilon^2}\, do \\ &\leq \frac{\pi}{\sqrt{\varepsilon}} \sup_{\partial B_{1/\sqrt{\varepsilon}}(0)} \Big\{ \big|1-\varrho_\varepsilon^2\big|^2 + \frac{1}{\varrho_\varepsilon^2}\big|\nabla_{B_\varepsilon} v_\varepsilon\big|^2 \Big\} \\ &\leq C\sqrt{\varepsilon} \to 0 \qquad \text{as } \varepsilon \to 0. \end{aligned}$$

But on the other hand, by Stokes' theorem

$$\int_{\partial B_{1/\sqrt{\varepsilon}}(0)} B_\varepsilon(\tau)\,do = \int_{B_{1/\sqrt{\varepsilon}}(0)} F_\varepsilon\,dx.$$

Hence we find that

$$E_1\left(v_\varepsilon, B_\varepsilon; B_{1/\sqrt{\varepsilon}}(0)\right) \geq \int_{\partial B_{1/\sqrt{\varepsilon}}(0)} d\psi\,do - C\sqrt{\varepsilon} = 2\pi d_i - C\sqrt{\varepsilon}.$$

Similary, if $d_i < 0$ we find the asymptotic lower bound $2\pi\,|d_i|$, concluding the proof. □

From the estimate

$$\begin{aligned} 2\pi d &= E_\varepsilon(u_\varepsilon, A_\varepsilon; \Sigma) \geq \sum_i E_\varepsilon\left(u_\varepsilon, A_\varepsilon; B_{\sqrt{\varepsilon}}(x_i^0)\right) \\ &\geq 2\pi \sum_i |d_i| - o(1) \end{aligned}$$

and using the fact that

$$\sum_i d_i = d$$

we then deduce that $I_0 \leq d$ and $d_i > 0$ for each i. Moreover, for any compact subset Σ' of $\Sigma \setminus \Sigma_0$, the energy

$$E'_\varepsilon\left(u_\varepsilon, A_\varepsilon; \Sigma'\right)$$

becomes arbitrarily small for sufficiently small $\varepsilon > 0$. Using Theorem 3.2 (observing that $|u| \geq \frac{1}{2}$ by Lemma 2.3) and scaling back, we then obtain that

$$|\nabla_{A_\varepsilon} u_\varepsilon|^2 + \frac{1}{4\varepsilon^2}\left(1 - |u|^2\right)^2 + \varepsilon^2\,|dA_\varepsilon|^2 \to 0 \tag{4.2}$$

uniformly on Σ' as $\varepsilon \to 0$. This implies already

$$|u_\varepsilon|^2 \to 1,\ \nabla_{A_\varepsilon} u_\varepsilon \to 0$$

as $\varepsilon \to 0$.

In order to show that the curvature of A_ε tends to 0 as well, we use the equation (1.21) and (1.26) to show the following estimate.

For $x_0 \in \Sigma'$ and $0 < \varrho < \operatorname{dist}(x_0, \partial\Sigma')$ let

$$\phi_\varepsilon(\varrho) = \fint_{B_\varrho(x_0)} \frac{1 - |u_\varepsilon|^2}{\varepsilon^2}\, dx,$$

where $\fint_{B_\varrho(x_0)} \cdots$ denotes the average over $B_\varrho(x_0)$. We may assume that $\varepsilon \le \sqrt{\varepsilon} \le \varrho_0 = \frac{1}{2}\operatorname{dist}(x_0, \partial\Sigma')$ and that $|u_\varepsilon|^2 \ge \frac{1}{2}$ on Σ'.

Lemma 4.3. $\sup_{\varepsilon \le \varrho \le \varrho_0} \phi_\varepsilon(\varrho) \to 0$ *as* $\varepsilon \to 0$.

Proof. Multiplying (1.21) by $\bar{u}$, we obtain the equation

$$\frac{1}{2}\Delta(1 - |u|^2) + |\nabla_A u|^2 = \frac{|u|^2}{2\epsilon^2}(1 - |u|^2) \ge \frac{1 - |u|^2}{4\epsilon^2} \tag{4.3}$$

Here and in the following we write u for u_ε etc.

Let $\varphi \in C_0^\infty(B_2(0))$ satisfy $0 \le \varphi \le 1$, $\varphi = 1$ on $B_1(0)$. For $\varepsilon \le \varrho \le \varrho_0$ we scale $\varphi_\varrho(x) = \varphi\left(\frac{x - x_0}{\varrho}\right)$.

Multiplying (4.3) by φ_ϱ and integrating by parts, we then obtain the estimate

$$\int_{B_\varrho} \frac{1 - |u|^2}{\epsilon^2}\, dx \le 4 \int_{B_{2\varrho}(x_0)} |\nabla_A u|^2\, dx + C_0^2 \int_{B_{2\varrho} \setminus B_\varrho} \frac{1 - |u|^2}{\varrho^2}\, dx$$

where $B_\varrho = B_\varrho(x_0)$, etc., and with a constant C_0 independent of ϱ and ε. Multiplying by $\frac{\varepsilon^2}{\varrho^2}$ and "filling the hole" on the right à la Widman [33] in a different context, we find

$$\begin{aligned}
(1 + C_0^2 \frac{\epsilon^2}{\varrho^2}) \int_{B_\varrho} \frac{1 - |u|^2}{\varrho^2}\, dx &\le 16\pi\epsilon^2 \sup_{\Sigma'} |\nabla_A u|^2 + C_0^2 \frac{\epsilon^2}{\varrho^2} \int_{B_{2\varrho}} \frac{1 - |u|^2}{\varrho^2}\, dx \\
&= \epsilon^2 o(1) + 4C_0^2 \frac{\epsilon^2}{\varrho^2} \int_{B_{2\varrho}} \frac{1 - |u|^2}{4\varrho^2}\, dx,
\end{aligned}$$

where $o(1) \to 0$ as $\epsilon \to 0$ independent of ϱ.

For $\epsilon \le \varrho \le \varrho_0$ let

$$\psi_\epsilon(\varrho) = \int_{B_\varrho} \frac{1 - |u|^2}{\varrho^2}\, dx.$$

Then we have

$$\psi_\varepsilon(\varrho) \le \varepsilon^2 o(1) + \frac{4C_0^2\varepsilon^2}{\varrho^2 + C_0^2\varepsilon^2} \psi_\varepsilon(2\varrho).$$

Note that for $2C_0\varepsilon \leq \varrho$ there holds

$$\frac{4C_0^2\varepsilon^2/\varrho^2}{1+C_0^2\varepsilon^2/\varrho^2} \leq \frac{4}{5} = \theta$$

and hence

$$\psi_\varepsilon(\varrho) \leq \varepsilon^2 o(1) + \theta\psi_\varepsilon(2\varrho).$$

By iteration, for $2C_0\varepsilon \leq \varrho \leq \bar{\varrho} \leq \frac{\varrho_0}{2}$ therefore we obtain that

$$\psi_\varepsilon(\varrho) \leq \varepsilon^2 o(1) + \psi_\varepsilon(\bar{\varrho}). \tag{4.4}$$

Moreover, for $\varrho = \varrho_0$ we have

$$\begin{aligned} \psi_\varepsilon(\varrho_0) &\leq \int_{B_{\varrho_0}} \frac{1-|u|^2}{\varrho_0^2}\,dx = \pi\varepsilon \fint_{B_{\varrho_0}} \frac{1-|u|^2}{\varepsilon}\,dx \\ &\leq \pi\varepsilon \sup_{\Sigma'} \frac{1-|u|^2}{\varepsilon} \leq \varepsilon\, o(1) \leq \varrho_0^2 o(1). \end{aligned}$$

Hence for $\bar{\varrho} = \frac{1}{2}\varrho_0$ there holds

$$\psi_\varepsilon(\bar{\varrho}) \leq \varepsilon^2 o(1) + 4C_0^2\frac{\varepsilon^2}{\bar{\varrho}^2}\psi_\varepsilon(\varrho_0) \leq \varepsilon^2 o(1). \tag{4.5}$$

Combining estimates (4.4) and (4.5), we find

$$\sup_{2C_0\varepsilon\leq\varrho\leq\varrho_0} \phi_\varepsilon(\varrho) \to 0 \qquad \text{as } \varepsilon \to 0.$$

Finally, observe that for $\varepsilon \leq \varrho \leq 2C_0\varepsilon$ there holds

$$\phi_\varepsilon(\varrho) \leq 4C_0^2\phi_\varepsilon(2C_0\varepsilon).$$

The claim follows. □

Lemma 4.4. *$\sup_{\Sigma'} \frac{1-|u_\varepsilon|^2}{\varepsilon^2} \to 0$ as $\varepsilon \to 0$.*

Proof. From (4.3) we have for $u = u_\varepsilon$, etc.,

$$-2\varepsilon^2\Delta\left(\frac{1-|u|^2}{\varepsilon^2}\right) + \frac{1-|u|^2}{\varepsilon^2} \leq 4|\nabla_A u|^2.$$

Fix $x_0 \in \Sigma'$ and suppose $\varepsilon \leq \varrho_0 = \frac{1}{2}\operatorname{dist}(x_0, \partial\Sigma')$.

Let

$$g = \frac{1 - |u|^2}{\varepsilon^2} : B_\varepsilon(x_0) \to \mathbb{R},$$

and for $\delta \geq \sup_{\Sigma'} |\nabla_A u|^2$ define

$$\bar{g}(x) = g(x_0 + \varepsilon x) + \delta |x|^2,$$

satisfying the equation

$$-2\Delta\bar{g} + \bar{g} \leq 4\big(|\Delta_A u|^2 - \delta\big) \leq 0 \quad \text{in } B_1(0).$$

That is, $\bar{g} > 0$ is a sub-solution for the operator $(-2\Delta + 1)$ on $B_1(0)$ and Moser's sup-estimate and Lemma 4.2 imply that

$$\left(\frac{1 - |u|^2}{\varepsilon^2}\right)(x_0) = \bar{g}(0) \leq C \int_{B_1(0)} \bar{g}\, dx$$
$$\leq C \fint_{B_\varepsilon(x_0)} \frac{1 - |u|^2}{\varepsilon^2}\, dx + C\delta \leq o(1) + C\delta,$$

where $o(1) \to 0$ as $\varepsilon \to 0$.

Since we may let $\delta \to 0$ as $\varepsilon \to 0$, the claim follows. □

Proof of Theorem 1.1 (completed): From Lemma 4.3 and equation (1.26) we now immediately deduce the asserted convergence

$$\sup_{\Sigma'} |\Lambda F_\varepsilon| \leq C \sup_{\Sigma'} \frac{1 - |u_\varepsilon|^2}{\varepsilon^2} \to 0$$

as $\varepsilon \to 0$.

Finally, since the total topological charge

$$\int_\Sigma \Lambda F_\varepsilon = 2\pi \deg L$$

is independent of ε, the preceding blow-up analysis and convergence result imply that

$$\Lambda F_\varepsilon \rightharpoonup 2\pi \sum_{j=1}^{l} d_j \delta(x_j)$$

in the sense of measure, as claimed.

This concludes the proof of Theorem 1.1. □

References.

[1] M.S. Berger and Y.Y. Chen, *Symmetric vortices for the Ginzburg-Landau equations of superconductivity and the nonlinear desingularization phenomenon,* Journal Functional Analysis, **82** (1989), 259–295.

[2] F. Bethuel, H. Brézis, and F. Hélein, *Asymptotics for the minimization of a Ginzburg-Landau functional,* Calc. Var., **1** (1993), 123–148.

[3] F. Bethuel, H. Brézis, and F. Hélein, *Ginzburg-Landau vortices,* Birkhäuser, (1994).

[4] F. Bethuel and T. Riviére, *Vortices for a variational problem related to superconductivity,* to appear.

[5] S. Bradlow, *Vortices in holomorphic line bundles over closed Kähler manifolds,* Comm. Math. Phys., **135** (1990), 1–17.

[6] S. Bradlow, *Special metrics and stability for holomorphic bundles with global sections,* J. Diff. Geom., **33** (1991), 169–214.

[7] Y.Y. Chen, *Vortices for the Ginzburg-Landau equations in the nonsymmetric case in bounded domain,* Contemp. Math., **108** (1990), 19–32.

[8] Y.M. Chen and M. Struwe, *Existence and partial regularity results for the heat flow for harmonic maps,* Math. Z., **201** (1989), 83–103.

[9] Y.M. Chen and F.-H. Lin, *Remarks on approximate harmonic maps,* Comm. Math. Helv., **70** (1995), 161–169.

[10] C. Cronström, *A simple and complete Lorentz-covariant gauge condition,* Phys. Lett., **90 B** (1980), 267–269.

[11] M. del Pino and P. Felmer, *Local minimizers for the Ginzburg-Landau energy,* Preprint, (1995).

[12] O. García-Prada, *Invariant connections and vortices,* Comm. Math. Phys., **156** (1993), 527–546.

[13] O. García-Prada, *A direct existence proof for the vortex equations over a compact Riemann surface,* Bull. London Math. Soc., **26** (1994), 88–96.

[14] O. García-Prada, *Dimensional reduction of stable bundles, vortices and stable pairs,* Intern. J. Math., **5** (1994), 1–52.

[15] V. Ginzburg and L. Landau, *On the theory of superconductivity,* Zh. Eksp. Teor. Fiz., **20** (1950), 1064–1082.

[16] R. Hardt and F.-H. Lin, *Singularities for p-energy minimizing unit vectorfields on planar domains,* Preprint,

[17] E. Heinz, *On certain nonlinear elliptic differential equations and univalent mappings,* Journ. d' Anal., **5** (1956/57), 197-272.

[18] A. Jaffe and C. Taubes, *Vortices and monopoles,* Birkhäuser, Basel, (1980)

[19] F.-H. Lin, *Some dynamical properties of Ginzburg-Landau vortices,* Preprint, (1994),

[20] L. Modica, *The gradient theory of phase transitions and the minimal interface criterion,* Arch. Rat. Mech. Anal., **98** (1986), 123-142.

[21] L. Modica and S. Mortola, *Un esempio di* Γ*-convergenza,* Boll. U.M.I. (5), **14-B** (1977), 285-299.

[22] G. Orlandi, *Asymptotic behavior of the Ginzburg-Landau functional on complex line bundles over compact Riemann surfaces,* Preprint, (1995).

[23] J. Qing, *Renormalized energy for Ginzburg-Landau vortices on closed surfaces,* Preprint, (1995).

[24] R. Schoen, *Analytic aspects of the harmonic map problem,* Math. Sci. Res. Inst. Publ., **2** Springer, Berlin (1984)

[25] M. Struwe, *On the asymptotic behavior of minimizers of the Ginzburg-Landau model in 2 dimensions,* Journal Differential Integral Equations **7**, (1994), 1062-1082.

[26] M. Struwe, *Une estimation asymptotique pour le modéle de Ginzburg-Landau,* C.R. Acad. Sci. Paris, **317** (1993), 677-680.

[27] M. Struwe, *Singular pertubations of geometric variational problems,* to appear

[28] C. Taubes, *Arbitrary n-vortex solutions to the first order Ginzburg-Landau equations,* Comm. Math. Phys., **72** (1980), 277–292.

[29] C. Taubes, *On the equivalence of the first and second order equation for gauge theories,* Comm. Math. Phys., **75** (1980), 207–227.

[30] C. Taubes, *The Seiberg-Witten and the Gromov invariants,* Preprint, (1995).

[31] C.Taubes, *From the Seiberg-Witten equations to pseudo-holomorphic curves,* Preprint, (1995).

[32] K. Uhlenbeck, *Connections with* L^p *bounds on curvature,* Comm. Math. Phys., **83** (1982), 31–42.

[33] K.-O. Widman, *Hölder continuity of solutions of elliptic systems,* Manuscripta Math., **5** (1971), 299–308.

RECEIVED MARCH 15, 1995.

The Dirichlet Problem for the Prescribed Curvature Quotient Equations with General Boundary Values[1]

NINA M. IVOCHKINA, MI LIN, NEIL S. TRUDINGER

In honour of the sixtieth birthday of Stefan Hildebrandt

1. Introduction.

In the paper [9], Lin and Trudinger consider the solvability of the classical Dirichlet problem for prescribed curvature equations of the form,

$$F[u] \equiv f(\kappa) = \Psi(x, u) \tag{1.1}$$

in domains Ω in Euclidean n-space, R^n, and constant boundary values. In equation (1.1), the function $u \in C^2(\Omega)$, $\kappa = (\kappa_1, \dots, \kappa_n)$ denotes the principal curvatures of the graph of u over Ω, Ψ is a given positive function on $\Omega \times R$ and f is a symmetric function of the form,

$$f = S_{k,l} \equiv \frac{S_k}{S_l} \tag{1.2}$$

where $0 \leq l < k \leq n$ and S_k denotes the kth elementary symmetric function,

$$S_k(\kappa) = \sum \kappa_{i_1} k_{i_2} \dots \kappa_{i_k}, 1 \leq k \leq n, \tag{1.3}$$

the sum being taken over all increasing k-tuples $i_1, i_2, \dots, i_k \subset \{1, \dots, n\}$ and $S_0 = 1$. The classical Dirichlet problem for the k-mean curvature equations, corresponding to the cases $l = 0$, was treated by Caffarelli, Nirenberg and Spruck [1] and Ivochkina [3], [4], (see also [12]), with the solvability for general boundary values under reasonable geometric conditions being established in [4]. In this paper, we adapt the treatment of curvature quotient equations in [9] to cover general boundary values under analogous geometric conditions to those in [4], [9], [12].

[1]Research supported by Australian Research Council and Russian Foundation for Fundamental Research

We shall adopt the same terminology as [9]. A function $u \in C^2(\Omega)$ is called *admissible* for the operator F of the form (1.1), if

$$f(\kappa + \eta) \geq f(\kappa) \quad \text{for all} \quad \eta_i \geq 0, i = 1, \dots, n. \tag{1.4}$$

Letting Γ_k denote the open cone in R^n, with vertex at the origin, given by

$$\Gamma_k = \{\kappa \in R^n | S_j(\kappa) > 0, \quad j = 1, \dots, k\}, \tag{1.5}$$

it follows that u is admissible for $f = S_{k,l}$ provided $\kappa \in \bar{\Gamma}_k$, ($\bar{\Gamma}_l$ if $k = l + 1$). Also, for $k = 1, \dots, n-1$, we say that the domain Ω with boundary $\partial\Omega \in C^2$ is *k-convex* (*uniformly k-convex*) if the principal curvatures of $\partial\Omega$, $\kappa' = (\kappa'_1, \dots, \kappa'_{n-1})$ satisfy $S_j(\kappa) \geq 0 (> 0)$ for $j = 1, \dots, k$, that is $\kappa' \in \bar{\Gamma}_k(\Gamma_k)$. It follows that non-degenerate level surfaces of ($S_{k,l}$) admissible functions bound $(k-1)$-convex domains.

The following theorem extends the basic result in [9], (Theorem 1.1), to general boundary values.

Theorem 1.1. *Let* $0 \leq l < k < n$, $0 < \alpha < 1$. *Assume that*

(i) Ω *is a bounded domain in* R^n *with boundary* $\partial\Omega \in C^{4,\alpha}$ *and* $\phi \in C^{4,\alpha}(\bar{\Omega})$ *is a given function;*

(ii) $\Psi \in C^{2,\alpha}(\Omega \times R)$, $\Psi > 0$, $\frac{\partial \psi}{\partial z} \geq 0$ *in* $\bar{\Omega} \times R$;

(iii) Ω *is* $(k-1)$*-convex with*

$$\Psi(x, \phi(x)) \leq S_{k,l}(\kappa') \quad on \quad \partial\Omega.$$

Then, provided there exists any bounded admissible subsolution of equation (1.1) *in* Ω, *there exists a unique admissible solution* $u \in C^{4,\alpha}(\bar{\Omega})$ *of* (1.1) *satisfying* $u = \phi$ *on* $\partial\Omega$.

As remarked in [9], the existence of a bounded subsolution, (which can be taken in the viscosity sense of [12]) can be replaced by conditions guaranteeing *a priori* solution bounds [13]. For convenience the smoothness conditions in Theorem 1.1 are expressed in Hölder spaces; they may be relaxed to $\partial\Omega \in C^{3,1}(\Omega)$, $\phi \in C^{3,1}(\bar{\Omega})$, $\Psi \in C^{1,1}(\bar{\Omega} \times R)$, with resultant solution $u \in C^{3,\beta}(\bar{\Omega})$ for all $\beta < 1$.

As in [9], the case $k = n$ is omitted from Theorem 1.1, as the corresponding extension of the Serrin condition (iii) would imply $\Psi = 0$ on $\partial\Omega$, contradicting (ii). To embrace this case for non-degenerate equations, we need to assume the existence of a subsolution taking on the given boundary values. Since our second derivative estimates will be derived under weaker conditions than (iii), we will consequently be able to infer more general results than Theorem 1.1. In particular we note the following improvement of Theorem 1.2 in [9].

Theorem 1.2. *Let $0 < l < k \leq n$, $0 < \alpha < 1$. Assume that hypotheses* (i) *and* (ii) *of Theorem* 1.1 *hold with ϕ constant. Then, provided there exists an admissible subsolution $u_0 \in C^{0,1}(\bar{\Omega})$ with $u_0 = \phi$ on $\partial\Omega$, there exists a unique admissible solution $u \in C^{4,\alpha}(\bar{\Omega})$ of equation* (1.1) *satisfying $u = \phi$ on $\partial\Omega$.*

Some extensions of Theorem 1.2 to general boundary values will be formulated in conjunction with our treatment of second derivative estimates. When ϕ is constant, the boundary $\partial\Omega$ is a level surface of the solution u, whence Ω is necessarily uniformly $(k-1)$-convex, with $\Psi(x, \phi) \leq S_{k-1,l-1}(\kappa')$ on $\partial\Omega$, and the pertinent second derivative estimates are already remarked in [9]. Theorem 1.2 extends to the case $l = 0$, provided we assume Ω is uniformly $(k-1)$-convex at the outset.

As with previous studies, ([1], [2], [4], [9]), the proofs of the above existence theorems utilize the method of continuity which reduces the problem of existence to that of *a priori* estimates, for a related family of Dirichlet problems in the Hölder space $C^{2,\beta}(\bar{\Omega})$ for some $\beta > 0$. For the curvature quotient equations, the crucial estimates are those of second derivatives on the boundary $\partial\Omega$. These were established for constant boundary values in [9], following ideas of Ivochkina [4] and Trudinger [15]. In this paper, we extend these estimates to cover more general boundary values through a more thorough analysis of the barrier construction. The other components of the solution estimate follow the constant boundary values case [9]. Our assumption of the existence of a bounded subsolution, u_0, automatically ensures, by the comparison principle, a solution bound

$$u_0 + \inf_{\partial\Omega}(\phi - u_0) \leq u \leq \sup_{\partial\Omega} \phi, \tag{1.6}$$

as well as a boundary gradient estimate, in the case of Theorem (1.2),

$$\sup_{\partial\Omega} |Du| \leq \sup_{\partial\Omega} |Du_0|. \tag{1.7}$$

Estimates for the first and second derivatives of solutions in terms of their boundary values are provided by Caffarelli, Nirenberg and Spruck [1], while the boundary gradient estimate for Theorem 1.1, under the Serrin condition (iii) is given in Trudinger [12]. Note that although the curvature quotient equations are excluded by the hypotheses in [1], the above estimates remain valid under more general hypotheses. The Hölder estimates for second derivatives follow from the Krylov theory for nonlinear, uniformly elliptic equations, (see [2, 7, 14]).

The plan of this paper is as follows. In the following section, we provide the barrier analysis enabling us to extend the estimates for mixed tangential normal derivatives at the boundary in [9] to general boundary values. In Section 3 we establish the double normal second derivative estimates at the boundary using an extension of the techniques in [9] and [15]. An alternative approach to this estimation, along the original lines of the case $l = 0$, is given in [5]. Finally in Section 4, we complete the proof of Theorem 1.1 and consider, in accordance with our discussion above, more general results which dispense with hypothesis (iii). We also remark here that the results of [9] have also been extended to general functions Ψ depending also on the gradient of u, [10]. Throughout this paper, we shall draw extensively on the notation and result in [9], so that this paper should be read as an extension of [9].

2. Mixed second derivative boundary estimates.

As in [4], [9], it is convenient to write equation (1.1), for admissible solutions $u \in C^2(\Omega)$, in the form

$$F[u] = F(Du, D^2u) = f(\lambda) = \bar{\Psi}(x, u, Du) \tag{2.1}$$

Where $\lambda = (\lambda_1, \dots, \lambda_n)$ denotes the eigenvalues of the matrix

$$C = (I - \nu \otimes \nu)d^2u, \qquad \nu = \frac{Du}{\sqrt{1 + |Du|^2}}. \tag{2.2}$$

and f and $\bar{\Psi}$ are given respectively by

$$f(\lambda) = [S_{k,l}]^{1/m}, \qquad m = k - l \tag{2.3}$$

$$\bar{\Psi}(x, z, p) = \Psi^{1/m}(x, z)\sqrt{1 + |p|^2}. \tag{2.4}$$

Fixing a point y on the boundary $\partial\Omega$ of the domain Ω, we choose the coordinate axes so that the x_n axis is directed along the inner normal at y. Let ξ be a C^2 vector field in some neighborhood $\mathcal{N}$ of y and consider the function,

$$w = \xi \cdot Du - \frac{1}{2}\sum_{s=1}^{n-1}(D_s u - D_s u(y))^2. \tag{2.5}$$

If $u \in C^3(\mathcal{N} \cap \Omega)$ is an admissible solution of equation (2.1), it is shown in [9], that w satisfies an elliptic differential inequality of the form,

$$F^{i,j} D_{ij} w - \Psi^i D_i w \leq C(|\tilde{D}\tilde{\Psi}| + F^{ii} + F^{ij} D_i w D_j w) \tag{2.6}$$

with coefficients F^{ij}, Ψ^i given by

$$F^{ij} = \frac{\partial F}{\partial r_{ij}}(Du, D^2 u), \quad \Psi^i = \frac{\partial \tilde{\Psi}}{\partial p_i}(x, u, Du) \tag{2.7}$$

and constant C depending on $\mathcal{N}, |\xi|_2$ and $|Du|_0$ where $\tilde{D}$ denotes the gradient in $R^{2n+1}(x, z, p)$. The function F is extended to $R^n \times S^n(p, r)$ as follows. For $p \in R^n$, we define

$$\Gamma_k(p) = \{r \in S^n | \, \lambda(p, r) \in \Gamma_k\} \tag{2.8}$$

where $\lambda = (\lambda_1, \dots, \lambda_n)$ denotes the eigenvalues of the matrix

$$\left(I - \frac{p \otimes p}{1 + |p|^2} \right) r. \tag{2.9}$$

Writing

$$S_k(p, r) = S_k(\lambda), \quad S_{k,l}(p, r) = \frac{S_k}{S_l}(p, r) \tag{2.10}$$

we then have

$$F(p, r) = f(\lambda) = [S_{k,l}(p, r)]^{1/m} \tag{2.11}$$

defined whenever $r \in \Gamma_k(p)$. For $r \in \partial\Gamma_k(p), S_k(p, r) = 0$ and we set $S_{k,l}(p, r) = 0$. As in [9], we conclude from (2.6), that a further function $\tilde{w}$ given by

$$\tilde{w} = 1 - e^{-\alpha w} - b|x - y|^2, \tag{2.12}$$

satisfies the simpler inequality,

$$F^{i,j} D_{ij} \tilde{w} \leq \Psi^i D_i \tilde{w} \tag{2.13}$$

for constants a, b depending on n, $|\xi|_2, |\tilde{D}\Psi|_0$ and $|Du|_0$ provided $diam\mathcal{N}$ is sufficiently small in terms of the same quantities.

Barrier Construction

In turns out that the barrier (3.25) in [9] which was similar to that used by Ivochkina [3, 4] suffices under condition (iii) in Theorem 1.1, but the inequality (2.17) in [9] is too coarse to show this and must be considerably refined. For this purpose, we introduce further notation. First, for $p \in R^n, i = 1, \dots, n$, we let $p(i)$ designate the vector obtained by setting $p_i = 0$ and similarly for $r \in R^n, i - 1, \dots, n$ we let $r(i)$ designate the symmetric matrix obtained by setting the ith row and column to zero and $r(i,i)$ the symmetric matrix obtained by setting $r_{ii} = 0$. For the elementary symmetric functions, we denote

$$\left\{ \begin{array}{r} S_{k;i}(k) = S_k(k(i)), \quad S_{k,l;i}(k) = S_{k,l}(k(i)), \\ S_{k;i}(p,r) = S_k(p(i), r(i)), \quad S_{k,l;i}(p,r) = S_{k,l}(p(i), r(i)). \end{array} \right. \tag{2.14}$$

In accordance with [4], we also let

$$r\begin{bmatrix} i_1 & \dots & i_k \\ j_1 & \dots & j_k \end{bmatrix}, \tag{2.15}$$

$$1 \le i_1 < \dots < i_k \le n, \quad 1 \le j_1 < \dots < j_k \le n,$$

denote the $k \times k$ minor corresponding to the $i_1, \dots, i_k$ rows and $j_1, \dots, j_k$ columns of $r \in S^n$. We then have the formulae,

$$\begin{aligned} S_k(p,r) &= \sum \left\{ r\begin{bmatrix} i_1 & \dots & i_k \\ i_1 & \dots & i_k \end{bmatrix} \right. \\ &\quad \left. - \frac{p_i p_j}{1+|p|^2} r\begin{bmatrix} i_1 & \dots & i & \dots & i_{k-1} \\ i_1 & \dots & j & \dots & i_{k-1} \end{bmatrix} \right\} \\ &= \sum \left\{ \frac{1+|p(1)|^2}{1+|p|^2} r\begin{bmatrix} 1 & i_2 & \dots & i_k \\ 1 & i_2 & \dots & i_k \end{bmatrix} \right. \\ &\quad - \frac{p_i(1)p_j(1)}{1+|p|^2} r\begin{bmatrix} 1 & i_2 & \dots & i & \dots & i_{k-1} \\ 1 & i_2 & \dots & j & \dots & i_{k-1} \end{bmatrix} \\ &\quad + r\begin{bmatrix} i_2 & \dots & i_{k+1} \\ i_2 & \dots & i_{k+1} \end{bmatrix} \\ &\quad - \frac{p_1 p_i(1)}{1+|p|^2} r\begin{bmatrix} i_2 & \dots & i & \dots & i_k \\ 1 & i_2 & \dots & \dots & i_k \end{bmatrix} \\ &\quad \left. - \frac{p_1 p_j(1)}{1+|p|^2} r\begin{bmatrix} 1 & i_2 & \dots & \dots & i_k \\ i_2 & \dots & j & \dots & i_k \end{bmatrix} \right\} \\ &= \frac{1+|p(1)|^2}{1+|p|^2} r_{11} S_{k-1;1}(p,r) + \bigcirc(|r(1,1)|^k). \end{aligned} \tag{2.16}$$

From (2.16), we deduce (replacing 1 by i),

$$\frac{\partial S_k(p,r)}{\partial r_{ii}} = \frac{1+|p(i)|^2}{1+|p|^2} S_{k-1;i}(p,r), \tag{2.17}$$

whence, for $1 \leq l < k \leq n$,

$$\begin{aligned} \lim_{r_{ii}\to\infty} S_{k,l}(p,r) &= S_{k-1,l-1;i}(p,r) \\ &> S_{k,l}(p,r) \end{aligned} \tag{2.18}$$

for $r \in \Gamma_k(p)$, by virtue of the ellipticity of $S_{k,l}$[9]. As a further application of (2.16), we have for $p = (p_1, \cdots, 0)$,

$$\begin{aligned} S_k(p,r) &= \sum \frac{1}{1+|p|^2} r \begin{bmatrix} 1 & i_2 & \dots & i_k \\ 1 & i_2 & \dots & i_k \end{bmatrix} + r \begin{bmatrix} i_2 & \dots & i_{k+1} \\ i_2 & \dots & i_{k+1} \end{bmatrix} \\ &= \frac{1}{1+|p|^2}(S_k(r) + |p|^2 S_{k;1}(r)), \end{aligned} \tag{2.19}$$

so that

$$\begin{aligned} S_{k,l}(p,r) &= \frac{S_k(r) + |p|^2 S_{k;1}(r)}{S_l(r) + |p|^2 S_{l;1}(r)} \\ &\geq \min\{S_{k,l}(r), S_{k,l;1}(r)\} \text{ provided } S_l, S_{l;1} > 0, \\ &> S_{k+1,l+1}(r), \text{ if } r \in \Gamma_{k+1}(0). \end{aligned} \tag{2.20}$$

Here we have abbreviated $S_k(r) = S_k(0,r)$ etc. Furthermore, if $r \in S^n$ has eigenvalues $\lambda = (\lambda_1, \cdots, \lambda_n)$,then

$$S_{k,l;1}(r) \geq \min_i S_{k,l;i}(\lambda) \tag{2.21}$$

if $r \in \Gamma_{k+1}$,and

$$S_{k,l}(r) = S_{k,l}(\lambda) > \min_i S_{k,l;i}(\lambda) \tag{2.22}$$

if $r \in \Gamma_k$, whence we obtain from (2.2),

$$\begin{aligned} S_{k,l}(p,r) &\geq \min_i S_{k,l;i}(\lambda), \text{ if } r \in \Gamma_k, \\ &> S_{k+1,l+1}(r) \end{aligned} \tag{2.23}$$

if $r \in \Gamma_{k+1}$. We remark also that

$$\min_i S_{k,l;i}(\lambda) = S_{k,l;j}(\lambda) \tag{2.24}$$

for $\lambda_j = \max \lambda_i$. An alternative estimate for $S_{k,l}(p,r)$, in terms of $S_{k,l}(r)$ also follows from (2.20), namely

$$\begin{aligned} S_{k,l}(p,r) &\geq \frac{S_k(r)}{S_l(r) + |p|^2 S_{l,1}(r)} \quad \text{if } r \in \Gamma_{k+1}, \\ &\geq \frac{S_k(\lambda)}{S_l(\lambda) + |p|^2 \max_i S_{l,i}(\Lambda)} \\ &\geq \frac{1}{1+(n-l)|p|^2} S_{k,l}(r). \end{aligned} \tag{2.25}$$

The desired barrier for inequality (2.13) now arises from the following lemma

Lemma 2.1. *Suppose* $\Psi \in C^{0,1}(\mathcal{N} \cap \bar{\Omega})$ *satisfies*

$$0 \leq \Psi \leq \min_i S_{k-1,l-1;i}(\kappa_i'), \qquad \kappa' \in \Gamma_{k-1}. \tag{2.26}$$

Then for any constant $\epsilon, M > 0$ *and function* $\theta \in C^{1,1}(\mathcal{N})$ *there exist a further neighborhood of* $y, \tilde{\mathcal{N}} \subset \mathcal{N}$, *and a function* $g \in C^2(\Omega \cap \tilde{\mathcal{N}})$ *with* $D^2 g \in \Gamma(p)$ *for any* $p : \bar{\Omega} \cap \tilde{\mathcal{N}} \to R^n$, *such that*

$$\begin{cases} S_{k,l}(p, D^2 g) \geq (1-\epsilon)\Psi |Dg|^m & \text{in } \Omega \cap \tilde{\mathcal{N}} \\ g(y) = \theta(y), \quad g \leq \theta & \text{on } \partial\Omega \cap \tilde{\mathcal{N}} \\ g \leq -M & \text{on } \Omega \cap \partial\tilde{\mathcal{N}}. \end{cases}$$

Proof. Let us fix a ball $B_R(y) \in \mathcal{N}$ so that each point $x \in \Omega \cap B_R(y)$ has a unique closet boundary point $\bar{x} \in \partial\Omega \cap \mathcal{N}$ with the distance function $d \in C^2(\bar{\Omega} \cap B_R(y))$, (see [2]). We may then take $\tilde{\mathcal{N}} = B_R(y)$ and g given by

$$\begin{cases} g(x) = -a_0|x-y|^2 + h(d(x)), \\ h(d) = b_0(e^{c_0 d} - 1), \end{cases} \tag{2.27}$$

with constants a_0, b_0 and c_0 to be determined. Without loss of generality, we may assume $\theta(y) = D_i\theta(y) = 0$, $i = 1, \ldots, n-1$, and fix $a_0 \geq M/R$ so that

$$-a_0|x-y|^2 \leq \theta \qquad \text{on } \partial\Omega \cap \tilde{\mathcal{N}}.$$

To use the preceding properties of symmetric functions, we fix a point $x \in \tilde{\mathcal{N}} \cap \Omega$ and a principal coordinate system at $\bar{x} \in \mathcal{N} \cap \partial\Omega$, (see [2]). Then we have the formulae

$$\begin{cases} D_i g = -2a_0(x_i - y_i) + \delta_{in} h', \\ D^2 g = -2a_0 I + \operatorname{diag}\left[\dfrac{-h'\kappa_1'}{1 - d\kappa_1'}, \ldots, h''\right], \end{cases} \tag{2.28}$$

where the boundary curvatures $\kappa' = (\kappa'_1, \dots, \kappa'_{n-1})$ are evaluated at $\bar{x} \in \mathcal{N} \cap \partial\Omega$. Lemma 2.1 follows from (2.16), with 1 replaced by n, and (2.22), (2.25) with k, l and r replaced by $k-1$, $l-1$ and $r(n)$, by taking b_0 and c_0 sufficiently large. Note that by (2.23) the geometric hypothesis in Lemma 2.1 is weaker than the Serrin condition (iii) in Theorem 1.1. □

By taking $p = Du$ in Lemma 2.1, we obtain

$$\begin{aligned} F^{ij} D_{ij} g &\geq [S_{k,l}(Du, D^2 g)]^{1/m}, \text{ by concavity,} \\ &\geq ((1-\epsilon)\Psi)^{1/m} |Dg| \\ &\geq \Psi^i D_i g \end{aligned} \tag{2.29}$$

for $\epsilon \leq 1 - |\nu|^m$. Consequently, with $\theta = \tilde{w}$ on $\mathcal{N} \cap \partial\Omega$ and $M = -\inf_{\mathcal{N}\cap\Omega} \tilde{w}$, we infer, from the maximum principle,

$$D_n \tilde{w}(y) \geq D_n g(y) \tag{2.30}$$

and hence, taking tangential vector fields ξ in (2.5), we obtain the *a priori* estimates for the mixed second derivatives $D_{in} u(y)$, $i = 1, \dots, n-1$.

For the case of constant ϕ, we assume $\kappa' \in \Gamma_{k-1}$ on $\mathcal{N} \cap \partial\Omega$, and revisiting the proof of Theorem 3.1 in [9], we select as a barrier,

$$w^* = Au + Bu^2 + g, \tag{2.31}$$

where A and B are positive constants to be determined and g is given by (2.27). The constant a_0 is chosen as before, while b_0 and c_0 are chosen to ensure $D^2 g \in \Gamma_k(0)$ in $\tilde{\mathcal{N}} \cap \Omega$. By subtracting a constant, we may also assume $u = 0$ on $\mathcal{N} \cap \partial\Omega$. The constants A and B are initially selected to satisfy

$$-2B \inf_{\mathcal{N}\cap\Omega} u < A \tag{2.32}$$

so that in particular $w^* \leq h$ in $\mathcal{N} \cap \Omega$. For sufficiently small p, say $|p| < \eta$, it follows that $D^2 g \in \Gamma_k(p)$ in $\tilde{\mathcal{N}} \cap \Omega$, with

$$S_{k,l}(p, D^2 g) \geq (\epsilon |Dg|)^m$$

for some $\epsilon > 0$, depending on η and g. Since

$$\begin{aligned} F^{ij} D_{ij} w^* &= (A + 2Bu) F^{ij} D_{ij} u + 2B F^{ij} D_i u D_j u + F^{ij} D_{ij} g \\ &= (A + 2Bu)\tilde{\Psi} + 2B F^{ij} D_i u D_j u + F^{ij} D_{ij} g, \end{aligned} \tag{2.33}$$

we then obtain

$$F^{ij} D_{ij} w^* \geq \Psi^i D_i w^*, \tag{2.34}$$

provided $|Du| < \eta$ and $\Psi|\nu|^m \leq \epsilon^m$. If these inequalities are not satisfied, then $|Du| \geq \eta^*$ for some positive η^* depending on η, Ψ and ϵ. By taking B sufficiently large, in terms of g, it follows from (2.19), that the matrix

$$2BDu \otimes Du + D^2 g \in \Gamma_k(Du)$$

and hence $w^* \in \Gamma_k(Du)$. By taking A sufficiently large we then conclude (2.34) again, similarly to the proof of Theorem 3.1 in [9]. As for the previous case of Lemma 2.1, we again obtain from the maximum principle

$$D_n w^* \leq D_n \tilde{w}, \tag{2.35}$$

and mixed second derivative estimates follow as before. Accordingly, we have proved the following extension of [9], Theorem 3.1.

Theorem. *Let Ω be a uniformly $(k-1)$-convex domain with boundary $\partial\Omega \in C^{2,1}$, Ψ a non-negative function in $\Omega \times R$, with $\Psi^{1/m} \in C^{0,1}(\bar{\Omega} \times R)$, and ϕ a function in $C^{2,1}(\bar{\Omega})$. Suppose either*

$$\min_i S_{k-1,l-1;i}(\kappa') \geq \Psi, \qquad (\geq 1), \tag{2.36}$$

on $\partial\Omega$ or ϕ is constant. Then if $u \in C^3(\Omega) \cap C^2(\bar{\Omega})$ is an admissible solution of equation (1.1) in Ω, satisfying $u = \phi$ on $\partial\Omega$, we have at any point $y \in \partial\Omega$, the estimate,

$$|D_{in} u(y)| \leq C, \quad i = 1, \ldots, n-1. \tag{2.37}$$

with constant C depending on n, $\partial\Omega$, $|D\Psi^{1/m}|_0$, $|u|_1$ and $|\phi|_{2,1}$.

Remarks.

(i) From [9] and also (2.8) we have

$$|\nu|^m S_{k-1,l-1}(\kappa') > \Psi \tag{2.38}$$

as a necessary condition for an admissible solution $u \in C^2(\Omega)$ with constant boundary values. Consequently if $\inf_{\partial\Omega} \Psi > 0$, there is no need to assume Ω is $(k-1)$-convex in Theorem 2.2 when $l > 0$.

(ii) In the special case $k = l+1$, the proof of Theorem 2.2 for constant ϕ, simplifies as the matrix F^{ij} is bounded.

(iii) As remarked in the introduction the case $k = n$ is covered by Theorem 2.2 when ϕ is constant but excluded by condition (2.36) when Ψ is positive. Using the inequality

$$\begin{aligned} F^{ij} D_{ij} g &\geq F^{ij} \lambda_{\min}(D^2 g) \\ &\geq [S_{k,l}(1, \dots, 1)]^{1/m} \lambda_{\min}(D^2 g) \end{aligned} \tag{2.39}$$

by [9], (2.9), where $\lambda_{\min}(r)$ denotes the minimum eigenvalue of the symmetric matrix r, we can embrace the case $k = n$, for general boundary values ϕ, by replacing conditions (2.26), (2.38) by the condition

$$\binom{n}{k} (\kappa'_{\min})^m \geq \binom{n}{l} \Psi, \tag{2.40}$$

with Ω being assumed uniformly convex. For other results, see [6].

3. Normal second derivative boundary estimates.

To complete the estimation of the second derivatives at the boundary, we need to estimate the double normal derivatives $D_{nn}u$ in term of the other second derivatives. We follow the technique of [9], making use of the more elaborate barrier constructions in the preceding section. Let $u \in C^3(\bar{\Omega})$ be an admissible solution of the equation (1.1) in Ω satisfying $u = \phi$ on $\partial\Omega$, with $\partial\Omega \in C^{3,1}$ and $\phi \in C^{3,1}(\bar{\Omega})$. As before the coordinate system is chosen at a point $y \in \partial\Omega$ so that the x_n axis is directed along the inner normal to $\partial\Omega$ at y. From (2.12) we may write equation (1.1), $l > 0$, in the form

$$\frac{A_k D_{nn} u + B_k}{A_l D_{nn} u + B_l} = (\tilde{\Psi})^m, \tag{3.1}$$

where

$$\begin{aligned} A_k &= S_{k-1;n}(Du, D^2 u), \quad 1 < k \leq n \\ &= S_{k-1}(Du(n), D^2 u(n)) \end{aligned} \tag{3.2}$$

and B_k depends only on Du, $D^2 u(n, n)$. For admissible u, we must have

$$A_k - A_l (\tilde{\Psi})^m \geq 0, \quad A_k, A_l \geq 0, \tag{3.3}$$

(with strict inequality if $\Psi > 0$), and a bound for $D_{nn}u$ will follow if the quantity

$$\eta = \frac{A_k}{A_l} - (\tilde{\Psi})^m = S_{k-1,l-1;n}(Du, D^2u) - (\tilde{\Psi})^m \tag{3.4}$$

is bounded away from zero. Following [9], we now let $y \in \partial\Omega$ be a point in $\partial\Omega$ where the function η is minimized on $\partial\Omega$. To proceed further we express the function η in terms of a fixed orthonormal frame, as in [15]. Fixing a principal coordinate system at y and a corresponding neighborhood $\mathcal{N}$ of y with $\gamma_n < 0$ in $\mathcal{N} \cap \partial\Omega$, where γ denotes the unit outer normal vector field on $\partial\Omega$, we let $\xi^{(1)}, \dots, \xi^{(n-1)} \in C^{\mathbf{2,1}}(\mathcal{N} \cap \partial\Omega)$, be an orthonormal vector field on $\partial\Omega$, which is tangential and agrees with our coordinate system at y. Writing

$$\begin{cases} \partial_i u = \xi_k^{(i)} D_k u, & \nabla_{ij} u = \xi_l^{(i)} \xi_k^{(j)} D_{kl} u, \\ \mathcal{C}_{ij} = \xi_{l^{(i)}\xi_k^{(j)}} D_l \gamma_k, & \partial u = (\partial_i u), \\ \nabla^2 u = |\nabla_{ij} u|, & \mathcal{C} = [\mathcal{C}_{ij}], \quad i, j = 1, \dots, n-1, \end{cases} \tag{3.5}$$

we obtain from the boundary condition, $u = \phi$ on $\partial\Omega$,

$$\begin{aligned} \nabla_u^2 &= (D_\gamma u)\mathcal{C} + \nabla^2 \phi - (D_\gamma \phi)\mathcal{C} \\ &= (D_\gamma u)\mathcal{C} + \tilde{\phi}. \end{aligned} \tag{3.6}$$

Hence we may write the function η in (3.4) as

$$\eta(x) = \tilde{\eta}(x, D_\gamma u(x)), \tag{3.7}$$

where

$$\tilde{\eta}(x, t) = S_{k-1,l-1}(\partial\phi, t\mathcal{C} + \tilde{\phi}) - \Psi\sqrt{t^2 + |\partial\phi|^2 + 1}. \tag{3.8}$$

Now suppose $\mathcal{C} \in \Gamma_{k-1}(\partial\phi)$ in $\mathcal{N} \cap \Omega$. Then for $t \geq D_\gamma u$, the matrix $t\mathcal{C} + \tilde{\phi} \in \Gamma_{k-1}(\partial\phi)$ and furthermore

$$\begin{aligned} \tilde{\eta}(x, t) &= \frac{\partial}{\partial r} S_{k-1,l-1}(\partial\phi, t\mathcal{C} + \tilde{\phi}) \cdot \mathcal{C} - m\Psi t(t^2 + |\partial\phi|^2 + 1)^{\frac{m}{2}-1} \\ &\geq m\{S_{k-1,l-1}^{1/m}(\partial\phi, \mathcal{C}) S_{k-1,l-1}^{1-1/m}(\partial\phi, t\mathcal{C} + \tilde{\phi}) \\ &- \Psi t(t^2 + |\partial\phi|^2 + 1)^{\frac{m}{2}-1}\} \\ &\geq m\{S_{k-1,l-1}^{1/m}(\partial\phi, \mathcal{C}) \\ &- \frac{t\Psi^{2/m}}{\sqrt{t^2 + |\partial\phi|^2 + 1}}\Big\} \Psi^{1-1/m}(t^2 + |\partial\phi|^2 + 1)^{\frac{m-1}{2}} \end{aligned} \tag{3.9}$$

by virtue of the concavity and homogeneity of $S^{1/m}_{k-1,l-1}$ on Γ_{k-1}, as well as (3.3). In particular we obtain

$$\begin{aligned}\tilde{\eta}_t(x, D_\gamma u) &\geq m\Psi^{1-1/m}\left\{S^{1/m}_{k-1,l-1}(\partial\phi, C) - \frac{\Psi^{1/m}D_\gamma u}{\sqrt{1+|Du|^2}}\right\} \\ &\geq m\Psi(1-|\nu|)\end{aligned} \tag{3.10}$$

if either (2.26) holds or ϕ is constant. From here, we may proceed by extending the function η into $\mathcal{N}\cap\Omega$, by extension of the vector fields, $\gamma, \xi^{(i)} \in C^{1,1}(\mathcal{N}\cap\Omega)$ or otherwise. In place of (2.5), we define

$$w = \tilde{\eta}(x, \gamma\cdot Du) - K\sum_{s=1}^{n-1}|D_s u - D_s u(y)|^2. \tag{3.11}$$

From the derivation of (2.6) in [9] (see Remark 4(ii) in [9]), it follows that the function w satisfies similar inequality with constant C depending on $\mathcal{N}$, $|Du|_0$ and $D^2\tilde{\eta}$ provided the constant K is chosen sufficiently large in terms of $\tilde{\eta}_{tt}$. The details are carried out in [10]. From the barrier constructions in the preceding section, we infer a lower bound for $D_n\eta(y)$, and hence by (3.11) an upper bound for $D_{nn}u(y)$ that is

$$D_{nn}u(y) \leq C \tag{3.12}$$

for some constant C depending on n, $\partial\Omega$, $|\phi|_4$, $|\Psi|_2$, $\min_{\partial\Omega}\Psi$ and $|Du|_0$. From (3.12) we conclude, as in [9], a lower bound for η, namely

$$\eta \geq \eta(y) \geq \delta \tag{3.13}$$

where δ is a positive constant depending on n, $\partial\Omega$, $|\phi|_4$, $|\Psi|_2$, $\min\Psi$ and $|Du|_0$. Note that since η is minimized at y, $D_s\eta = 0$, $s = 1, \ldots, n-1$, whence

$$D_{sn}u(y) = [\tilde{\eta}_{x_s}/\tilde{\eta}_t - \kappa_s D_s u](y) \tag{3.14}$$

so that B_k, B_l in (3.1) are independent of the mixed tangential-normal second derivatives at the point y. From (3.1) and (3.13), we finally obtain an upper bound for $D_{nn}u$ on all of the boundary $\partial\Omega$, in terms of the quantities in (3.12), (3.13) together with the mixed second derivatives. A lower bound clearly follows from the admissibility of u. On combination with Theorem 2.2 we thus have the following boundary second derivative estimate.

Theorem 3.1. *Let Ω be a uniformly $(k-1)$-convex domain with boundary $\partial\Omega \in C^{3,1}$, Ψ a positive function in $C^{1,1}(\bar{\Omega}\times R)$ and ϕ a function in $C^{3,1}(\bar{\Omega})$. Suppose either* (2.36) *is satisfied or ϕ is constant. Then if $u \in C^3(\Omega)\cap C^2(\bar{\Omega})$ is an admissible solution of equation* (1.1) *in Ω, satisfying $u = \phi$ on $\partial\Omega$, we have the estimate*

$$\max_{\partial\Omega} |D^2 u| \leq C \tag{3.15}$$

where the constant C depends on n, $\partial\Omega$, $|\Psi|_2$, $\min_{\partial\Omega} \Psi$, $|\phi|_4$ and $|Du|_0$.

Remarks. **(i)** The esimate (3.15) can be localized by fixing any point $y \in \partial\Omega$ with either (2.36) satisfied or ϕ constant on $\partial\Omega \cap \mathcal{N}$ for some neighborhood $\mathcal{N}$ of y. An estimate for the mixed tangential normal derivatives of u at y is already provided by the arguments in Section 2. The double normal derivative $D_{nn}u(y)$ is estimated by replacing the function η by η^* given by

$$\eta^*(x) = \eta(x) + L|x-y|^2 \tag{3.16}$$

where $B_d(y) \subset \mathcal{N}$ and $L = \sup_{\partial\Omega\cap\mathcal{N}} \eta/d^2$. The minimum of η on $\partial\Omega \cap \mathcal{N}$ must then be taken on at some point $z \in B_d(y)\cap\partial\Omega$ and we proceed as before to deduce an upper bound for $D_{nn}u(z)$, from which follows lower bounds for $\eta(z)$ and hence $\eta(y)$, (see also [5, 14]). Consequently we conclude

$$|D^2 u(y)| \leq C, \tag{3.17}$$

with constant C now depending on n, $\mathcal{N}\cap\partial\Omega$, $|\Psi|_{2;\mathcal{N}\cap\partial\Omega}$, $\inf_{\mathcal{N}\cap\Omega} \Psi$, $|\phi|_{4;\mathcal{N}\cap\partial\Omega}$ and $|Du|_{0;\mathcal{N}\cap\Omega}$.

(ii) The estimates (3.15), (3.17) embrace the case $l = 0$, by considering the function

$$\eta = S_{k-1}(\partial\phi, D_\gamma uC + \tilde{\phi}), \tag{3.18}$$

under the hypothesis that Ω is uniformly $(k-1)$-convex if ϕ is constant and additionally

$$\min_i S_{k-1;i}(\kappa') \geq 0 \tag{3.19}$$

on $\mathcal{N} \cap \partial\Omega$ in the general case $\phi \in C^{3.1}$. Note that, in accordance with Remark (i) in Section 2, the hypothesis that Ω is $(k-1)$-convex is not necessary in the case when ϕ is constant and $l > 0$.

(iii) The estimation of the double normal derivative can also be achieved by replacement of $\tilde{\eta}$ by the function

$$\tilde{\eta}(x,t) = S^{1/m}_{k-1,l-1}(\partial\phi, tC + \tilde{\phi}) - \Psi \tag{3.20}$$

and using the concavity of $\tilde{\eta}$ with respect to t, for $tC + \tilde{\phi} \in \Gamma_{k-1}(\partial\phi)$, to follow more closely the proofs in [9] or [15]. Indeed this approach is simpler in the case when ϕ is constant, as the function (2.5) can be used directly with $\xi = \gamma$.

4. Completion of existence proofs.

As in [9], to apply the method of continuity to the proofs of Theorems 1.1 and 1.2, we need to designate suitable families of problems. For Theorem 1.2 we select any smooth admissible function g such that

$$0 < F[g] \leq \Psi. \tag{4.1}$$

For example, $g(x) = c|x|^2$ for sufficiently small c. Then we consider the problems

$$\begin{cases} F[u] = t\Psi + (1-t)F[g] & \text{in } \Omega, \\ u = t\phi + (1-t)g & \text{on } \partial\Omega \end{cases} \tag{4.2}$$

for $0 \leq t \leq 1$. For Theorem 1.2 we may take $g = u_0$, (with Ω replaced by $\{u_0 < \phi - \varepsilon\}$, for some $\varepsilon > 0$, if u_0 is not smooth enough at $\partial\Omega$). Clearly all out preceding estimates are independent of the parameter t and we conclude an *a priori* estimate of the form

$$|u|_{2,\alpha;\Omega} \leq C \tag{4.3}$$

with constant C depending on n, Ω, Ψ, ϕ, and u_0. The unique solvability of the Dirichlet problems (4.2), for all $0 \leq t \leq 1$, then follows.

As remarked in the introduction, the hypothesis (2.36), used for the derivation of the second derivative boundary estimates, permits a more general version of Theorem 1.1.

Theorem 4.1. *Suppose hypothesis* (iii) *of Theorem* 1.1 *is replaced by,* **(iii)′** Ω *is* $(k-1)$*-convex with*

$$\Psi(x,\phi(x)) \leq \min_i S_{k-1,l-1;i}(\kappa') \qquad \text{on } \partial\Omega.$$

Then, provided there exists an admissible subsolution $u_0 \in C^{0,1}(\bar{\Omega})$ *with* $u_0 = \phi$ *on* $\partial\Omega$, *there exists an admissible solution* $u \in C^{4,\alpha}(\bar{\Omega})$ *of equation* (1.1) *satisfying* $u = \phi$ *on* $\partial\Omega$.

References.

[1] Caffarelli, L., Nirenberg, L., and Spruck, J., *Nonlinear second-order elliptic equations V. The Dirichlet problem for Weingarten hypersurfaces.* Comm. Pure Appl. Math. 41 (1988), 47–70.

[2] Gilbarg, D. and Trudinger, N.S., *Elliptic Partial Differential Equations of Second Order.* Second Edition, Springer-Verlag, 1983.

[3] Ivochkina, N.M., *Solution of the Dirichlet problem for curvature equations of order m*, English tran.: Math USSR Sbornik 67 (1990), No.2, 317–339.

[4] Ivochkina, N.M., *The Dirichlet problem for the equations of curvature of order m*, English trans.: Leningrad Math. J. 2 (1991), No.3, 631–645.

[5] Ivochkina, N.M., *Local estimates of normal second derivatives for solutions of the Dirichlet problem for the prescribed quotients curvature equations*, Zap. Nauchn. Sem. St. Petersburg. 221 (1995), 114–126.

[6] Ivochkina, N.M. and F. Tomi (in preparation).

[7] Krylov, N.V., *Nonlinear elliptic and parabolic equations of the second order.* Reidel, Dordrecht, 1987.

[8] Lin, M. and Trudinger, N.S., *On some inequalities for elementary symmetric functions*, Bull. Aust. Math. Soc. 50 (1994), 317–326.

[9] Lin, M. and Trudinger, N.S., *The Dirichlet problem for the prescribed curvature quotient equations*, Topological Methods in Nonlinear Analysis, 3 (1994) 1–17.

[10] Lin, M. and Trudinger, N.S., *The Dirichlet problem for the prescribed curvature quotient equations* 11 (in preparation).

[11] Serrin, J., *The problem of Dirichlet for quasilinear elliptic differential equations with many independent variables.* Philos. Trans. Roy. Soc. London Ser. A 264 (1969), 413–496.

[12] Trudinger, N.S., *The Dirichlet problem for the prescribed curvature equations*, Arch. Rat. Mech. Anal. 111 (1990), No.2, 153–179.

[13] Trudinger, N.S., *Maximum principles for curvature quotient equations*, J. Math. Sci. Univ. Tokyo 1 (1994), 551–566.

[14] Trudinger, N.S., *Lectures on Nonlinear Elliptic Equations of Second Order*, Univ. Tokyo, 1995.

[15] Trudinger, N.S., *On the Dirichlet problem for Hessian equations*, Acta Math. **175** (1995), 151–164.

RECEIVED NOVEMBER 13, 1995.

ST. PETERSBURG ACADEMY OF CIVIL ENGINEERING,
ST. PETERSBURG, RUSSIA.

DEPARTMENT OF MATHEMATICS,
NORTHWESTERN UNIVERSITY, EVANSTON, USA.

AND

CENTRE FOR MATHEMATICS AND ITS APPLICATIONS,
AUSTRALIAN NATIONAL UNIVERSITY, CANBERRA, AUSTRALIA.

Generalized harmonic maps between metric spaces

JÜRGEN JOST

Dedicated to Stefan Hildebrandt for his sixtieth birthday

Contents.

Introduction.

Harmonic maps between Riemannian manifolds M, N are critical points of the energy integral

$$E(f) = \int_M \|df(x)\|^2 \, \mathrm{dvol}_M(x)$$

(see § 1 for the notation involved).

They have been introduced by Bochner and studied by Al'ber, Eells-Sampson, Hartman, and Hamilton in case N has nonpositive curvature, and by Hildebrandt-Kaul-Widman, Schoen-Uhlenbeck, Giaquinta-Giusti and many others in the general case (see §1 for more precise references). Important links have been developed with the calculus of variations and elliptic and parabolic PDE, inspiring also much research in those fields, and many important geometric applications have been found, for example to rigidity questions (see the survey articles quoted in § 1).

For questions in the context of Margulis' superrigidity (see [Ma1]), it was found necessary, however, to consider more general target spaces than Riemannian manifolds. For this reason, Gromov-Schoen [GS] developed a theory of generalized harmonic maps with values in certain metric spaces, for example Euclidean Tits buildings. Those target spaces have nonpositive curvature in the sense of Alexandrov (see §3). A general theory of harmonic maps between metric spaces was developed by Jost [J2], [J3] and Korevaar-Schoen [KS] independently, with results usually also requiring nonpositive curvature of the target. In particular, in [J2], [J3], the existence problem was solved in a very general setting. In the present paper, we survey and extend those constructions and point out some further directions. Important themes are the connection with convexity properties of the target and mean value properties of harmonic maps (in this direction, see also the interesting work of Kendall [Ke1].)

In §1, we briefly recall the theory of harmonic mappings between Riemannian manifolds. §2 introduces the notion of generalized harmonic maps between metric spaces. As an important tool, we use Γ-convergence in the sense of de Giorgi. In §3, we explain upper curvature bounds in the sense of Alexandrov and their geometric consequences. In §4, we demonstrate a general result on the existence of minimizers of convex functionals on metric spaces with certain convexity properties. This result is used in §5 to obtain existence results for generalized harmonic maps between metric spaces, including those of Jost [J2], [J3] and Korevaar-Schoen [KS].

The regularity question for generalized harmonic maps even with values in spaces of nonpositive curvature requires additional assumptions on the domain. If the domain is a Riemannian manifold, Korevaar-Schoen [KS] have shown the Lipschitz continuity of such maps. In §6, we indicate a somewhat different proof of this Lipschitz continuity that links it with mean value properties. Higher regularity is known only under more specialized assumptions (see [GS] and results announced in [KS]), and this topic is not addressed here.

Kendall [Ke3], [Ke4] explored the connection of harmonic maps with Brownian motion. In §7, we present a stochastic interpretation of harmonic maps not involving Brownian motion. This interpretation also suggests a fully parallel numerical approach for the computation of harmonic maps.

The author has had interesting discussions about generalized harmonic maps with many people, including Scott Adams, Stephanie Alexander, Gabriele Anzellotti, David Berg, Richard Bishop, Pat Eberlein, Wilfrid Kendall,

Igor Nikolaev, Pierre Pansu, Rick Schoen, Karl-Theodor Sturm, and Shing-Tung Yau.

The author thanks Prof. Jean-Pierre Ezin for organizing a visit to the IMSP in Porto Novo (Benin) where parts of this paper have been written. The author acknowledges generous financial support from the DFG (Leibniz programme, SFB 237) of the research underlying this paper.

1. Harmonic mappings.

Let M and N be Riemannian manifolds. For (sufficiently regular) maps $f : M \to N$, we consider the energy integral

$$E(f) = \int_M ||df(x)||^2 \, \mathrm{dvol}_M(x) \tag{1.1}$$

where the norm of $df(x) : T_x M \to T_{f(x)} N$ is defined with the help of the Riemannian metrics on M and N. If the metrics of M and N are given in local coordinates by $(\gamma_{\alpha\beta})_{\alpha,\beta=1,\dots,m}$ ($m = \dim M$) and $(g_{ij})_{i,j=1,\dots,n}$ ($n = \dim N$), resp., we have with the usual tensor notation $(\gamma^{\alpha\beta}) = (\gamma_{\alpha\beta})^{-1}$,

$$e(f)(x) := ||df(x)||^2 = \gamma^{\alpha\beta}(x) g_{ij}(f(x)) \frac{\partial f^i}{\partial x^\alpha} \frac{\partial f^j}{\partial x^\beta} \tag{1.2}$$

with the standard summation convention.

Critical points, e.g. minimizers of E, are called harmonic maps if they are smooth. They satisfy the Euler-Lagrange equations for E, i.e.

$$\frac{1}{\sqrt{\gamma}} \frac{\partial}{\partial x^\alpha} \left(\gamma^{\alpha\beta} \sqrt{\gamma} \frac{\partial f^i}{\partial x^\beta} \right) + \gamma^{\alpha\beta} \Gamma^i_{jk}(f(x)) \frac{\partial f^i}{\partial x^\alpha} \frac{\partial f^k}{\partial x^\beta} = 0 \quad \text{for } i = 1, \dots, n, \tag{1.3}$$

with $\gamma := \det(\gamma_{\alpha\beta})$,

$$\Gamma^i_{jk} := \frac{1}{2} g^{il} \left(\frac{\partial}{\partial f^j} g_{kl} + \frac{\partial}{\partial f^k} g_{jl} - \frac{\partial}{\partial f^l} g_{jk} \right) \quad \text{(Christoffel symbols of } N\text{)} .$$

(1.3) constitutes a nonlinear system of elliptic partial differential equations and since the nonlinearity is quadratic in the gradient of f and hence analytically of the same weight as the linear part of (1.3), the Laplace-Beltrami operator of M applied to f, the question of regularity of solutions cannot be solved by general principles.

The first existence and regularity results for harmonic maps are due to Morrey who solved the Dirichlet problem for harmonic maps if the domain M is two-dimensional and who showed that in this case, minimizers of E are regular [Mo]. He did not specify a topological class for the maps produced by his method, but he applied them to construct minimal surfaces in Riemannian manifolds. These results were later extended by Lemaire [L] and Sacks-Uhlenbeck [SkU] who showed the existence of a harmonic map in a prescribed homotopy class for a two-dimensional domain under the topological assumption $\pi_2(N) = 0$.

Using quite different methods, namely the parabolic system associated with (1.3) (consider maps $f : M \times [0, \infty) \to N$ and replace the right hand side of (1.3) by $\frac{\partial f^i}{\partial t}$), Eells-Sampson [ES], Al'ber [A1], [A2], and Hartman [Hr] solved the problem of existence, regularity, and uniqueness affirmatively for maps between compact Riemannian manifolds in a prescribed homotopy class under the assumption that the image N has nonpositive sectional curvature. (For extensions, see [DO], [D], [C], [JY1], [La]). This curvature assumption leads to a useful differential inequality. Namely, one has the so-called Bochner formula

$$\begin{aligned}\frac{1}{2}\Delta \left\|df\right\|^2 &= \left\|\nabla df\right\|^2 + \left\langle \mathrm{Ric}^M\left(df(e_\alpha)\right), df(e_\alpha)\right\rangle \\ &\quad - \left\langle R^N\left(df(e_\alpha), df(e_\beta)\right) df(e_\beta), df(e_\alpha)\right\rangle \qquad (1.4)\end{aligned}$$

where ∇df is the Hessian of f, (e_α) is an orthonormal frame at the point x under consideration, $< \cdot, \cdot >$ is the product determined by the metrics of M and N, Ric^M is the Ricci tensor of M, and R^N is the curvature tensor of N.

Thus, if N has nonpositive sectional curvature, we obtain

$$\Delta \left\|df\right\|^2 \geq -c_1 \left\|df\right\|^2 \qquad (1.5)$$

for some constant c_1 (depending only on a lower bound for the Ricci curvature of M), and from this inequality one may deduce the estimate

$$\left\|df(x_0)\right\|^2 \leq \frac{c_2}{R^2}\int_{B(x_0,R)} \left\|df(x)\right\|^2 \mathrm{dvol}_M(x) = \frac{c_2}{R^2} E\left(f|_{B(x_0,R)}\right), \qquad (1.6)$$

where $B(x_0, R) := \{x \in M : d(x, x_0) < R\}$ is a distance ball ($d(\cdot,\cdot) =$ Riemannian distance function of M).

A similar consideration is possible in the parabolic situation, and the resulting estimates are crucial for the existence and regularity proofs just indicated.

Hamilton [Hm] then was able to solve the Dirichlet problem for harmonic maps under the assumption that the image has nonpositive curvature again.

For the general case where N may also have positive curvature, fundamental progress was achieved by Hildebrandt-Kaul-Widman [HKW1], [HK-W2]. Employing and developing [HW] powerful methods for elliptic PDE, they could solve the Dirichlet problem for harmonic maps if the image is contained in a strictly convex ball, i.e. a ball $B(p, R) := \{q \in N : d(p,q) < R\}$ that is disjoint to the cut locus of p and satisfies $R < \frac{\pi}{2\sqrt{K}}$, where $K \geq 0$ is an upper curvature bound. (For the unit spheres S^n of curvature 1, this means that the image has to lie in an open hemisphere.) Their work linked the existence and regularity problem for harmonic maps to convexity properties of the image. They also showed by an example that the restriction on R is sharp. Namely, the map

$$f : B(0, \varrho) \subset \mathbb{R}^m \to S^{m-1} \subset \mathbb{R}^m$$

$$x \mapsto \frac{x}{|x|}$$

for $m \geq 3$ is a weak solution of (1.3) with a singularity at 0. One may also compose this map with the totally geodesic equatorial embedding

$$i : S^{m-1} \to S^m$$

in order to obtain a singular weakly harmonic map $i \circ f$ with image contained in a closed hemisphere of S^m, but not in an open one. (For the open hemisphere, [HKW] implies regularity!).

Subsequently, the structure of the singular set of energy minimizing maps was investigated by Schoen-Uhlenbeck [SU1], [SU2], and in a somewhat different context, by Giaquinta-Giusti [GG1], [GG2] (see also Steffen [St] for a comprehensive account including later developments), and it was found that any possible singularity after a blow-up becomes the one discovered by Hildebrandt-Kaul-Widman. Thus, they had found the prototype for the singularities of energy minimizing maps.

Here, we refrain from giving a more detailed account of the theory of harmonic maps between Riemannian manifolds and their geodesic applications and quote the survey articles [EL1], [EL2], [EL3], [H], [J1], [JY2], [JY3] instead.

2. Energy functionals for maps between metric spaces.

We let M by a metric space, with metric denoted by $d(\cdot,\cdot)$, and equipped with a Radon measure μ.

We let N be another metric space, with metric likewise denoted by $d(\cdot,\cdot)$. We shall usually assume that N is complete. For a continuous curve $\gamma : [0,a] \to N$, we may define its length as

$$L(\gamma) := \sup \sum_{i=i}^{m} d\left(\gamma(t_i), \gamma(t_{i+1})\right) \in \mathbb{R}^{\geq 0} \cup \{\infty\}$$

where the supremum is taken over all partitions

$$0 = t_0 < t_1 < \ldots < t_m < t_{m+1} = a$$

of $[0,a]$.

We say that γ is parametrized proportionally to arclength if there exists a constant λ with

$$L\left(\gamma|_{[0,t]}\right) = \lambda t \quad \text{for all } t \in [0,a].$$

In case $\lambda = 1$, we say that γ is parametrized by arclength. We say that γ is geodesic if

$$\exists \eta > 0 \forall t_1, t_2 \in [0,a] \quad \text{with } 0 < t_2 - t_1 \leq \eta :$$
$$L\left(\gamma|_{[t_1,t_2]}\right) = d\left(\gamma(t_1), \gamma(t_2)\right).$$

In particular, any shortest connection of its endpoints, i.e. any curve $\gamma : [0,a] \to N$ with $\gamma(0) = p$, $\gamma(a) = q$ and

$$L(\gamma) = d(p,q)$$

is geodesic as follows from the triangle inequality.

In the sequel, all geodesics will be assumed to be parametrized proportionally to arclength.

We shall also assume that N is geodesically complete, i.e. that any $p, q \in N$ can be connected by a shortest geodesic γ. (In particular, N has to be connected). A metric space with this property is sometimes called geodesic or length space. Given some for example continuous map $g : M \to N$, we may define the space $L^2(M,N)$ as the completion of the metric space of those maps $f : M \to N$ for which $d^2(f(x), g(x))$ is μ-measurable and

$$\int d^2(f(x), g(x))\, d\mu(x) < \infty.$$

For technical reasons, however, it is preferable to lift the maps f, g to the universal covers, e.g.

$$\tilde{f} : \tilde{M} \to \tilde{N},$$

and use the lifted metric on $\tilde{N}$, again denoted by $d(\cdot,\cdot)$, instead of the metric on N and consider

$$\int d^2(\tilde{f}(x), \tilde{g}(x))d\mu(x),$$

where the integration now is carried out on a fundamental region for M in $\tilde{M}$.

In the sequel, we shall always implicitly perform this lift to universal covers without explicit mention.

Usually, we shall consider more generally the situation where we have a homomorphism

$$\varrho : \pi_1(M) \to I(\tilde{N}),$$

$I(\tilde{N})$ denoting the group of isometries of $\tilde{N}$, and where all maps $f : \tilde{M} \to \tilde{N}$ are required to be ϱ-equivariant, i.e.

$$f(\gamma x) = \varrho(\gamma) f(x) \quad \text{for all } x \in \tilde{M}, \gamma \in \pi_1(M).$$

Still more generally, $\pi_1(M)$ may be replaced here by an arbitrary subgroup Γ of the isometry group $I(\tilde{M})$, and we suppose that the measure μ is defined on the quotient M/Γ. In other words, with a slight change of notation, we consider simply connected spaces X_1 and X_2, X_2 carrying a geodesically complete metric $d(\cdot,\cdot)$, with isometry group $I(X_2)$, a group Γ operating on X_1, a homomorphism $\varrho : \Gamma \to I(X_2)$, and a Γ-equivariant measure μ on X_1 (i.e. $\gamma_{\#}\mu = \mu$ for all $\gamma \in \Gamma$) which then induces a measure μ_Γ on X_1/Γ. $L^2_\varrho(X_1, X_2)$ then is the metric completion of the space of ϱ-equivariant maps $f : X_1 \to X_2$ with metric defined by

$$d^2(f,g) := \int d^2(f(x), g(x))d\mu_\Gamma(x).$$

We also suppose that the nonnegative, symmetric function $h(\cdot,\cdot) : X_1 \times X_1 \to \mathbb{R}^{\geq 0}$ is Γ-invariant, i.e.

$$h(\gamma x, \gamma y) = h(x,y) \quad \text{for all } \gamma \in \Gamma, x, y, \in X_1.$$

For $f \in L^2_\varrho(X_1, X_2)$, we define the energy density as

$$e_h(f)(x) := \int h(x,y) d^2(f(x), f(y)) d\mu_\Gamma(y),$$

and the energy as

$$E_h(f) := \int e_h(f)(x) d\mu_\Gamma(x) = \int h(x,y) d^2(f(x), f(y)) d\mu_\Gamma(y) d\mu_\Gamma(x).$$

The energy functional

$$E : L^2_\varrho(X_1, X_2) \to \mathbb{R}^{\geq 0} \cup \{\infty\}$$

can then be obtained as the Γ-limit in the sense of de Giorgi of a suitable family of such functionals $(E_{h^n})_{n \in \mathbb{N}}$ (possibly after selecting a subsequence). Because of the symmetry of h, f minimizes E_h iff $f(x)$ is the $h(x,\cdot)$ weighted mean value of f for a.a. $x \in M$, i.e. $f(x)$ minimizes

$$\varphi(q) := \int h(x,y) d^2(q, f(y)) d\mu_\Gamma(y).$$

For example, assume that X_1 is also a metric space with a Γ-invariant metric $d_1(\cdot,\cdot)$, put

$$B(x,r) := \{y \in X_1 : d_1(x,y) < x\}$$

and

$$h^n(x,y) := \frac{\chi|_{B(x,1/n)}(y)}{\text{normalization factor}} \quad (\chi \text{ denotes the characteristic function}),$$

where the normalization factor is chosen in such a way as to get a nontrivial limit for the functionals for $n \to \infty$ (if X_1 is a Riemannian manifold of dimension m, one may choose the normalization factor as $n^{-(m+2)}$). In this case, E_{h^n} is minimized iff for a.a. $x \in M$, $f(x)$ is the mean value of f on the ball $B(x, 1/n)$.

In general, if

$$F_n : Y \to \mathbb{R} \cup \{\pm\infty\}$$

is a sequence of functionals defined on some metric space Y, a functional $F : Y \to \mathbb{R} \cup \{\pm\infty\}$ is called Γ-limit of $(F_n)_{n \in \mathbb{N}}$,

$$F = \Gamma - \lim_{n \to \infty} F_n$$

if

(i) For all convergent sequences $x_n \to x$ in Y,

$$F(x) \leq \liminf_{n \to \infty} F_n(x_n).$$

(ii) For every $x \in Y$, there exists some sequence $y_n \to x$ in Y with

$$F(x) \geq \limsup_{n\to\infty} F_n(y_n).$$

(There also exist more general versions of this definition, e.g. in terms of filters.)

Here, the first condition is the more important one. Γ-limits have many important variational properties some of which we are going to mention (The general reference for all relevant results is dal Maso's book [dM]):

- Γ-limits are lower semicontinuous
- Γ-limits of convex functionals are convex ($F : Y \to \mathbb{R} \cup \{\pm\infty\}$ is called convex if the restriction of F to any geodesic in Y is a convex function of the arclength parameter)
- If x_n minimizes F_n is some fixed class, and if $x_n \to x \in Y$, then x minimizes F in this class. Thus, if one wants to minimize F, it suffices to minimize the approximating functionals F_n and find a convergent sequence of these minimizers.
- Provided one is allowed to take a subsequence of $(F_n)_{n\in\mathbb{N}}$, the Γ-limit exists under very general circumstances; for example, it suffices to assume that Y is second countable.

Sometimes it is convenient to define $H^1_\varrho(X_1, X_2)$ as the space of those maps of class $L^2_\varrho(X_1, X_2)$ with

$$E(f) < \infty.$$

In example 1) below, this is a Sobolev space of maps. More generally, Sobolev spaces of maps between metric spaces have been studied by Ambrosio [Am] and Korevaar-Schoen [KS].

Minimizers of E_h are called h-harmonic or h-equilibrium maps, minimizers of E harmonic or equilibrium maps.

Examples.

1) M and N compact Riemannian manifolds (with universal covers X_1 and X_2), μ the Riemannian volume of M (lifted to X_1),

$$\mu^\varepsilon_x = \mu \llcorner B(x, \varepsilon),$$

with

$$\begin{aligned} B(x,\varepsilon) &= \{y \in X_1 : d(y,x) < \varepsilon\}, \\ E_\varepsilon &:= \frac{1}{\text{normalization factor}} \cdot \int\int d^2(f(x), f(y)) d\mu_x^\varepsilon(y) d\mu(x), \\ E &:= \Gamma - \lim_{\varepsilon \to 0} E_\varepsilon . \end{aligned}$$

In this case, one may show that $E(f)$ is the standard energy functional (up to a constant factor), and this example in fact motivates much of the preceding constructions.

Minimizers of E_ε are called ε-harmonic and characterized by the mean value property on balls of radius ε. We recall that harmonic functions on Euclidean domains satisfy this mean value property on balls of any radius contained in their domain. For harmonic functions on Riemannian domains, or for harmonic maps between Riemannian manifolds, however, in general this is not true anymore. Nevertheless if we consider harmonic functions or maps as limits of ε-harmonic functions or maps, then they still satisfy an infinitesimal mean value property ($\varepsilon \to 0$).

2) M, N, μ as in 1), $m := \dim M$,

$$\mu_x^\varepsilon(y) = \frac{1}{(4\pi\varepsilon)^{\frac{m}{2}}} e^{-\frac{d(x,y)}{4\varepsilon}} \mu(y) \quad \text{(heat kernel or Gaussian measure)} ,$$

similar results as in 1).

3) M, N, μ as in 1),

$$\mu_x^\varepsilon = \mu \llcorner B\left(x, \frac{1}{\varepsilon}\right) .$$

In this case, in the limit $\varepsilon \to 0$, we obtain a global mean value property.

In 1)–3), nothing essential changes if we let N be a general complete geodesic metric space.

4) $M = \mathbb{R}^d$, μ =Lebesgue measure,

$$\mu_x^\varepsilon = \mu \llcorner \left\{ y = (y^1, \dots, y^d) : \sum_{i=1}^{c} (y^i)^2 \le \varepsilon^2, \sum_{j=c+1}^{d} (y^j)^2 \le \varepsilon^4 \right\}$$

for some $c \in \mathbb{N}$ with $1 \leq c < d$.
This construction can be generalized to any (not necessarily) smooth foliation of a Riemannian manifold (or a metric space), in order to obtain in the limit $\varepsilon \to 0$ objects that are harmonic in the directions tangential to the leaves of the foliation while still preserving some regularity properties in the normal direction (obviously, ε^4 may be replaced by any function of ε going to zero faster than quadratically). Thus, the construction can be interpreted as a viscosity method for leafwise or foliated harmonic mappings, for example in the sense considered by Gromov [G].

The construction of generalized harmonic maps presented here follows [J2]. A somewhat similar construction was independently developed in [KS].

Generalizations.

1) In the definition of E_h or E_ε, $d^2(f(x), f(y))$ may be replaced by $\psi(f(x), f(y))$, where ψ is any smooth function with an asymptotic expansion

$$\psi(z, w) = d^2(z, w) + \text{higher order terms}$$

near $z = w$.

In the above example 1), this generalization will still reproduce the standard energy functional E in the limit $\varepsilon \to 0$. This generalization will be important below for establishing the existence of harmonic maps with values in convex balls.

2) One may replace $d^2(f(x), f(y))$ in the definition of the ε-energy by $d^p(f(x), f(y))$ for $p \geq 1$; or, more generally by $\psi(d(f(x), f(y)))$ for any convex function with $\psi(0) = 0$ as the unique minimum. The existence theory presented in this paper is not affected by this generalization, but the question of higher differentiability properties (not addressed here) becomes more difficult. It is presently not clear whether this generalization will lead to new applications beyond those for $\psi(d) = d^2$, with the possible exception of $\psi(d) = d$.

3) Measure valued equilibrium maps
We let M, N, μ, μ_x^ε etc. as before. Of course, everything below may again be carried out on universal covers in an equivariant manner.

Let

$$\begin{aligned} V &:= \{\nu \text{ Radon measure on } N\}, \\ V_1 &:= \left\{\nu \in V : \int_N d\nu(q) = 1\right\}. \end{aligned}$$

Let

$$F := \{f : M \to V\}, \quad F_1 := \{f : M \to V_1\}.$$

We make V and V_1 into metric spaces by putting

$$d^2(\nu^1, \nu^2) = \int\int d^2(p,q) d\nu^1(p) d\nu^2(q),$$

and similarly for F and F_1

$$d^2(f_1, f_2) = \int_M \int_N \int_N d^2(p,q) d\nu_x^1(p) d\nu_x^2(q) d\mu(x)$$

with $\nu_x^i := f_i(x)$.

We also define weak convergence

$$f_n \rightharpoonup f :\Leftrightarrow \int \varphi(p)\eta(x) d\nu_x^n(p) dx \to \int \varphi(p)\eta(x) d\nu_x(p) dx$$

$$\text{for all } \varphi \in C^0(N), \eta \in C_0^0(M)$$

(with $\nu_x^n = f_n(x)$, $\nu_x = f(x)$) and take $\bar{F}_1$ as the smallest subset of F that contains F_1 and is closed under weak convergence.

For $f \in \bar{F}_1$ and $\varepsilon > 0$, we put

$$e_\varepsilon(f)(x) := \frac{\int\int\int d^2(p,q) d\nu_x(q) d\nu_y(p) d\mu_x^\varepsilon(y)}{\text{normalization}}$$

and define the ε-energy as

$$E_\varepsilon(f) := \int e_\varepsilon(f)(x) d\mu(x),$$

and, provided this exists (possibly after taking a subsequence $\varepsilon_n \to 0$), the energy as

$$E = \Gamma - \lim_{\varepsilon \to 0} E_\varepsilon.$$

One attempts to minimize $E(f)$ in the class $f \in \bar{F}_1$, or some subclass defined e.g. by a ρ-equivariance condition. If we lift to universal covers

X of M, Y of N and look at maps that are equivariant w.r.t. some representation

$$\rho : \pi_1(M) \to I(Y),$$

and if we assume that Y has nonpositive curvature in the sense of Alexandrov (see §3), then for minimizing the energy, it suffices to look at ordinary maps $f : X \to Y$, because energy will be decreased if each measure ν_x is replaced by the Dirac measure associated with its center of gravity. A similar remark applies for maps with values in convex balls (see below). In general, however, phenomena like the splitting off of minimal 2-spheres for energy minimizing sequences of ordinary maps make even the passage from F_1 to $\bar{F}_1$ necessary. (One starts with the space F_1 in order to exclude maps whose images consist of trivial measures.) Thus, measure valued maps may be a useful tool for understanding the formation of singularities of energy minimizing maps.

There should also exist interesting relations with the theory of Cartesian currents of Giaquinta-Modica-Souček [GMS].

4) Nonlinear cocycles

Let M be a metric space with measures μ, μ_x^ε, Y a complete 1-connected metric space. Let $\mathcal{U}$ be an acyclic open covering of M. Suppose that for any $U_1, U_2 \in \mathcal{U}$ with $U_1 \cap U_2 \neq \emptyset$, there is given $i_{U_1U_2} \in I(Y)$ (isometry group of Y). We then look for a collection of harmonic maps

$$f_U : U \to Y \ , \quad U \in \mathcal{U}$$

s.t. for $U_1, U_2 \in \mathcal{U}$ with $U_1 \cup U_2 \neq \emptyset$,

$$f_{U_2|U_1 \cap U_2} = i_{U_1U_2} f_{U_1|U_1 \cap U_2}.$$

Since the energy E remains invariant under isometries of the image, such a collection $\{f_U\}_{U \in \mathcal{U}}$ can be found by minimizing as in §5 below if Y has nonpositive curvature (under some reductivity assumption), provided that the class under consideration is nonempty. For f to be well defined, one needs to assume the nonlinear cocycle conditions:

$$\begin{aligned} i_{U_1U_1} &= \text{id} \\ i_{U_1U_2} &= (i_{U_2U_1})^{-1} \quad \text{if } U_1 \cap U_2 \neq \emptyset \\ i_{U_1U_2} i_{U_2U_3} i_{U_3U_1} &= \text{id} \quad \text{if } U_1 \cap U_2 \cap U_3 \neq \emptyset. \end{aligned}$$

Since $I(Y)$ is noncommutative in general, this is more general than giving a homomorphism $\varrho : \pi_1(M) \to I(Y)$; it corresponds to a homomorphism defined only up to conjugation; a corresponding ϱ-equivariant map then satisfies

$$f(\gamma x) = \lambda^{-1}\varrho(\gamma)\lambda f(x) \quad \text{for some } \lambda \in I(Y),$$

i.e.

$$\lambda f(\gamma x) = \varrho(\gamma)\lambda f(x).$$

Thus $\lambda f(x)$ satisfies the equivariance condition, or f satisfies it up to some isometry λ of Y.

3. Geometry of spaces with upper curvature bounds.

Let N be a metric space with metric $d(\cdot,\cdot)$. In order to prove the existence of equilibrium maps with values in N, we shall need to impose a geometric restriction on N, namely an upper curvature in the generalized sense of Alexandrov.

Definition 1. We say that N has curvature $\leq K$ in the sense of Alexandrov if

(i) each $p \in N$ has a neighborhood any two points of which can be connected by a shortest curve

and

(ii) for each sufficiently small geodesic triangle (i.e. whose sides are geodesics $\gamma_i : [0, a_i] \to N$ $(i = 1, 2, 3)$ which we assume to be parametrized by arclength), there exists a geodesic triangle in the sphere S_K^2 of curvature K in case $K > 0$, the Euclidean plane $\mathbb{R}^2$ in case $K = 0$, or the hyperbolic plane H_K^2 of constant curvature K in case $K < 0$, with sides $\bar{\gamma}_i$ of the same lengths as γ_i for $i = 1, 2, 3$, and if p_i and $\bar{p}_i$ denote the vertices of the triangles opposite to γ_i and $\bar{\gamma}_i$, resp., we have

$$d(\gamma_i(t), p_i) \leq d_K(\bar{\gamma}_i(t), \bar{p}_i)$$

for $0 \leq t \leq a_i$ where we assume that $\bar{\gamma}_i$ is also parametrized by arclength on $[0, a_i]$, and where $d_K(\cdot,\cdot)$ is the distance function of the above metric of constant curvature K.

The formulation given here is somewhat different from the original formulation of Alexandrov. The equivalence of the two versions is a consequence of the work of Reshetnyak. For an exposition, we refer to Berestovskij-Nikolaev [BN].

Every Riemannian manifold with sectional curvature $\leq K$ is a space of curvature $\leq K$ in Alexandrov's sense.

If one wants to bring out the global aspects, one requires the comparison property for every geodesic triangle of perimeter $\leq \frac{2\pi}{\sqrt{K}}$ in case $K > 0$, not only for sufficiently small ones. (The universal cover of a space of curvature ≤ 0 for example satisfies this global property.) For simplicity of notation, we say in this case that N is a space of global curvature $\leq K$.

From the work of Reshetnyak [Re], important comparison properties for spaces of global curvature $\leq K$ can be deduced. We quote the results here only in case $K = 0$ for which proofs can also be found in [KS].

Theorem 3.1. *Let X be a space of global nonpositive curvature. Then*

$$d^2 : X \times X \to \mathbb{R}$$

is a strictly convex function.

More precisely, if we have two geodesics $\gamma_1, \gamma_2 : [0,1] \to X$ (parametrized proportionally to arclength); then we have for $0 \leq t \leq 1$ and a parameter $0 \leq \lambda \leq 1$

$$\begin{aligned} d^2(\gamma_1(t), \gamma_2(t)) \quad \leq \quad & (1-t)d^2(\gamma_1(0), \gamma_2(0)) + td^2(\gamma_1(1), \gamma_2(1)) \\ & -t(1-t)\big(\lambda(d(\gamma_1(0), \gamma_1(1)) - d(\gamma_2(0), \gamma_2(1)))^2 \\ & +(1-\lambda)(d(\gamma_1(0), \gamma_2(0)) - d(\gamma_1(1), \gamma_2(1)))^2\big) \end{aligned}$$

and also

$$\begin{aligned} & d^2(\gamma_1(0), \gamma_2(t)) + d^2(\gamma_1(1), \gamma_2(1-t)) \\ & \quad \leq d^2(\gamma_1(0), \gamma_2(0)) + d^2(\gamma_1(1), \gamma_2(1)) + 2t^2 d^2(\gamma_2(0), \gamma_2(1)) \\ & \qquad + t\Big\{d^2(\gamma_1(0), \gamma_1(1)) - d^2(\gamma_2(0), \gamma_2(1)) - \lambda(d(\gamma_1(0), \gamma_1(1)) \\ & \qquad - d(\gamma_2(0), \gamma_2(1)))^2 - (1-\lambda)(d(\gamma_1(0), \gamma_2(0)) - d(\gamma_1(1), \gamma_2(1)))^2\Big\} \end{aligned}$$

Note that these inequalities are sharp in the Euclidean case for certain special quadrilaterals. In case of an upper curvature bound $K > 0$, we have a similar result by comparison with the geometry on a sphere S^2_K of curvature K. We need to assume in addition that

$$d(\gamma_1(0), \gamma_1(1)) + d(\gamma_1(0), \gamma_2(0)) + d(\gamma_2(0), \gamma_2(1)) + d(\gamma_1(1), \gamma_2(1)) < \frac{2\pi}{\sqrt{K}}$$

and that the geodesic arcs γ_1 and γ_2 are contained in a geodesic ball $B(p,R)$ with the property that each $q \in B(p,R)$ has a unique shortest (geodesic) connection with p. We shall call such a ball simple.

In the sequel, we shall need the following result of W. Kendall [Ke2]:

Theorem 3.2. *Let B be a simple geodesic ball of radius R and center p with curvature $\leq K$ $(K>0)$*

$$R < \frac{\pi}{2\sqrt{K}}.$$

Then

$$\phi : B \times B \to \mathbb{R}$$

$$\phi(z,w) = \left(\frac{1-\cos\sqrt{K}d(z,w)}{\cos\sqrt{K}d(z,p)\cos\sqrt{K}d(w,p) - \lambda^2} \right)^{\nu+1}$$

is strictly convex whenever $\nu \geq 1$; $0 < \lambda < \cos\sqrt{K}R$, and

$$2\nu\lambda^2\left(\cos^2\sqrt{K}R - \lambda^2\right) \geq 1.$$

The asymptotic expansion of ϕ in terms of $d(z,w)$ near $z = w$ starts with the exponent $2(\nu+1) > 2$. We therefore choose $\sigma > 0$ such that

$$\phi(z,w) \geq \sigma d^2(z,w) \quad \text{whenever } d(z,w) \geq \frac{\pi}{4\sqrt{K}}$$

and take

$$\psi(z,w) := \max\left(\phi(z,w), \sigma d^2(z,w)\right).$$

Since $d^2(z,w)$ is strictly convex for $d(z,w) < \frac{\pi}{2\sqrt{K}}$ and since the maximum of two strictly convex functions is again strictly convex, ψ is a strictly convex function on $B \times B$ with asymptotic expansion near $z = w$ starting with $d^2(z,w)$.

For a map $f : M \to B$, with M a metric space with measures μ and $\mu_x^\varepsilon := \mu\llcorner B(x,\varepsilon)$ for $x \in M$, $\varepsilon > 0$, we can therefore define the ε-energy as

$$E_\varepsilon(f) := \int_M \frac{\int \psi(f(x), f(y))d\mu_x^\varepsilon(y)}{\int d^2(x,y)d\mu_x^\varepsilon(y)} d\mu(x)$$

and the energy as

$$E = \Gamma - \lim_{\varepsilon \to 0} E_\varepsilon \quad \text{(possibly taking a subsequence)}.$$

In the Riemannian case, this again reproduces the usual energy functional, because of the behavior of ψ near $z = w$.

Let us now consider mapping spaces of the type

$$L^2(M, N)$$

as introduced in §2. Let $f_0, f_1 \in L^2(M, N)$. A shortest geodesic between f_0 and f_1 is given by

$$\gamma : [0, 1] \to L^2(M, N)$$

where $\gamma(t)(x) = \gamma_x(t)$ for each $x \in M$ is a shortest geodesic from $f_0(x)$ to $f_1(x)$. We also write

$$f_t(x) := \gamma(t)(x),$$

and thus the family of maps $f_t : M \to N$ constitutes a shortest geodesic from f_0 to f_1. Of course, one has to check whether $f_t \in L^2(M, N)$ for all t, but this holds if the curvature of N is bounded from above by some K. In fact, if we have a global curvature bound K for N, the preceding consideration implies that the corresponding space

$$Y := L^2(M, N)$$

also has this global curvature bound K. This follows from the simple observation that the comparison properties required for the curvature bound are not affected by an integration over M.

In particular, we record

Lemma 3.3. *In the situation considered in §2, let*

$$\varrho : \Gamma \to I(X_2)$$

be a homomorphism where Γ operates isometrically on the metric space X_1, and where the complete geodesically connected metric space X_2 has global nonpositive curvature. Then $L^2_\varrho(X_1, X_2)$, the space of ϱ-equivariant L^2-maps between X_1 and X_2 (w.r.t. a Γ-invariant measure μ as in §2) also has global nonpositive curvature. A corresponding result holds if X_2 is contained in a simple ball in a space of global curvature $\leq K$.

4. Minimizers of convex functionals.

Let Y be a complete geodesic metric space, "geodesic" meaning that any two points p, q in Y can be connected by a geodesic arc γ of length $d(p,q)$, the distance between p and q. A function

$$F : Y \to \mathbb{R} \cup \{\infty\}$$

is called convex if its restriction to any geodesic arc is convex. We suppose that Y satisfies the following convexity assumption.

(C) There exists a lower semicontinuous function

$$\psi : Y \to \mathbb{R}^+$$

which is strictly convex in the following quantitative sense:
For any geodesic $\gamma : [0,1] \to Y$, and any $\varepsilon > 0$, there exists $\delta > 0$ such that if

$$\psi(\gamma(\frac{1}{r})) \geq \frac{1}{2}\psi(\gamma(0)) + \frac{1}{2}\psi(\gamma(1)) - \delta, \tag{4.1}$$

then

$$d(\gamma(0), \gamma(1)) < \varepsilon.$$

Let $F : Y \to \mathbb{R} \cup \{\infty\}$ be a function. For $\lambda > 0$, we define the Moreau-Yosida approximation F^λ of F by

$$F^\lambda := \inf_{y \in Y} (\lambda F(y) + \psi(y)).$$

Lemma 4.1. *Suppose $F : Y \to \mathbb{R}^+ \cup \{\infty\}$ is convex, lower semicontinuous, bounded below, and not identically ∞. Then for every $\lambda > 0$, there exists a unique $y_\lambda \in Y$ with*

$$F^\lambda = \lambda F(y_\lambda) + \psi(y_\lambda). \tag{4.2}$$

Proof. Let $(y_n)_{n \in \mathbb{N}}$ be a minimizing sequence, i.e.

$$\lambda F(y_n) + \psi(y_n) \to F^\lambda = \inf_{y \in Y} (\lambda F(y) + \psi(y)) \tag{4.3}$$

For $i, j \in \mathbb{N}$, we let $y_{i,j}$ be a mid point of y_i and y_j, i.e. we connect y_i and y_j by a geodesic arc $\gamma : [0,1] \to Y$ with $\gamma(0) = y_i$, $\gamma(1) = y_j$, and put $y_{i,j} = \gamma(\frac{1}{2})$. Then

$$F^\lambda \leq \lambda F(y_{i,j}) + \psi(y_{i,j}) \tag{4.4}$$

By convexity of F, $F(y_{i,j}) \leq \frac{1}{2}F(y_i) + \frac{1}{2}F(y_j)$. Assumption (C), (4.3), and (4.4) then imply that $(y_n)_{n\in\mathbb{N}}$ is a Cauchy sequence. Since Y is complete, it has a unique limit y_λ, and since F and ψ are lower semicontinuous, $\lambda F(y_\lambda) + \psi(y_\lambda) = F^\lambda$.

□

Theorem 4.2. *Let Y be a complete geodesic metric space satisfying assumption (C). Let $F : Y \to \mathbb{R}^+ \cup \{\infty\}$ be convex, lower semicontinuous, bounded below, $\not\equiv \infty$. Define y_λ as in Lemma 4.1.*
If $(\psi(y_{\lambda_n}))_{n\in\mathbb{N}}$ is bounded for some sequence $\lambda_n \to \infty$, then $(y_\lambda)_{\lambda>0}$ converges to a minimizer of F as $\lambda \to \infty$.

Proof. **1)** y_{λ_n} minimizes $F(y) + \frac{1}{\lambda_n}\psi(y)$. Since $\psi(y_{\lambda_n})$ is bounded, $(y_{\lambda_n})_{n\in\mathbb{N}}$ therefore is a minimizing sequence for F.
2) Let $0 < \mu_1 < \mu_2$. From the definition of y_{μ_1}, we have

$$F(y_{\mu_2}) + \frac{1}{\mu_1}\psi(y_{\mu_2}) \geq F(y_{\mu_1}) + \frac{1}{\mu_1}\psi(y_{\mu_1}),$$

hence

$$F(y_{\mu_2}) + \frac{1}{\mu_2}\psi(y_{\mu_2}) \geq F(y_{\mu_1}) + \frac{1}{\mu_2}\psi(y_{\mu_1}) + \left(\frac{1}{\mu_1} - \frac{1}{\mu_2}\right)(\psi(y_{\mu_1}) - \psi(y_{\mu_2}))$$

The definition of y_{μ_2} then implies

$$\psi(y_{\mu_1}) \leq \psi(y_{\mu_2}),$$

i.e. $\psi(y_\lambda)$ is a nondecreasing function of λ.
3) Since $\psi(y_\lambda)$ is nondecreasing and bounded on the sequence $(\lambda_n)_{n\in\mathbb{N}}$, it is a bounded function of λ.
4) We have

$$F(y_\lambda) = \inf\{F(y) : \psi(y) \leq \psi(y_\lambda)\}.$$

Since $\psi(y_\lambda)$ is nondecreasing by 2), $F(y_\lambda)$ is a nonincreasing function of λ, with limit $\inf_{y\in Y} F(y)$ by 1).

5) Given $\varepsilon > 0$, we choose $\delta > 0$ as in (4.1), and by 2) and 3), we may find $\lambda_0 > 0$ such that for $\lambda, \mu \geq \lambda_0$

$$|\psi(y_\lambda) - \psi(y_\mu)| < 2\delta.$$

If $\lambda_0 \leq \lambda \leq \mu$, we then have $F(y_\lambda) \geq F(y_\mu)$ by 4), and letting $y_{\lambda,\mu}$ be the mid point of y_λ and y_μ, we obtain from the definition of y_λ

$$F(y_\lambda) + \frac{1}{\lambda} \leq F(y_{\lambda,\mu}) + \frac{1}{2}\psi(y_{\lambda,\mu}). \tag{4.5}$$

Also, by convexity of F,

$$F(y_{\lambda,\mu}) \leq F(y_\lambda).$$

Therefore, we must have

$$\begin{aligned} \psi(y_{\lambda,\mu}) &\geq \psi(y_\lambda) \\ &\geq \frac{1}{2}\left(\psi(y_\lambda) + \psi(y_\mu)\right) - \delta. \end{aligned}$$

Assumption (C) then implies

$$d(y_\lambda, y_\mu) < \varepsilon.$$

Therefore $(y_\lambda)_{\lambda>0}$ satisfies the Cauchy property for $\lambda \to \infty$.
6) Since Y is complete, the Cauchy property of $(y_\lambda)_{\lambda>0}$ implies the existence of a unique limit point $\bar{y}$ for $\lambda \to \infty$. 4) and the lower semicontinuity of F imply that $\bar{y}$ minimizes F.

□

5. Existence theorems for generalized harmonic maps.

The first main result of this § will be an existence result for ϱ-equivariant harmonic maps $X_1 \to X_2$, where $\varrho : \Gamma \to I(X_2)$ is a homomorphism as explained in §2. This homomorphism shall need to satisfy an (essentially necessary) technical condition whose formulation is based on [JY1], [La] and suggestions of Scot Adams.

Let X_2 be a complete, simply connected metric space of nonpositive curvature in the sense of Alexandrov as defined in §3. A subgroup G of $I(X_2)$ now is called reductive if there exists a complete convex (hence also nonpositively curved) subspace X of X_2 that is stabilized by G (i.e. $\gamma(X) \subset$

X for all $\gamma \in G$) with the following property:
Whenever there is an unbounded sequence $(p_n)_{n\in\mathbb{N}} \subset X$ with

$$d(p_n, \gamma p_n) \le \text{const}$$

for all $\gamma \in G$ (with a constant depending on γ, but independent of n), then G stabilizes a finite-dimensional complete, flat, totally geodesic subspace of X.

A homomorphism $\varrho : \Gamma \to I(X_2)$ then will be called reductive if its image $\varrho(\Gamma)$ is.

We let X_1 be a metric space with measures μ, μ_x^ε as above. Let Γ be a subgroup of the isometry group $I(X_1)$ of X_1. Assume that the measures μ and μ_x^ε are Γ-equivariant, i.e. $\gamma_*\mu = \mu$ and $\gamma_*\mu_x^\varepsilon = \mu_{\gamma x}^\varepsilon$ for all $\gamma \in \Gamma$. μ then induces a measure μ_Γ on X_1/Γ.

Let X_2 be a complete metric space with isometry group $I(X_2)$, and let

$$\varrho : \Gamma \to I(X_2)$$

be a homomorphism. We define the following functionals on the class of ϱ-equivariant maps $f : X_1 \to X_2$, for a Γ-invariant, nonnegative, symmetric h as in §2,

$$E_\varepsilon^\varrho(f) := \int h(x,y) d^2(f(x), f(y)) d\mu_\Gamma(y) d\mu_\Gamma(x)$$

and

$$E^\varrho := \Gamma - \lim_{\varepsilon\to\infty} E_{h^n}^\varrho \quad \text{(for a suitably normalized sequence } h^n\text{)}$$

Theorem 5.1. *Suppose X_2 is a complete metric space of global nonpositive curvature in the sense of Alexandrov. Suppose*

$$\varrho : \Gamma \to I(X_2)$$

is a reductive homomorphism.

Suppose there exists some ϱ-equivariant map $f : X_1 \to X_2$ with

$$E^\varrho(f) < \infty.$$

Assume also that

$$\int d^2(f(x), f(\gamma x)) \mu_\Gamma(dx) \le c(\gamma)\,(E^\varrho(f) + 1) \quad \textit{for all } \varrho\textit{-equivariant } f, \textit{ with a constant } c(\gamma) \textit{ depending on } \gamma. \tag{5.1}$$

Then there exists a ϱ-equivariant harmonic map

$$f : X_1 \to X_2,$$

namely a minimizer for E^ϱ.

Proof. We consider the functionals $E^\varrho_{h^n}$ and E^ϱ on the space of ϱ-equivariant maps from X_1 to X_2 of class L^2,

$$Y = L^2_\varrho(X_1, X_2),$$

with metrics

$$d^2(f,g) = \int_{X_1/\Gamma} d^2(f(x), g(x))d\mu_\Gamma(x).$$

Y is a complete metric space of nonpositive curvature, because X_2 is (see Lemma 3.1), and since $d^2 : X_2 \times X_2 \to \mathbb{R}$ is convex (see Theorem 3.1), the functionals E^ϱ_h are convex, and consequently also E^ϱ by a general result about Γ-limits (see §2). Moreover, for any $f_0 \in L^2_\varrho(X_1, X_2)$, $d^2(f_0, \cdot)$ is strictly convex and satisfies (C) of §4.1. In order to apply Theorem 4.1, we therefore only have to show that the maps f_n minimizing $nE^\varrho(f) + d^2(f_0, f)$ (see Lemma 4.1) are bounded for $n \in \mathbb{N}$.

If f_n is unbounded, i.e. if $f_n(x)$ is unbounded on a set of positive μ_Γ measure, by our reductivity assumption and (5.1), $\varrho(\Gamma)$ stabilizes a totally geodesic finite-dimensional flat subspace L. Let $p_n = f_n(x)$ be an unbounded sequence in L. By (5.1), we may assume

$$d^2(p_n, \varrho(\gamma)p_n) \leq c(\gamma) \quad \text{independent of } n \text{, for every } \gamma \in \Gamma\ . \tag{5.2}$$

Since L is Euclidean, (5.2) implies that p_n stays within bounded distance from some axis of $\varrho(\gamma)$, for every $\gamma \in \Gamma$. Therefore, there must exist some line l in L such that every $\varrho(\gamma)$ has an axis that is parallel to l. Therefore, $\varrho(\Gamma)$ commutes with translations in the direction of l, and we may compose f_n with a suitable such translation in order to bring $f_n(x)$ back to some bounded region in X_2 without destroying the ϱ-equivariance. In this manner, we may decrease $d^2(f_0, f_n)$ without affecting $E^\varrho(f_n)$. This contradicts the definition of f_n, and therefore $(f_n)_{n\in\mathbb{N}}$ must be bounded. Thus, Theorem 4.1 applies. □

Remark. The argument given in [J3] claiming that $d^2(f_n, f_n\circ\gamma) \leq d^2(f_0, f_0 \circ \gamma)$ is incorrect, because Lemma 4 of that paper is not applicable to $f_n \circ \gamma$.

Examples.

1) Consider the negatively curved surface

$$N_1 := \left\{ \left(x, \cos\vartheta e^{-x}, \sin\vartheta e^{-x}\right) : x \in \mathbb{R}, \vartheta \in [0, 2\pi] \right\} \subset \mathbb{R}^3.$$

Then, in no homotopically nontrivial class of maps $\gamma : S^1 \to N_1$, the infimum of energy is achieved. The reason is that the associated homomorphism

$$\varrho : \mathbb{Z} \left(= \pi_1(S^1)\right) \to I\left(\tilde{N}_1\right)$$

is not reductive.

2) Consider the negatively curved surface

$$N_2 := \{(x, \cos\vartheta \cosh x, \sin\vartheta \cosh x) : x \in \mathbb{R}, \vartheta \in [0, 2\pi]\} \subset \mathbb{R}^3.$$

This time, the infimum of energy is achieved in every homotopy class of maps $\gamma : S^1 \to N_2$, although N_2 is noncompact, because the reductivity condition is satisfied. Namely, the associated homomorphism

$$\varrho : \mathbb{Z} \to I(\tilde{N}_2)$$

has an invariant line in $\tilde{N}_2$.

3) One may relax the condition in the reductivity assumption that the invariant flat subspace L is finite dimensional. For example, the result also holds if $\varrho(\Gamma)$ operates via translations only, i.e. in a commutative manner on a Hilbert space L. Again, one may bring an unbounded sequence $f_n(x)$ as in the proof of Theorem 5.1 back to some bounded region by a suitable translation that commutes with all elements of $\varrho(\Gamma)$ and therefore respects the ϱ-equivariance. Without some additional commutativity assumption, however, the result ceases to be true for infinite dimensional L. For example, one may choose an unbounded sequence $(q_n)_{n\in\mathbb{N}} \subset L$ and an isometry σ that operates by rotations with center q_n in mutually orthogonal two-planes. The cyclic group generated by σ then is not representable by an equivariant harmonic map from $\mathbb{R}$, the universal cover of S^1, to L.

Remarks.

1) Theorem 5.1 was obtained in [J3], with an earlier version in [J2], and in the special case where Γ is the fundamental group of a smooth compact Riemannian manifold essentially the same result was obtained by Korevaar-Schoen [KS].

2) We should also mention the work of Margulis [Ma2], where he obtained an existence result for some class of generalized harmonic maps by a method that has some similarity with the one presented above, as well as a very interesting rigidity theorem for commensurability subgroups as an application.

3) Finally, we should point out that the idea of averaging maps in order to decrease energy was already found in [F].

We now extend the existence theorem of Hildebrandt-Kaul-Widman [HKW2]:

Theorem 5.2. *Let B be a simple geodesic ball of radius R and center p with curvature $\leq K$ and $R < \frac{\pi}{2\sqrt{K}}$.*

Let M be a compact Riemannian manifold with boundary ∂M, and let

$$\varphi : M \to B$$

be given. Suppose that there exists an L^2-map $g : M \to B$ with energy $E(g) < \infty$ as defined above and with $g|_{\partial M} = \varphi$ in the sense of traces (this means that $d(g(x), \varphi(x))$ is a (real-valued) function of Sobolev class $H_0^{1,2}(M)$; see Korevaar-Schoen [KS] or Ambrosio [Am]). Then there exists an energy minimizing map $f : M \to B$ with $f|_{\partial M} = \varphi$.

Proof. By the properties of ψ established as a consequence of Kendall's theorem 3, E is a convex functional on

$$Y := \{g \in L^2(M, B), g|_{\partial M} = \varphi\},$$

and Y satisfies condition (C) because of Theorem 3 again. Therefore, Theorem 4.1 produces the desired energy minimizing map.

□

Remark. We do not address the issue of boundary regularity here. See Serbinowski [Se1] in this regard. We have required the domain M to be a Riemannian manifold rather than a more general metric space in order to be able to formulate the boundary condition $g|_{\partial M} = \varphi$. Presumably, one can generalize the setting as follows: Let M be a metric space with boundary ∂M. Define $H^{1,2}(M)$ to be the space of L^2-functions with finite energy on M. $(f_n)_{n\in\mathbb{N}} \subset H^{1,2}(M)$ is said to converge to $f \in H^{1,2}(M)$ if it converges in L^2 and if $E(f_n - f)$ converges to 0. We let $H_0^{1,2}(M)$ be the closure w.r.t. this topology of the space of continuous $H^{1,2}(M)$ functions that vanish on the boundary ∂M. We say that $g|_{\partial M} = \varphi|_{\partial M}$ in the sense of traces if $d(g(x), \varphi(x)) \in H_0^{1,2}(M)$.

6. Regularity.

Generalized harmonic maps $f : M \to N$ need not be continuous, even if N has nonpositive curvature. For example, let

$$M = M_1 \cup M_2,$$

where M_1 and M_2 are compact smooth Riemannian manifolds of dimension ≥ 2, and

$$M_1 \cap M_2$$

is a single point p. Let $f : M \to N$ be harmonic (with $E(f) < \infty$), N a space of nonpositive curvature. Then f restricts to a harmonic map

$$f : M_i \setminus \{p\} \to N \quad \text{for } i = 1, 2,$$

and one may apply regularity theory as outlined in §1 to conclude that $f|_{M_i \setminus \{p\}}$ is smooth and extends as a smooth harmonic map f_i to M_i. Using the uniqueness results for harmonic maps into spaces of nonpositive curvature, one may easily construct examples where the harmonic maps f_i have to satisfy

$$f_1(p) \neq f_2(p).$$

This means that f is discontinuous at p. If M, however, is a smooth Riemannian manifold, Korevaar-Schoen [KS] have shown that any generalized harmonic map (of finite energy) into a space of nonpositive curvature is Lipschitz continuous, extending earlier results of Gromov-Schoen. It is plausible that the Lipschitz constant will depend only on a lower bound for the Ricci curvature of M, the dimension of M, and the energy of f. Here, we want to indicate an estimate that is somewhat different from the one in [KS] and that gives explicit bounds, in terms of the sectional rather than the Ricci curvature, however. In the spirit of the present paper, this estimate will exploit mean value properties of ε-harmonic maps.

Let $x_1, x_2 \in M$, $\varepsilon > 0$. Let $d(x_1, x_2) <$ injectivity radius of M. We suppose that M has bounded Ricci curvature.

We introduce normal coordinates on the balls $B(x_1, \varepsilon)$ and $B(x_2, \varepsilon)$, and we use these coordinates to define a diffeomorphism

$$\varphi : B(x_1, \varepsilon) \to B(x_2, \varepsilon)$$

by mapping an orthonormal frame at x_1 into such a frame at x_2 through parallel transport along the shortest geodesic from x_1 to x_2. Let $d\nu_i$ be the volume form of $B(x_i, \varepsilon)$, $i = 1, 2$. Put $V_i := \mathrm{vol}(B(x_i, \varepsilon))$, $i = 1, 2$. We then have

$$|d\nu_1 - \varphi^* d\nu_2| \leq c\varepsilon^2, \tag{6.1}$$

where c depends on upper and lower bounds for the Ricci curvature of M. (Actually, if φ happens to be an isometry, e.g. for M a quotient of hyperbolic space, we get 0 on the right hand side of (6.1). Thus, in certain special cases, (6.1) can be considerably improved.)

Let $f : M \to Y$ be an ε-equilibrium map into a space Y of global non-positive curvature in the sense of Alexandrov, i.e. $f(x)$ is the mean value of f on the ball $B(x, \varepsilon)$ for every $x \in M$. The following inequality was derived by Korevaar-Schoen [KS] as a consequence of the comparison theorem of Reshetnyak (Theorem 3.1)

$$\begin{aligned} & d^2\left(f(x_1), f(x_2)\right) \\ \leq\ & \frac{1}{V_1} \int_{B(x_1,\varepsilon)} d^2\left(f(y), f(\varphi(y))\right) d\nu_1(y) \\ & - \frac{1}{V_1} \int_{B(x_1,\varepsilon)} \left(d(f(y), f(\varphi(y)) - d(f(x_1), f(x_2))\right)^2 d\nu_1(y) \\ & + 2d\left(f(x_1), f(x_2)\right) \int_{B(x_2,\varepsilon)} d\left(f(y), f(y_2)\right) \left| \frac{d\nu_2(y)}{V_2} - \varphi * \frac{d\nu_1(y)}{V_1} \right| \end{aligned} \tag{6.2}$$

This implies

$$\begin{aligned} & d(f(x_1), f(x_2)) \\ \leq\ & \frac{1}{V_1} \int_{B(x_1,\varepsilon)} d(f(y), f(\varphi(y))) d\nu_1(y) \\ & + \int_{B(x_2,\varepsilon)} d\left(f(y), f(x_2)\right) \left| \frac{d\nu_2(y)}{V_2} - \varphi * \frac{d\nu_1(y)}{V_1} \right|. \end{aligned} \tag{6.3}$$

We note that

$$d(y,\varphi(y)) \leq d(x_1,x_2)\cosh\omega\varepsilon^2 \quad \text{for all } y \in B(x_1,\varepsilon) \tag{6.4}$$

if $-\omega^2 \leq K$ is a lower bound for the sectional curvature of M. If we now iterate (6.3), i.e. we estimate the quantities $d(f(y), f(\varphi(y)))$ and $d(f(y), f(x_2))$ appearing in the integrals on the r.h.s. of (6.3) by applying (6.3), and use (6.4), we obtain in case $d(x_1,x_2) \leq \varepsilon$

$$\frac{d(f(x_1), f(x_2))}{\varepsilon} \leq c_1(K)\int_M\int_{B(x,\varepsilon)} \frac{d(f(y), f(x))}{\varepsilon}\mathrm{dvol}(y)\mathrm{dvol}(x), \tag{6.5}$$

where the constant $c_1(K)$ depends on the aforementioned curvature bounds. Applying Hölder's inequality to the r.h.s. of (6.5), we get in case $d(x_1,x_2) \leq \varepsilon$

$$\begin{aligned}\frac{d(f(x_1), f(x_2))}{\varepsilon} &\leq c_2(K)\left(\int_M \frac{\int_{B(x,\varepsilon)} d^2(f(y),f(x))\mathrm{dvol}(y)}{\int_{B(x,\varepsilon)} d^2(y,x)\mathrm{dvol}(y)}\mathrm{dvol}(x)\right)^{\frac{1}{2}} \\ &\leq c_3(K)E_\varepsilon(f)^{\frac{1}{2}}.\end{aligned} \tag{6.6}$$

By the triangle inequality, we also conclude that in case $d(x_1,x_2) \leq \nu\varepsilon$ for $\nu \in \mathbb{N}$

$$d(f(x_1), f(x_2)) \leq c_2(K)E_\varepsilon(f)^{\frac{1}{2}}\nu\varepsilon. \tag{6.7}$$

If we let $\varepsilon \to 0$, we then obtain a Lipschitz bound for a harmonic map $f: M \to Y$:

$$d(f(x_1), f(x_2)) \leq c_2(K)E(f)^{\frac{1}{2}}d(x_1,x_2). \tag{6.8}$$

Higher regularity requires additional assumptions in the target, see the results in Gromov-Schoen [GS] and the results announced in Korevaar-Schoen [KS].

Without the assumption of nonpositive image curvature partial regularity results in the sense of Schoen-Uhlenbeck have been obtained by Kourouma [K2] and Serbinowski [Se2].

7. A stochastic construction.

As in §2, let a Radon measure μ and a nonnegative, symmetric function $h(\cdot,\cdot)$ on the metric space M be given. Assuming a finiteness condition, we may normalize the situation so that

$$\begin{aligned} \int h(x,y)d\mu(y) &= 1 \quad \text{for all } x \in M \\ \mu(M) &= 1. \end{aligned}$$

We let P be the probability distribution associated with μ.

We also assume that N has nonpositive curvature and lift to universal covers locally as explained in §2, in order to have geodesics uniquely defined. In order to construct a generalized harmonic map

$$f : M \to N,$$

we define the following iterative procedure:

We start with an arbitrary map f_0 in the given class. Having iteratively defined $f_{\nu-1}$ for $\nu \in \mathbb{N}$, we randomly select $x \in M$ according to the probability distribution $P(x)$. For $y \in M$, we let $\gamma : [0,1] \to N$ be the geodesic arc from $f_{\nu-1}(y)$ to $f_{\nu-1}(x)$ (parametrized proportionally to arclength), and we put

$$f_\nu(y) := \gamma\left(\eta(\nu)h(x,y)\right) \tag{7.1}$$

where $(\eta(\nu))_{\nu\in\mathbb{N}}$ is a sequence of positive real numbers converging to 0 in a suitable manner as explained below.

Our prescription has the effect of moving the value of the map at y in the support of h closer to the value of the map at x, according to the probability distribution defined by $h(x,\cdot)\mu$. In the limit $\nu \to \infty$, the effects on the value at y of all the points $x \in M$ will balance out with probability 1, and because of the symmetry assumption for h this will mean that an equilibrium map will be achieved. This requires of course the right choice of the rate of convergence of $\eta(\nu)$ to 0. If $\eta(\nu)$ converges too slowly, the iteration will oscillate forever without asymptotic equilibrium. If it converges too fast, the iteration may freeze the process asymptotically in a nonequilibrium state.

For purposes of numerical implementation, it is useful to consider the case where M is a discrete space $\{z_j : j = 1, \ldots, m\}$. We assume that each $z_j \in M$ has neighbors in M that are denoted by z_j^μ, $\mu = 1, \ldots, k$ (in principle, k may depend on j, but we suppress this possibility for simplicity). μ just is normalized counting measure. The points z_j should be considered as the processing units of a parallel computer, and at each step, one of them is randomly activated, and we then change the values at its neighbors according to the rule (7.1), i.e.

$$f_\nu(z_j^\mu) = \gamma\left(\frac{\eta(\nu)}{k}\right) \quad \text{for } \mu = 1, \ldots, k, \tag{7.2}$$

where $\gamma : [0,1] \to N$ again is the geodesic from $f_{\nu-1}(z_j^\mu)$ to $f_{\nu-1}(z_j)$. Again, $\eta(\nu)$ has to converge to 0 at the right rate for $\nu \to \infty$.

Of course, the process may be modified by letting the units z_j activate themselves independently according to some probability distribution $p(t)$, where $t \geq 0$ stands for time, and where we now have to use a decay function $\eta(t)$ in (7.2) that converges to 0 (at the right rate) as $t \to \infty$.

This process then represents a fully parallel algorithm for constructing harmonic mappings, and thus in particular harmonic functions. It is also easy to incorporate Dirichlet boundary conditions as in numerical difference quotient schemes for computing harmonic functions. Likewise, if we wish to approximate harmonic functions or mappings on a connected space M, we may approximate M by discrete spaces M^ε, perform the preceding constructions on the M^ε and pass to the limit $\varepsilon \to 0$.

References.

[A1] Al'ber, S.I., *On n-dimensional problems in the calculus of variations in the large*, Sov. Math. Dokl. 5, (1964), 700-704

[A2] Al'ber, S.I., *Spaces of mappings into a manifold with negative curvature*, Sov. Math. Dokl. 9, (1967), 6-9

[Am] Ambrosio, *Metric space valued functions of bounded variation*, Ann. Sc. Norm. Sup. Pisa(IV) 17 (1990), 439–478

[BN] Berestovskij, V.N., and Nikolaev, I.G., *Multidimensional generalized Riemannian spaces*, in: Encyclopaedia Math. Sciences Vol. 70, Geometry IV (ed. Yu. G. Reshetnyak), pp. 246–362, Springer, 1993

[C] K. Corlette, *Flat G-Bundles with canonical metrics*, J. Diff. Geom. 28 (1988), 361–382

[D] S. Donaldson, *Twisted harmonic maps and the self-duality equations*, Proc. London Math. Soc. 55 (1987), 127–131

[dM] G. dal Maso, *An introduction to Γ-convergence*, Birkhäuser, 1993

[DO] K. Diederich and T. Ohsawa, *Harmonic mappings and disk bundles over compact Kähler manifolds*, Publ. Res. Inst. Math. Sci. 21 (1985), 819–833

[EL1] Eells, J., and Lemaire, L., *A report on harmonic maps*, Bull. Londen Math. Soc. 10(1978), 1–68

[EL2] Eells, J., and Lemaire, L., *Selected topics in harmonic maps*, CBMS Reg. Conf. Series 50, AMS, 1983

[EL3] Eells, J., and Lemaire, L., *Another report on harmonic maps*, Bull. London Math. Soc. 20 (1988), 385–524

[ES] Eells, J., and Sampson, J., *Harmonic mappings of Riemannian manifolds*, Am. J. Math. 85 (1964), 109-160

[F] Frankel, S., *Locally symmetric and rigid factors for complex manifolds via harmonic maps*, Ann. Math. **141** (1995), 285-300.

[G] Gromov, M., *The foliated Plateau problem*, Part I, GAFA 1 (1991), 14–79, Part II, GAFA

[GG1] Giaquinta, M., and Giusti, E., *On the regularity of the minima of variational integrals*, Acta Math. 148 (1982), 31–46

[GG2] Giaquinta, M., and Giusti, E., *The singular set of the minima of certain quadratic functionals*, Ann. Sc. Norm. Sup. Pisa (IV) 11 (1984), 45–55

[GMS] Giaquinta, M., Modica, G., and Souček, J., *Cartesian currents and variational problems for mappings into spheres*, Ann. Sc. Norm. Sup. Pisa (IV) 16 (1989), 393–485

[GS] Gromov, M., and Schoen, R., *Harmonic maps into singular spaces and p-adic superrigidity for lattices in groups of rank one*, Publ. Math. IHES 76 (1992), 165–246

[H] Hildebrandt, S., *Nonlinear elliptic systems and harmonic mappings*, Proc. Beijing Symp. Diff. Geom. 8 Diff. Eq. 1980, Science Press, Beijing, 1982

[HKW1] Hildebrandt, S., Kaul, H., and Widman, K.O., *Dirichlet's boundary value theorem for harmonic mappings of Riemannian manifolds*, Math. Z. 147 (1976), 225–236

[HKW2] Hildebrandt, S., Kaul, H., and Widman, K.O., *An existence theorem for harmonic mappings of Riemannain manifolds*, Acta Math. 138 (1977), 1–16

[Hm] Hamilton, R., *Harmonic maps of manifolds with boundary*, Springer L.N.M. 471 (1975)

[Hr] P. Hartman, *On homotopic harmonic maps*, Can. J. Math. 19, (1967), 673-687

[HW] Hildebrandt, S., and Widman, K.O., *On the Hölder continuity of weak solutions of quasilinear elliptic systems of second order*, Ann. Sc. N. Sup. Pisa IV (1977), 145–178

[J1] Jost, J., *Unstable solutions of two-dimensional geometric variational problems*, Proc. Symp. Pure Math. 54 (1993), Part 1, 205–244

[J2] Jost, J., *Equilibrium maps between metric spaces*, Calc. Var. 2 (1994), 173–204

[J3] Jost, J., *Convex functionals and generalized harmonic maps into spaces of nonpositive curvature*, Comment. Math. Helv. 70 (1995), 659–673

[JY1] Jost, J., and Yau, S.T., *Harmonic maps and group representations*, in: Differential Geometry and Minimal Submanifolds (B. Lawson and K. Tenenblat, eds.), Longman Scientific, 1991, pp. 241–260

[JY2] Jost, J., and Yau, S.T., *Harmonic maps and Kähler geometry*, in: Prospects in Complex Geometry (eds. J. Noguchi, T. Oshawa), Springer LNM 1468, 1991, pp.340–370

[JY3] Jost, J., and Yau, S.T., *Appications of quasilinear PDE to algebraic geometry and arithmetic lattices*, in Algebraic geometry and related topics (eds. J.H. Yang, Y. Namikawa, K. Ueno), International Press, 1994, pp. 169-190

[K1] Kourouma, M., *Harmonic sections of Riemannian fiber bundles*, Preprint

[K2] Kourouma, M., *About the regularity of energy minimizing maps with values in a metric space of curvature bounded above*, Preprint

[Ka] Karcher, H., *Riemannian center of mass and mollifier smoothing*, CPAM 30 (1977), 509–541

[Ke1] Kendall, W., *Brownian motion and partial differential equations from the heat equation to harmonic maps*, Proc. 151 49th session, Firenze 1993, pp. 85–101

[Ke2] Kendall, W., *Convexity and the hemisphere*, J. London Math. Soc. (2) 43 (1991), 567–576,

[Ke3] Kendall, W., *Probability, convexity, and harmonic maps with small image I: uniqueness and fine existence*, Proc. London Math. Soc. (3) 61 (1990), 371-406

[Ke4] Kendall, W., *Probability, convexity, and harmonic maps II: Smoothness via probabilistic gradient inequalities*, J. Funct. Anal., to appear

[KS] N. Korevaar and R. Schoen, *Sobolev spaces and harmonic maps for metric space targets*, Comm. Anal. Geom. 1 (1993), 561–569

[L] Lemaire, L., *Applications harmoniques de surfaces riemanniénnes*, J. Diff. Geom. 13 (1978), 51–78

[La] F. Labourie, *Existence d'applications harmoniques tordues à valeurs dans les variétés à courbure négative*, Proc. AMS 111 (1991), 877–882

[Ma1] Margulis, G.A., *Discrete subgroups of semisimple Lie groups*, Springer, 1991

[Ma2] Margulis, G.A., *Superrigidity for commensurability subgroups and generalized harmonic maps*, preprint.

[Mo] Morrey, C., *The problem of Plateau on a Riemannian manifold*, Ann. Math. 49 (1948), 807–851

[N] Nikolaev, I., *Synthetic methods in Riemannian geometry*, Lecture Notes

[Re] Reshetnyak, Y.G., *Inextensible mappings in a space of curvature no greater than K*, Siberian Math. Journ. 9 (1968), 683–689

[Se1] Serbinowski, T., *Boundary regularity of harmonic maps to nonpositively curved metric spaces*, Comm. Anal. Geom. 2 (1994), 139–153

[Se2] Serbinowski, T., *Regularity of small energy minimizing maps into locally compact metric spaces with the upper curvature bound*, Preprint

[SkU] Sacks, J., and Uhlenbeck, K., *The existence of minimal immersions of 2-spheres*, Ann, Math. 113 (1981), 1–24

[St] Steffen, K., *An introduction to harmonic mappings*, Vorlesungsreihe SFB256, Vol. 18, Bonn, 1991

[SU1] Schoen, R., and Uhlenbeck, K., *A regularity theory for harmonic maps*, J. Diff. Geom. 17 (1982), 307–335

[SU2] Schoen, R., and Uhlenbeck, K., *Boundary regularity and miscellaneous results on harmonic maps*, J. Diff. Geom. 18 (1983), 253–268

Received March 10, 1995.

Equations of Mean Curvature Type with Contact Angle Boundary Conditions

N. KOREVAAR[1] AND L. SIMON[2]

Given a domain $\Omega \subset \mathbf{R}^n$, we are interested in regularity near the boundary $\partial\Omega$ for solutions $u \in C^2(\Omega)$ of the mean-curvature equation

$$\textbf{0.1} \qquad \sum_{i=1}^{n} D_i \nu^i = H(x,u) \quad \text{on} \quad \Omega,$$

subject to the "contact angle" boundary condition

$$\textbf{0.2} \qquad \nu \cdot \eta = \cos\theta,$$

where $\nu = (\nu^1, \dots, \nu^n)$ is given by $\nu^i = (1 + |Du|^2)^{-1/2} D_i u$, $i = 1, \dots, n$, and where η is the outward pointing unit normal of $\partial\Omega$. Thus geometrically we are looking for a function u on $\overline{\Omega}$ whose graph has the prescribed mean curvature H and which meets the boundary cylinder in the prescribed angle θ. $H = H(x,z)$ is assumed to be a given locally Lipschitz function on $\overline{\Omega} \times \mathbf{R}$ satisfying the structural conditions

$$\textbf{0.3} \qquad H_z(x,z) \geq 0, \quad |H_z(x,z)| + \sum_{i=1}^{n} |H_{x^i}(x,z)| \leq \Lambda$$

for $x \in \Omega$, $z \in [-M, M]$, where Λ, M are constants, and where the subscripts denote partial derivatives of H with respect to the indicated variables; thus,

$$H_z(x,z) = \partial H(x,z)/\partial z, \qquad H_{x^i}(x,z) = \partial H(x,z)/\partial x^i.$$

For the first part of our discussion we also assume the *a-priori* estimate

$$\textbf{0.4} \qquad \sup |u| \leq M.$$

Our main interest is in the mathematically most difficult case, when $\theta = 0$, in which case the solution graph at the boundary is required to be tangent to the boundary cylinder at each point. This case is mathematically

[1]Partially supported by NSF grant DMS-9208666 at University of Utah
[2]Partially supported by NSF grant DMS-9207704 at Stanford University

more subtle than the case when $\theta \in (0, \pi)$, for the obvious reason that $\theta = 0$ implies that the gradient is unbounded on approach to $\partial\Omega$.

Theorem 1, in §1 below, is the key result here; it gives a local gradient estimate of special type near the boundary for solutions of 0.1–0.4, including uniform estimates for solutions which have θ close to zero. As explained in Remark 1.2, the result of Theorem 1 guarantees that it is possible to "tilt" coordinates slightly to ensure that the graph of the solution gives a Lipschitz graph with uniformly bounded gradient, even in cases when, in the original coordinate system, the solution has unbounded gradient.

By using Theorem 1 in combination with standard PDE methods, in §2 we prove that for zero contact angle solutions the unit normal function $\nu = (1 + |Du|^2)^{-1/2}(Du, -1)$ (considered as a function defined on Ω), is automically continuous up to the boundary and that, as a function on the *graph*, ν is Lipschitz continuous. The exact statement is discussed in §2.

Some, but by no means all, of the results here can also be obtained by the methods of Geometric Measure Theory; in any case the methods we are proposing here are of a more elementary character, and apply to quite a large class of quasilinear problems for which GMT methods are not relevant. Indeed in §§3, 4 we discuss extension of the results of §§1, 2 to other classes of quasilinear elliptic equations.

1. The Main Gradient Estimate.

Here Ω is any domain in $\mathbf{R}^n$ and $x_0 \in \partial\Omega$ is arbitrary; we assume that Ω is at least a Lipschitz domain, so that there is $\rho > 0$ such that, in a suitable coordinate system, $\partial\Omega \cap B_\rho(x_0)$ is contained in the graph of a Lipschitz function.

For convenience of notation we shall assume in the following theorem that coordinates are set up so that $x_0 = 0$ and $e_n = (0, \ldots, 0, 1)$ points into Ω at 0; ψ_0 will denote a Lipschitz function on $\mathbf{R}^{n-1}$ representing the boundary of Ω near 0 in the sense that

1.1 $$\Omega \cap B_\rho = \{x = (x^1, \ldots, x^n) \in B_\rho \,:\, x^n > \psi_0(x^1, \ldots, x^{n-1})\},$$

where, here and subsequently, B_ρ is an abbreviation for $B_\rho(0)$.

We let $\nu = (1 + |Du|^2)^{-1/2}(Du, -1)$ be the unit normal function for graph u; this is naturally defined on Ω, because u is defined on Ω, but we shall also on occasion want to think of ν as being the unit normal function on the graph. In this latter point of view we think of ν as being a function of the $(n+1)$-variables $(x^1, \dots, x^n, x^{n+1})$ which happens to be independent of the $(n+1)^{\text{st}}$ variable x^{n+1}, and then ν restricted to the graph $x^{n+1} = u(x^1, \dots, x^n)$ is the unit normal of the graph.

Theorem 1. *Suppose* $0 < \mu < \mu_0 < 1$, $u \in C^2(\Omega \cap B_\rho) \cap C^1(\overline{\Omega} \cap B_\rho)$ *satisfies* 0.1, 0.3, 0.4 *and* $\mu_0 v - u_n \le 0$ *on* $\partial\Omega \cap B_\rho$. *Then*

$$\mu v - u_n \le C, \quad \text{on} \quad \Omega \cap B_{7\rho/8},$$

where $C = C(\Lambda\rho^2, M/\rho, \mu, \mu_0)$.

1.2 Remark. Notice that the theorem with $\mu = \frac{1}{2}$, $\mu_0 = \frac{3}{4}$ says that, provided $\nu_n \equiv e_n \cdot \nu \ge \frac{3}{4}$ on $\partial\Omega \cap B_\rho$, there is $C_0 = C_0(\Lambda\rho^2, M/\rho, n)$ such that

$$(\ddagger) \qquad \alpha \cdot \nu \ge \frac{1}{2}(1 + C^2)^{-1/2} \quad \text{in} \quad \Omega \cap B_{7\rho/8}$$

for any $C \ge C_0$, where $\alpha = (1+C^2)^{-1/2}(0, \dots, 0, 1, -C)$ and where ν denotes the downward-pointing unit normal $v^{-1}(Du, -1)$ for graph u. Now introduce a coordinate transformation $x \mapsto y = Q(x)$, where Q is an orthogonal transformation which fixes the first $n-1$ coordinates and which takes α to $-e_{n+1}$; thus

$$(*) \qquad \begin{cases} y^j = x^j, \ j = 1, \dots, n-1, \\ y^n = -\alpha_2 x^n + \alpha_1 x^{n+1}, \ y^{n+1} = -\alpha_1 x^n - \alpha_2 x^{n+1}, \end{cases}$$

where $(\alpha_1, \alpha_2) = (1+C^2)^{-1/2}(1, -C)$. Then near 0 the graph of u transforms to a new hypersurface with unit normal $\widetilde{\nu}$ satisfying $e_{n+1} \cdot \widetilde{\nu} \equiv \alpha \cdot \nu \ge \frac{1}{2}(1 + C^2)^{-1/2}$. In other words the new hypersurface is locally, near any of its points, the graph of a function $\widetilde{u}$ such that $(1 + |D\widetilde{u}|^2)^{1/2} \le 2(1 + C^2)^{1/2}$. As a matter of fact we have the following more precise result:

Corollary 1. *Suppose the hypotheses are as in the lemma with* $\mu = \frac{1}{2}$, $\mu_0 = \frac{3}{4}$, *and let* $\mathcal{G} \subset B_\rho \times \mathbf{R}$ *be the piecewise* C^1 *submanifold defined by*

$$\mathcal{G} = (\operatorname{graph} u \,|\, \Omega \cap B_\rho) \cup \{(x, t) \in (\partial\Omega \cap B_\rho) \times \mathbf{R} : t \le u(x)\}.$$

If $C \ge C_0$ ($C_0 = C_0(\Lambda\rho^2, M/\rho, n)$ *as above*), *and if* Q *is the transformation* $x \mapsto y$ *of* $(*)$, *then there are* $\beta = \beta(\Lambda\rho^2, M/\rho, n) > 0$ *and Lipschitz functions* $\widetilde{u}$, ψ *on* $B_{3\rho/4}$ *such that*

$$\begin{aligned} \overline{\mathcal{G}} \equiv Q(\mathcal{G}) \cap (B_{3\rho/4} \times \mathbf{R}) &= \operatorname{graph} \widetilde{u}, \qquad |D\widetilde{u}| \le \beta, \\ Q(\partial\Omega \times \mathbf{R}) \cap (B_{3\rho/4} \times \mathbf{R}) &= \operatorname{graph} \psi, \qquad \widetilde{u} \le \psi, \end{aligned}$$

and such that $Q^{-1}(\widetilde{\mathcal{G}}) \supset \mathcal{G} \cap (B_{\rho/2} \times \mathbf{R})$. Also, there is the additional property that

$$(**) \qquad \widetilde{u}(y) < \psi(y) \Rightarrow \widetilde{u}(y + ze_n) < \psi(y + ze_n)$$

for any $y \in B_{\rho/2}$ and $z \geq 0$ such that $y + ze_n \in B_{\rho/2}$.

Remarks. (1) The reader may check that ψ is given explicitly by

$$\psi(y) = -\frac{\alpha_2 y^n + \psi_0(y^1, \dots, y^{n-1})}{\alpha_1},$$

where ψ_0 (as above) is the function whose graph represents $\partial\Omega$ in the ball B_ρ.
(2) Notice the last property $(**)$ is a direct consequence of the fact that $\widetilde{\mathcal{G}}$ is obtained by rotation of graph u in the $x^n - x^{n+1}$ plane

Proof of Corollary 1. Notice that the inequality (‡) in Remark 1.2 guarantees that the $\widetilde{\mathcal{G}}$ is locally (in a neighbourhood of each of its points) expressible as the graph of a Lipschitz function with Lipschitz constant $\leq C$. Also, since $\sup|u| \leq M$, $\partial\,\mathrm{graph}\, u$ ($\equiv$ closure $\mathrm{graph}\, u \setminus \mathrm{graph}\, u$) is contained in $(\partial\Omega \times \mathbf{R}) \cup (\partial B_\rho \times \mathbf{R})$, so, if necessary taking a larger C (i.e. α in $(*)$ closer to $(0, \dots, 0, -1)$), we can arrange $\partial\widetilde{\mathcal{G}}$ ($=$ closure $\widetilde{\mathcal{G}} \setminus \widetilde{\mathcal{G}}$) $\subset \partial B_{3\rho/4} \times \mathbf{R}$. Then the orthogonal projection $(x^1, \dots, x^n, x^{n+1}) \mapsto (x^1, \dots, x^n)$ of $\mathbf{R}^{n+1}$ onto $\mathbf{R}^n$ provides a Lipschitz covering projection of $\widetilde{\mathcal{G}}$ onto $B_{3\rho/4}$, and hence $\widetilde{\mathcal{G}}$ is the union of finitely many pairwise disjoint Lipschitz graphs all defined over the whole of $B_{3\rho/4}$; it is now not difficult to check that in fact there can be just one such graph, so the corollary is proved.

With zero contact angle boundary conditions for u on $\partial\Omega \cap B_\rho$, we show in the next section that further regularity statements about u follow by using the lemma in combination with standard PDE estimates on $\widetilde{u}$.

Before we begin the proof of Theorem 1, we recall some important identities needed in the sequel. First note that by differentiating the equation with respect to the variable x^l, we get the identity

$$\textbf{1.3} \qquad \sum_{i,j=1}^{n} D_i(v^{-1} g^{ij} D_j u_l) = H_{x^l}(x, u) + H_z(x, u) u_l,$$

where, here and subsequently, we use the notation

$$\textbf{1.4} \qquad v = \sqrt{1 + |Du|^2}, \quad g^{ij} = \delta_{ij} - \nu^i \nu^j, \quad u_l = D_l u,$$

and where 1.3 (and corresponding subsequent identities) are to be interpreted in the weak (or distribution) sense, because u is only assumed to be C^2, whereas third derivatives appear in 1.3. Notice that by multiplying through by ν^l, rearranging terms, and summing on l, we obtain the identity

$$\mathbf{1.5} \qquad \sum_{i,j=1}^{n} D_i(v^{-1}g^{ij}D_j v) = \mathcal{C}^2 + \sum_{l=1}^{n} \nu^l H_{x^l}(x,u) + H_z(x,u)(v - v^{-1}),$$

where $\mathcal{C}^2 = v^{-2} \sum_{i,j,r,s=1}^{n} g^{ij}g^{rs}u_{ir}u_{js}$. (Geometrically $\mathcal{C}^2$ is just the sum of squares of principal curvatures of the graph of u.) Using 1.5 in combination with 1.3 (with $l = n$), we conclude

$$\sum_{i,j=1}^{n} D_i(v^{-1}g^{ij}D_j(\mu v - u_n)) = \mu\mathcal{C}^2 \qquad \mathbf{1.6}$$

$$+ H_z(x,u)(\mu v - u_n - \mu v^{-1}) + \mu\sum_{j=1}^{n} \nu^j H_{x^j}(x,u) - H_{x^n}(x,u)$$

for any real $\mu \in \mathbf{R}$. This identity will be of basic importance in the proof of Theorem 1.

Proof of Theorem 1. We use the notation

$$v = \sqrt{1+|Du|^2}, \quad w = \log v$$
$$\chi_\mu = (\mu - \nu_n)_+, \quad \psi_\mu = (\mu v - u_n)_+ \equiv v\chi_\mu, \quad h_\mu = \log(1+\psi_\mu),$$

where all functions f are thought of as functions of the variables $(x,z) \in \Omega \times \mathbf{R}$ which are constant on the vertical lines $\{x\} \times \mathbf{R}$; also we use the notation $f_+ = \max\{f, 0\}$.

The gradient of such functions f taken on $G = \operatorname{graph} u$ is denoted ∇f, and, in terms of the coordinates $(x^1, \dots, x^n, u(x)) \mapsto (x^1, \dots, x^n)$ for $\operatorname{graph} u$ (where $x = (x^1, \dots, x^n) \in \Omega$), is represented by

$$\nabla f = \sum_{i,j=1}^{n} g^{ij} D_j f D_i,$$

so that

$$(1) \qquad |\nabla f|^2 = \sum_{i,j=1}^{n} g^{ij} D_i f D_j f.$$

Notice that then for the quantity $\mathcal{C}^2$ introduced above we have

$$\mathcal{C}^2 = |\nabla\nu|^2 \equiv \sum_{j=1}^{n+1} |\nabla\nu_j|^2, \tag{2}$$

as one readily checks by direct computation.

Concerning the functions ψ_μ, note that, for $0 < \mu_2 < \mu_1$,

(3)
$$\begin{aligned}\psi_{\mu_2} = (\mu_2 v - u_n)_+ \equiv ((\mu_1 - (\mu_1 - \mu_2)v - u_n)_+ \le (\mu_1 v - u_n)_+ \equiv \psi_{\mu_1} \\ v\chi_{\mu_1} = \psi_{\mu_1} = (\mu_2 v - u_n + (\mu_1 - \mu_2)v)_+ \ge (\mu_1 - \mu_2)v \quad \text{on} \quad \Omega_{\mu_2}.\end{aligned}$$

Also $\psi_\mu = h_\mu = 0$ on $\partial\Omega \cap B_\rho$ for $\mu \le \mu_0$. We henceforth take $0 < \mu < \mu_1 < \mu_0$.

Note that multiplying by ψ_μ in the identity 1.6 leads directly to the inequality

$$D_i(\chi_\mu g^{ij} D_j \psi_\mu) \ge \mu\psi_\mu \mathcal{C}^2 + v^{-1}|\nabla\psi_\mu|^2 - C\Lambda\psi_\mu$$

in the weak sense on Ω, where $C = C(n,\mu)$. Thus since $\nu^n \ge \mu_0 > \mu_1 > \mu$ on $\partial\Omega \cap B_\rho$ (so that $\psi_\mu \equiv 0$ on $\partial\Omega \cap B_\rho$), we conclude that

$$\int_\Omega (\mu\chi_\mu \mathcal{C}^2 + v^{-2}|\nabla\psi_\mu|^2\zeta) v\, dx + \int_\Omega \chi_\mu \langle \nabla\psi_\mu, \nabla\zeta\rangle\, dx \le C\Lambda \int_\Omega \psi_\mu \zeta\, dx \tag{4}$$

for any $\zeta \in C_c^\infty(\Omega \cap B_\rho)$ with $\zeta \ge 0$.

Notice that if we replace ζ by $\zeta^2\Psi(\psi_\mu)$, where $\Psi : \mathbf{R} \to \mathbf{R}$ is an increasing C^1 function, then the above inequality implies

$$\int_{\Omega_\mu} (\chi_\mu \mathcal{C}^2 + v^{-2}|\nabla\psi_\mu|^2)\Psi(\psi_\mu)\zeta^2 v\, dx \le C \int_{\Omega_\mu} \Psi(\psi_\mu)(|\nabla\zeta|^2 + \Lambda\zeta^2) v\, dx, \tag{5}$$

where $C = C(n, \mu)$, and where $\Omega_\mu = \{x \in \Omega : \nu_n < \mu\}$.

The rest of the argument, beginning with a Moser-type iteration, closely parallels the Bombieri De Giorgi Miranda proof [BDM] of the interior gradient estimate for solutions of the minimal surface equation, using the inequality (5) at those places where the usual gradient bound argument uses the equation for the Laplacian of the unit normal (which is equivalent to the identity 1.5 above).

For this we need a Sobolev inequality

$$\left(\int_\Omega \zeta^{2\kappa} v\, dx\right)^{1/\kappa} \le C \left(\int_\Omega \zeta^2 v\, dx\right)^{1/2} \left(\int_\Omega (|\nabla\zeta|^2 + H^2\zeta^2) v\, dx\right)^{1/2}, \quad \kappa = \frac{n}{n-1},$$

which is valid for any $\zeta \in C_c^\infty(\Omega)$. (See e.g. [MS] or [SL1] for a discussion of this inequality.) Since H is the mean curvature (i.e., the sum of principal curvatures) and $\mathcal{C}$ is the sum of squares of principle curvatures, we have the inequality $H^2 \le n\mathcal{C}^2$, and hence the above inequality implies

$$(6) \quad \left(\int_\Omega \zeta^{2\kappa} v\,dx\right)^{1/\kappa} \le C\left(\int_\Omega \zeta^2 v\,dx\right)^{1/2}\left(\int_\Omega (|\nabla\zeta|^2 + \mathcal{C}^2\zeta^2) v\,dx\right)^{1/2}.$$

Now in (6) we replace ζ by $\chi_\mu^2\zeta h_\mu^k$, where $\zeta \in C_c^\infty(B_\rho)$ is non-negative, and where k is any positive integer. This gives

$$\left(\int_{\Omega_\mu} (\chi_\mu^2\zeta h_\mu^k)^{2\kappa} v\,dx\right)^{1/\kappa} \le \left(\int_{\Omega_\mu} (\chi_\mu^2\zeta h_\mu^k)^2 v\,dx\right)^{1/2}\left(\int_{\Omega_\mu} (|\nabla(\chi_\mu^2\zeta h_\mu^k)|^2 + \mathcal{C}^2\chi_\mu^2\zeta^2 h_\mu^{2k}) v\,dx\right)^{1/2}.$$

Since we have

$$|\nabla\chi_\mu| \le |\nabla\nu_n| \le \mathcal{C},$$

and

$$\chi_\mu|\nabla h_\mu| = \chi_\mu\frac{|\nabla\psi_\mu|}{1+\chi_\mu v} \le v^{-1}|\nabla\psi_\mu|,$$

by using the Cauchy-Schwarz inequality we see

$$\left(\int_{\Omega_\mu} (\chi_\mu^2\zeta h_\mu^k)^{2\kappa} v\,dx\right)^{1/\kappa} \le \tag{7}$$
$$C\rho\int_{\Omega_\mu} \left(v\chi_\mu^2(|\nabla\zeta|^2 + \rho^{-2}\zeta^2) + v\chi_\mu\mathcal{C}^2\zeta^2 + k^2v^{-1}|\nabla\psi_\mu|^2\zeta^2\right)\left(h_\mu^{2k} + h_\mu^{2k-2}\right)dx$$

for any $\mu \in (0, \mu_0)$. Notice that by using (5) (with $\Psi(t) = \log^{2k}(t) + \log^{2k-2}(t)$) we then conclude that

$$(8) \quad \left(\int_{\Omega_\mu} (\chi_\mu^2\zeta h_\mu^k)^{2\kappa} v\,dx\right)^{1/\kappa} \le C\rho\int_{\Omega_\mu} (|\nabla\zeta|^2 + \rho^{-2}\zeta^2)(h_\mu^{2k} + h_\mu^{2k-2}) v\,dx.$$

Now let $0 < \mu_2 < \mu_1 < \mu_0$, $\mu = (\mu_1+\mu_2)/2$ and note that $h_{\mu_2} \le h_\mu \le h_{\mu_1}$, and also, on Ω_μ, $h_{\mu_1} \ge \log(1+v(\mu_1-\mu)) \ge (\mu_1-\mu_2)/4$ and $\chi_{\mu_1} \ge \frac{1}{2}(\mu_1-\mu_2)$, while on Ω_{μ_2} we have $\chi_\mu \ge (\mu-\mu_2) = \frac{1}{2}(\mu_1-\mu_2)$. Then, taking any

$\eta \in (0,1)$, choosing ζ to be a cutoff function which is 1 on $B_{\eta\rho}$ and 0 outside B_ρ, we see that (8) gives

$$\left(\int_{\Omega_{\mu_2,\eta\rho}} \chi^2_{\mu_2} h^{2k\kappa}_{\mu_2} v\,dx\right)^{1/\kappa} \le Ck^2(1-\eta)^{-2}(\mu_1-\mu_2)^{-4}\rho^{-1}\int_{\Omega_{\mu_1,\rho}} \chi^2_{\mu_1} h^{2k}_{\mu_1} v\,dx,$$

where $\Omega_{\mu,\rho} = \Omega \cap B_\rho$. If we now let $k = \kappa^{q-1}$, where q is a positive integer, and if we replace η by $\eta_q \equiv 1 - \sum_{j=1}^{q} 2^{-j-3}$ (with $\eta_0 = 1$) and μ_1, μ_2 by μ_{q-1}, μ_q respectively, where $\mu_q \equiv \mu_1 - \sum_{j=1}^{q} 2^{-j}(\mu_1-\mu_2)$, we see that then

$$\Phi(\kappa^q) \le (C\rho^{-1})^{\kappa^{-(q-1)}} C^{(q-1)\kappa^{-(q-1)}} \Phi(\kappa^{q-1}), \quad q \ge 1,$$

where $\Phi(\kappa^q) = \left(\int_{\Omega_{\mu_q,\eta_q\rho}} \chi^2_{\mu_q} h^{2\kappa^q}_{\mu_q} v\,dx\right)^{1/\kappa^q}$, and by iteration this gives

$$\Phi(\kappa^q) \le (C\rho^{-1})^{\sum_{j=0}^{q-1}\kappa^{-j}} C^{\sum_{j=1}^{q-1} j\kappa^{-j}} \Phi(1).$$

Passing to the limit as $q \to \infty$ and using the facts that $\sum_{j=1}^{\infty} \kappa^{-j} = n$, $\sum_{j=1}^{\infty} j\kappa^{-j} < \infty$, we conclude

$$\sup_{\Omega_{\mu_2,3\rho/4}} h^2_{\mu_2} \le C\rho^{-n}\int_{\Omega_{\mu_1,\rho}} \chi^2_{\mu_1} h^2_{\mu_1} v\,dx, \tag{9}$$

where C depends only on n, $\Lambda\rho^2$, μ_1 and μ_2.

It remains to bound integrals like the one on the right of (9) in terms of $\rho^{-1}M$, n, μ_1, μ_0 and $\Lambda\rho^2$. This is done as follows:

The weak form of the equation is $\int_\Omega \nu^i D_i\zeta\,dx = \int_\Omega H(x,u)\zeta\,dx$, valid for any $\zeta \in C^2_c(\Omega \cap B_\rho)$. We replace ζ by $u\zeta^2$; this gives

$$\int_\Omega \sum_{i=1}^{n} \nu^i D_i u \zeta^2\,dx = \int_\Omega u(-2\zeta\nu^i D_i\zeta + \zeta^2 H)\,dx,$$

and since $\sum_{i=1}^{n} \nu^i D_i u = v^{-1}|Du|^2 = v - v^{-1} \ge v - 1$, we have

$$\int_\Omega v\zeta^2\,dx \le \int_\Omega \zeta^2\,dx + M\int_\Omega (2\zeta|D\zeta| + \zeta^2|H|)\,dx.$$

Now replace ζ by $\zeta\chi_\mu$, where ζ is no longer required to vanish along $\partial\Omega \cap B_\rho$. This then gives

$$\int_{\Omega_\mu} v\chi_\mu^2\zeta^2\,dx \le \int_{\Omega_\mu} \zeta^2\,dx + M\int_{\Omega_\mu} \left(2\zeta\chi_\mu^2|D\zeta| + \chi_\mu^2\zeta^2|H| + 2\zeta^2\chi_\mu|D\nu_n|\right)dx.$$

Now by direct computation one can check that

$$|D\nu|^2 (\equiv \sum_{j=1}^{n+1} |D\nu_j|^2) = v^{-2}\sum_{i,j,l=1}^{n} g^{ij}D_iu_lD_ju_l \equiv \mathcal{C}^2 + v^{-2}|\nabla v|^2, \tag{10}$$

and $|H| \le \sqrt{n}\,\mathcal{C}$ because H is the mean curvature and $\mathcal{C}^2$ is the sum of squares of principal curvatures, hence,

$$\int_{\Omega_\mu} \chi_\mu^2\zeta^2 v\,dx \le \int_{\Omega_\mu} \zeta^2\,dx + CM\int_{\Omega_\mu} \left(\chi_\mu(\mathcal{C} + v^{-1}|\nabla v|)\zeta^2 + \chi_\mu^2\zeta|D\zeta|\right)dx. \tag{11}$$

Finally we replace ζ by ζh_μ. The above inequality then gives

$$\int_\Omega \chi_\mu^2 h_\mu^2\zeta^2 v\,dx \le \int_{\Omega_\mu} (\zeta^2 + 2M\zeta|D\zeta|)h_\mu^2\,dx +$$
$$CM\int_\Omega \left(\chi_\mu(\mathcal{C} + v^{-1}|\nabla v|)h_\mu^2\zeta^2 + C\chi_\mu^2 h_\mu|Dh_\mu|\zeta^2\right)dx.$$

Since

$$\begin{aligned}\chi_\mu|Dh_\mu| &= \chi_\mu(1+\chi_\mu v)^{-1}|D\psi_\mu| \le v^{-1}|D\psi_\mu| \le |\nabla\psi_\mu| \\ &= |\chi_\mu\nabla v - v\nabla\nu_n| \le |\nabla v| + v\mathcal{C}\end{aligned} \tag{12}$$

on Ω_μ, we then have (after an application of Cauchy's inequality) that

$$\int_{\Omega_\mu} \chi_\mu^2 h_\mu^2\zeta^2 v\,dx \le 2\int_{\Omega_\mu} ((\zeta^2 + CM\zeta|D\zeta|)h_\mu^2)\,dx$$
$$+ CM^2\int_{\Omega_\mu} (\mathcal{C}^2 + v^{-2}|\nabla v|^2)\zeta^2 v\,dx.$$

Using the Cauchy inequality again, this gives

$$\int_{\Omega_\mu} \chi_\mu^2 h_\mu^2\zeta^2 v\,dx \le C\int_{\Omega_\mu} (\zeta^2 + M^2|D\zeta|^2)h_\mu^2\,dx + CM^2\int_{\Omega_\mu} \zeta^2(\mathcal{C}^2 + v^{-2}|\nabla v|^2)v\,dx.$$

Since $h_\mu \le C_\alpha v^\alpha$ for each $\alpha > 0$, this gives

(13)
$$\int_{\Omega_\mu} v\chi_\mu^2 h_\mu^2\zeta^2\,dx \le C\int_{\Omega_\mu} (\zeta^2 + M^2|D\zeta|^2)v\,dx + CM^2\int_{\Omega_\mu} (\mathcal{C}^2 + v^{-2}|\nabla v|^2)\zeta^2 v\,dx.$$

But by (5), with $\Psi(t)$ approximating the characteristic function of the interval $[\mu, \infty)$, we have

$$\int_{\Omega_\mu} (\chi_\mu \mathcal{C}^2 + v^{-2}|\nabla\psi_\mu|^2)v\,dx \le C\int_{\Omega_\mu} (|D\zeta|^2 + \Lambda\zeta^2)v\,dx. \tag{14}$$

Now if $\mu < \mu_1 < \mu_0$ we have $\mu_1 - \mu \le \chi_{\mu_1}$ and $\chi_{\mu_1}|\nabla v| \le |\nabla\psi_{\mu_1}| + v\mathcal{C}$ on $\Omega_\mu \subset \Omega_{\mu_1}$, and so using (14) with μ_1 in place of μ we have

$$\begin{aligned}\int_{\Omega_\mu} (\mathcal{C}^2 + v^{-2}|\nabla v|^2)v\,dx &\le C(\mu_1-\mu)^{-2}\int_{\Omega_{\mu_1}} (\chi_{\mu_1}\mathcal{C}^2 + v^{-2}|\nabla\psi_{\mu_1}|^2)v\,dx \\ &\le C(\mu_1-\mu)^{-2}\int_{\Omega_{\mu_1}} (|D\zeta|^2 + \Lambda\zeta^2)v\,dx.\end{aligned} \tag{15}$$

Using (13) together with (15) we obtain

$$\int_{\Omega_\mu} \chi_\mu^2 h_\mu^2 \zeta^2 v\,dx \le C(\mu_1-\mu)^{-2}\int_{\Omega_{\mu_1}} (M^2|D\zeta|^2 + (1+M^2\Lambda)\zeta^2)v\,dx,$$

which can of course be written

$$\int_{\Omega_\mu} \chi_\mu^2 h_\mu^2 \zeta^2 v\,dx \le C(\mu_1-\mu)^{-2}\int_{\Omega_{\mu_1}} (\rho^2|D\zeta|^2 + \zeta^2)v\,dx, \tag{16}$$

where now C depends on $\Lambda\rho^2$ and $\rho^{-1}M$. On the other hand (11) implies that

$$\begin{aligned}\int_{\Omega_\mu} \chi_\mu^2\zeta^2 v\,dx \le{}& C\int_{\Omega_\mu} (\zeta^2 + M^2|D\zeta|^2) \\ &+ C\left(\int_{\Omega_\mu} \zeta^2\,dx\right)^{1/2}\left(\int_{\Omega_\mu} (\mathcal{C}^2 + v^{-2}|\nabla v|^2)\zeta^2\right)^{1/2}\end{aligned}$$

and so by (15) we deduce that if $\mu_2 < \mu$ then

$$\begin{aligned}(\mu-\mu_2)^2\int_{\Omega_{\mu_2}} \zeta^2 v\,dx &\le \int_{\Omega_\mu} \chi_\mu^2\zeta^2 v\,dx \\ &\le C(\mu_1-\mu)^{-2}\rho^{n/2}\left(\int_{\Omega_{\mu_1}} (\rho^2|D\zeta|^2 + \zeta^2)v\,dx\right)^{1/2},\end{aligned}$$

where $C = C(\rho^{-1}M, \Lambda\rho^2, \mu, n)$. Then by virtue of a Moser-type iteration (as we used above in the argument to bound $\sup h_{\mu_2}$) we deduce that

$$\int_{\Omega_{\mu_2,\rho/2}} v\,dx \le C\rho^n, \tag{17}$$

where $C = C(\rho^{-1}M, \Lambda\rho^2, \mu_2, \mu_1, \mu_0, n)$.

Now by combining (9) (with a new μ_1, μ_2), (16) and (17) we have the required estimate. This completes the proof of Theorem 1.

2. Applications to the Zero Contact Angle Problem.

We illustrate the use of Theorem 1 with an application to the zero contact angle problem. We prove both a local regularity theorem (Theorem 2) and a global existence theorem (Theorem 3).

To describe the local regularity result, take $x_0 \in \partial\Omega$, $\rho > 0$; we shall work in $\Omega \cap B_\rho(x_0)$ or $\overline{\Omega} \cap B_\rho(x_0)$. In fact for convenience of notation we suppose without loss of generality that $x_0 = 0 \in \partial\Omega$.

For the regularity result, we assume no *a-priori* regularity at the boundary, so we are going to write the contact angle boundary condition and the equation in weak form, as follows:

Let η be any $C^0(\overline{\Omega})$ function which on $\partial\Omega$ coincides with the inward pointing unit normal of $\partial\Omega$, $\theta : \partial\Omega \cap B_\rho \to [0, \pi]$ is a given function, measurable with respect to $(n-1)$-dimensional Hausdorff measure $\mathcal{H}^{n-1}$ on $\partial\Omega \cap B_\rho$, and let $u \in C^2(\Omega \cap B_\rho) \cap W^{1,1}(\Omega \cap B_\rho)$. u is said to satisfy the contact angle problem

$$\textbf{2.1} \qquad \begin{cases} \sum_{i=1}^{n} D_i\big((1+|Du|^2)^{-1/2}D_iu\big) = H(x,u) & \text{in} \quad \Omega \cap B_\rho \\ \nu \cdot \eta \equiv \cos\theta \quad \text{on} \quad B_\rho \cap \partial\Omega, \end{cases}$$

in the weak sense provided

$$\textbf{2.1}' \qquad \int_{\Omega \cap B_\rho} \left(\sum_{i=1}^{n} \nu^i D_i\zeta + H(x,u)\zeta \right) = -\int_{\partial\Omega \cap B_\rho} \cos\theta\zeta \, d\mathcal{H}^{n-1}$$

for each $\zeta \in C^1(\overline{\Omega} \cap B_\rho)$. It is of interest to note that if one uses the Riesz representation theorem for the representation of the linear functional $\mathcal{F}(\zeta) = \int_{\Omega \cap B_\rho} \sum_{i=1}^{n} \nu^i D_i\zeta + H(x,u)\zeta$, then by the classical divergence theorem (applied over a sequence of smooth domains Ω_t which increase to Ω as $t \downarrow 0$) we have $\mathcal{F}(\zeta) \equiv \int_{\partial\Omega} \gamma\zeta \, d\mathcal{H}^{n-1}$ for some $\mathcal{H}^{n-1}$-measurable function γ with values in $[-1,1]$, and then 2.1′ is equivalent to $\gamma = \cos\theta$ $\mathcal{H}^{n-1}$-a.e. on $\partial\Omega \cap B_\rho$; thus $W^{1,1}(\Omega)$ solutions of 0.1 automatically satisfy a weak identity of the form 2.1′ with some measurable choice of θ. (Actually we shall not explicitly need this observation here.)

If Ω is connected, then a $W^{1,1}(\Omega)$ solution of 2.1′ is uniquely determined, up to an additive constant, by the function γ. Because if u_1 is another solution of 2.1′ then by taking differences (and noting by calculus that $(1+|Du_1|^2)^{-1/2}D_iu_1-(1+|Du|^2)^{-1/2}D_iu=\int_0^1\frac{d}{dt}((1+|Du_t|^2)^{-1/2}D_iu_t)=\sum_{j=1}^n a_{ij}(x)D_j(u_1-u)$, where $a_{ij}=\int_0^1(1+|Du_t|^2)^{-1/2}(\delta_{ij}-\nu_i^t\nu_j^t)\,dt$ with $\nu^t=(1+|Du_t|^2)^{-1/2}Du_t$ and $u_t=u+t(u_1-u)$) we obtain the identity

$$\int_\Omega\sum_{i=1}^n\left(\sum_{j=1}^n a_{ij}D_j(u_1-u)\right)D_i\zeta\,dx=\int_\Omega(H(x,u_1)-H(x,u))\zeta\,dx$$

for any $\zeta\in C^1(\overline{\Omega})$. Evidently, by approximation, since $\sum_{j=1}^n a_{ij}(x)D_j(u_1-u)\equiv(1+|Du_1|^2)^{-1/2}D_iu_1-(1+|Du|^2)^{-1/2}D_iu$ is bounded for each i, we deduce that this is in fact also holds for any $\zeta\in W^{1,1}(\Omega)$, so we can substitute $\zeta=(u_1-u)_+$. Since the right side is then ≤ 0 (because $H(x,z)$ is an increasing function of z), this gives

$$\int_{\Omega_+}\sum_{i,j=1}^n a_{ij}D_j(u_1-u)D_i(u_1-u)\,dx\leq 0,$$

where Ω_+ is the subset of Ω where $u_1\geq u$. On the other hand a_{ij} is a positive definite matrix a.e. in Ω, so we have $Du_1=Du$ a.e. in the set Ω_+. Similarly $Du_1=Du$ a.e. in the subset of Ω where $u_1\leq u$, so we deduce that $Du_1=Du$ a.e. in Ω, and hence u_1 and u differ at most by a constant as claimed.

We are mainly interested in the case when $\theta\equiv 0$ (i.e. "the zero contact angle problem"). This case is mathematically more difficult than the case $\theta\in(0,\pi)$ because $\theta=0$ implies that the graph is vertical ($|Du|=\infty$) at $\partial\Omega$.

The main regularity result is as follows:

Theorem 2. *Suppose $u\in C^2(\Omega\cap B_\rho)\cap W^{1,1}(\Omega\cap B_\rho)$ satisfies 0.3, 0.4, and 2.1′ with $\theta\equiv 0$, $\partial\Omega\cap B_\rho$ is of class C^2. Then $\nu\equiv\frac{(Du,-1)}{\sqrt{1+|Du|^2}}\in C^0(B_{\rho/2}\cap\overline{\Omega})$, and u satisfies the boundary condition classically (i.e. 2.1 holds with $\theta\equiv 0$ on $\Omega\cap B_{\rho/2}$). Furthermore, considered as a function on the graph of $u\,|\,\Omega\cap B_{\rho/2}$, ν is Lipschitz continuous. If $\kappa>0$ is a given constant, if $H_z(x,z)\geq\kappa$ on $B_\rho\times[-M,M]$, if $H\in C^{k-1}(\mathbf{R}^n\times\mathbf{R})$, and if $\partial\Omega\cap B_\rho$*

is C^k for some $k \ge 4$, then $u \mid \Omega \cap B_{\rho/2}$ has trace Γ on $\partial\Omega \cap B_{\rho/2}$ which is C^{k-1}, (closure graph u)$\cap(\Omega\cap B_{\rho/2})\times\mathbf{R}$ *is an embedded C^k submanifold with boundary Γ, and u is Hölder continuous with exponent $\frac{1}{2}$ up to $\partial\Omega \cap B_{\rho/2}$.*

Remark. Concerning the conclusion that ν is Lipschitz on graph $u \mid B_{\rho/2}$, we note that in fact we are going to prove the stronger result that $M =$ (closure graph u) $\cap$ ($B_{\rho/2} \times \mathbf{R}$) is contained in a properly embedded $C^{1,1}$ submanifold of $\mathbf{R}^{n+1}$.

The $C^{1,1}$ part of the proof involves using the estimate of Theorem 1 (and Corollary 1) on an approximating sequence of smooth solutions to show that the usual "obstacle problem" $C^{1,1}$ estimates [GC] can be applied.

We shall need the following general lemma, which is a consequence of the main theorem of [GC]. Here H is any locally Lipschitz function in $\Omega \times \mathbf{R}$, $A_i = A_i(x,p)$ is locally $C^{1,1}$ on $\Omega \times \mathbf{R}^n$ with $(\partial A_i(x,p)/\partial p_j)$ positive definite for each $(x,p) \in \Omega \times \mathbf{R}$.

Lemma. *Let $\psi \in C^{1,1}(\Omega)$, $u \in C^{0,1}(\Omega)$, $u \le \psi$ in Ω, and suppose that u is a weak solution of the equation $D_i[A_i(x,Du)] - B = 0$ in Ω, where B is locally bounded in Ω and $B|_{\{x\in\Omega : u(x)<\psi(x)\}} = H(x,u)|_{\{x\in\Omega : u(x)<\psi(x)\}}$, with H locally Lipschitz on $\Omega \times \mathbf{R}$. Then u is locally of class $C^{1,1}$ in Ω (i.e. $D_j u$ are locally Lipschitz functions in Ω, $j = 1, \ldots, n$), and in fact for any ball $\overline{B}_\sigma(y) \subset \Omega$ we have*

$$\|u\|_{C^{1,1}(B_{\sigma/2}(y))} \le C,$$

where C is determined by M, μ, ρ, and K, where $M =$ any upper bound for $|u| + |Du| + \|\psi\|_{C^{1,1}}$ on $B_\sigma(y)$, μ is any positive constant such that

$$\sum_{i,j=1}^{n} \xi^i \xi^j \partial A_i(x,p)/\partial p_j \ge \mu|\xi|^2$$

for all $(x,p,\xi) \in B_\sigma(y) \times B_M(0) \times \mathbf{R}^n$, and K is any upper bound for

$$\|B\|_{L^\infty(B_\sigma(y))} + \|H\|_{C^{0,1}(B_\sigma(y)\times[-M,M])} + \sum_{i=1}^{n} \|A_i\|_{C^{1,1}(B_\sigma(y)\times[-M,M]\times B_M(0))}.$$

Proof. The proof involves some manipulations to show that the hypotheses needed for the application of the argument of [GC, pp. 391–2] are satisfied, as follows:

Let $A(u) = \sum_{i=1}^{n} D_i[A_i(x, Du)]$, and note first that the standard $W^{2,p}$ estimates for uniformly elliptic equations give that $u \in C^{1,\alpha}(\Omega) \cap W^{2,p}_{\text{loc}}(\Omega)$ for each $\alpha \in (0,1)$ and each $p \geq 1$. Thus in particular $A(u) = A(\psi)$ a.e. in I, where I is the "coincidence set" $\{x \in \Omega : u(x) = \psi(x)\}$. Thus the equation can be written

$$(1) \qquad A(u) - H(x,u) = (A(\psi) - H(x,\psi))\chi_I \quad \text{a.e. in} \quad \Omega,$$

where χ_I denotes the characteristic function of I. Notice that in particular we then have

$$A(u) - A(\psi) - (H(x,u) - H(x,\psi)) = -(A(\psi) - H(x,\psi))\chi_{\Omega \setminus I}.$$

Since this can be written in the form

$$(2) \quad D_i(a_{ij}D_j(u-\psi)) + \sum_j b_j D_j(u-\psi) + c(u-\psi) = -(A(\psi) - H(x,\psi)\chi_{\Omega\setminus I},$$

where a_{ij}, b_j, c are locally Hölder continuous in Ω and where (a_{ij}) is positive definite, we can then apply the strong maximum principle for weak solutions ([GT, Th. 8.19]) to assert that

$$(3) \qquad A(\psi) - H(x,\psi) \geq 0 \quad \text{at each point of} \quad \Omega \cap \partial I,$$

otherwise (2) would imply

$$D_i(a_{ij}D_j(u-\psi)) + \sum_j b_j D_j(u-\psi) - c_-(u-\psi) \geq 0$$

($c_- = \max\{-c, 0\}$) in the neighbourhood of some point $x_0 \in \Omega \cap \partial I$ (where $u - \psi$ attains its zero maximum value), thus contradicting the strong maximum principle for weak solutions.

Now the equation (1) gives

$$(4) \qquad A(u) - H(x,u) = \chi_I Q_+ - \chi_I Q_-,$$

where $Q = A(\psi) - H(x,\psi)$ and $Q_\pm = \max\{\pm Q, 0\}$. Now evidently by (3) we have

$$\chi_I Q_- \quad \text{is locally Lipschitz on} \quad \Omega$$

(with Lipschitz constant $\leq$ the Lipschitz constant of Q), and hence (4) implies that

$$(5) \qquad A(u) + R = f(\leq 0),$$

where $R = -H(x,u) + \chi_I Q_-$ is Lipschitz on Ω and $f = \chi_I Q_+ \geq 0$. Now let $y \in \Omega \cap \partial I$, $\overline{B}_\sigma(y) \subset \Omega$, and let $\varphi \geq 0$ be C^1 in Ω and vanish on I. Then the second difference quotient $\delta_h^2 \varphi \equiv |h|^{-2}(\varphi(x+h) - 2\varphi(x) + \varphi(x-h))$ is non-negative everywhere on $I \cap B_\sigma(y)$ for $|h| < \operatorname{dist}(B_\sigma(y), \partial\Omega)$, so we have by (5)

$$(6) \quad (A(u) + R)(\delta_h^2 \varphi) \geq 0 \quad \text{a.e. on} \quad B_\sigma(y), \quad |h| < \operatorname{dist}(B_\sigma(y), \partial\Omega).$$

(We emphasize that it is not necessary for $\delta_h^2 \varphi$ to vanish on I; we only require that φ vanish on I.)

Now the argument of [GC], pp. 391–392 is applicable and gives the remaining conclusions of the lemma. (The rest of the proof involves applying Theorem 8.15 of [GT] to the inequality obtained for difference quotients $\delta_h^2 u$ by integrating (6) over Ω. The fact that $u \leq \psi$ with $\psi \in C^{1,1}$ gives a fixed upper bound λ, independent of h, on these second difference quotients at points of $\Omega \cap \partial I$, which means that $(\delta_h^2 u - \lambda - 1)_+$ is a weak subsolution of an inhomogeneous equation to which the relevant part of De Giorgi Nash theory ([GT, Th. 8.15]) can be applied, giving $\delta_h^2 u \leq C$ on $B_{\sigma/2}(y)$, with C independent of $|h|$. Letting $|h| \to 0$ we thus have in $B_{\sigma/2}(y) \setminus I$ that the eigenvalues of $D^2 u$ are all bounded above by a fixed constant, and then by using the equation again these eigenvalues are all bounded below also. For the details of this argument, we refer the reader to [GC].)

We can now give the proof of Theorem 2, but first we make a remark for later reference.

2.2 Remark. If $y \in \Omega \cap \partial I$ and $B_\sigma(y) \subset \Omega$ as in the above proof, if the function $\mathcal{K}(x) \equiv A(\psi) - H(x,\psi)$ has strictly negative partial derivative with respect to the variable x^n in the ball $B_\rho(y)$, and if property (**) of Corollary 1 holds, then we can actually prove that *strict* inequality holds in (3). Otherwise there would exist $x_0 \in \Omega \cap \partial I$ with $\mathcal{K}(x_0) = 0$, and then strict negativity of $D_n\mathcal{K}(x_0)$ and the fact that $\mathcal{K}$ is C^2 implies by Sard's theorem that there is an open U with $x_0 \in \partial U$, with e_n pointing into U at x_0, and with $\mathcal{K} < 0$ in U. But in view of equation (2) we can then apply the $C^{1,\alpha}$ version of Hopf boundary point lemma (for divergence-form equations with Hölder continuous coefficients—see [FG] and also [HS, p. 474]) at x_0 to assert that $u \equiv \psi$ in $U \cap B_\sigma(x_0)$ for some $\sigma > 0$. In view of (**) and the fact that e_n points into U at x_0, this implies that $B_\delta(x_0) \subset I$ for some $\delta > 0$, contradicting the fact that $x_0 \in \partial I$.

Proof of Theorem 2. Without loss of generality we can suppose that $e_n = (0,\dots,0,1)$ is the inward pointing unit normal for $\partial\Omega$ at 0.

First choose a C^2 domain $\widetilde{\Omega}$ with

$$B_{3\rho/4}\cap\Omega\subset\widetilde{\Omega}\subset B_\rho\cap\Omega;$$

we also assume that $\widetilde{\Omega}$ is selected so that $\partial\widetilde{\Omega}$ makes "clean" contact with $\partial\Omega$ in the sense that $\partial(\partial\widetilde{\Omega}\setminus\partial\Omega)$ is an $(n-2)$-dimensional submanifold Σ (or two points in case $n=2$). Notice that by replacing ζ by $\varphi\zeta$ in 2.1′, where φ is a C^2 approximation (from above) for the characteristic function of $\widetilde{\Omega}$, we have that u satisfies

$$\int_{\widetilde{\Omega}}\sum_{i=1}^n \nu^i D_i\zeta = \int_{\partial\widetilde{\Omega}}\gamma\zeta\, d\mathcal{H}^{n-1}, \tag{1}$$

where

$$\gamma = \begin{cases}\tilde{\eta}\cdot\nu & \text{on } \partial\widetilde{\Omega}\setminus\partial\Omega\\ 1 & \text{on } \partial\widetilde{\Omega}\cap\partial\Omega,\end{cases} \tag{2}$$

where $\tilde{\eta}$ denotes the inward pointing unit normal of $\partial\widetilde{\Omega}$. Now on $\widetilde{\Omega}$ we consider, for $\epsilon\in(0,1)$, the solution $w=w_\epsilon$ of the problem

$$\begin{cases}\sum_{i=1}^n D_i\big((1+|Dw|^2)^{-1/2}D_iw\big) = H_M(x,w) & \text{in } \widetilde{\Omega}\\ \tilde{\eta}\cdot(1+|Dw|^2)^{-1}Dw \equiv \gamma_\epsilon & \text{on } \widetilde{\Omega},\end{cases} \tag{3}$$

where $\gamma_\epsilon\in C^1(\partial\widetilde{\Omega})$ satisfies $|\gamma_\epsilon|\le 1-\epsilon$ on $\partial\Omega$ and $\gamma_\epsilon(x)\equiv(1-\epsilon)\gamma$ if $\operatorname{dist}(x,\Sigma)>\epsilon$, with $\Sigma=\partial(\partial\widetilde{\Omega}\setminus\partial\Omega)$, and where $H_M(x,z)=H(x,z)$ for $|z|\le M$, $H_M(x,z)=H(z,M)+z-M$ for $z>M$ and $H_M(x,z)=H(x,-M)+z+M$ for $z<-M$. Then H_M is increasing in z and $\to\pm\infty$ as $z\to\pm\infty$, so by [CF] there are *a-priori* bounds for $\sup|w_\epsilon|$ independent of ϵ; further, by e.g. [GC] or [UN2] or [SS] there exists a $C^{2,\alpha}$ solution of the problem (3) for each $\epsilon>0$.

The standard interior gradient estimates ([GT] or [SL2]) (and the consequent applicability of the De Giorgi Nash theory to give local $C^{1,\alpha}$ estimates and then Schauder theory to give $C^{2,\alpha}$ (see e.g. [GT]), we have for any $\alpha\in(0,1)$ and any compact $K\subset\widetilde{\Omega}$

$$|w_\epsilon|_{C^{2,\alpha}}\le C,$$

where C depends on K but not on ϵ. Thus taking a suitable sequence $\epsilon = \epsilon_j \downarrow 0$ we get $w_{\epsilon_j} \to w$ in $\widetilde{\Omega}$ with w satisfying

$$\int_{\widetilde{\Omega}} \sum_{i=1}^{n} (1 + |Dw|^2)^{-1/2} D_i w D_i \zeta \, dx = \int_{\widetilde{\Omega}} H_M(x, w) \zeta \, dx + \int_{\partial\widetilde{\Omega}} \gamma\zeta;$$

that is, w satisfies the same identity (1) as u.

Also, the estimate of Theorem 1 is applicable for each $\epsilon > 0$, giving

$$D_n w_\epsilon \geq \theta(1 + |Dw_\epsilon|^2)^{1/2} - C \tag{4}$$

on $B_{3\rho/4} \cap \widetilde{\Omega}$, where C is independent of ϵ. By construction w_ϵ satisfies the identity

$$\int_{\widetilde{\Omega}} \sum_{i=1}^{n} (1 + |Dw_\epsilon|^2)^{-1/2} D_i w_\epsilon D_i \zeta \, dx = \int_{\widetilde{\Omega}} H_M(x, w_\epsilon) \zeta \, dx + \int_{\partial\widetilde{\Omega}} \gamma_\epsilon \zeta \tag{5}$$

for any $\zeta \in C^1(\overline{\widetilde{\Omega}})$.

But now by the uniqueness discussion preceding the theorem (which is applicable because $H_M(x, u) \equiv H(x, u)$ by virtue of 0.4), we deduce that $u \equiv w + \delta$ for some constant δ, and hence the estimate (4) holds with u in place of w_ϵ:

$$D_n u \geq \theta(1 + |Du|^2)^{1/2} - C \quad \text{on} \quad B_{3\rho/4} \cap \Omega,$$

where C depends on $\Lambda\rho^2$, M/ρ, n.

Further, by Corollary 1 we know that, with $u_j = w_{\epsilon_j}$ there is a fixed orthogonal transformation Q as in 1.2 such that $Q(\operatorname{graph} u_j \cup \{(x,t) \in \partial\Omega \cap B_\rho \times \mathbf{R} : t \leq u_j(x)\}) \cap (B_{3\rho/4}(0) \times \mathbf{R}) = \operatorname{graph} \widetilde{u}_j$, where

$$|D\widetilde{u}_j| \leq \beta, \quad \widetilde{u}_j \leq \psi \quad \forall j,$$

where $\beta = \beta(\Lambda\rho^2, M/\rho, n)$ is fixed independent of j and ψ is as in Corollary 1.

Of course then we also have that $Q(\operatorname{graph} u) \subset \widetilde{\mathcal{G}}$, where $\widetilde{\mathcal{G}}$ is the graph of $\widetilde{u} = \lim \widetilde{u}_j$, so that

$$|D\widetilde{u}| \leq \beta, \quad \widetilde{u} \leq \psi.$$

Now let I_j be the set of $x \in B_{3\rho/4}$ such that $\widetilde{u}_j = \psi$; thus $I_j \times \mathbf{R}$ is the set of points $(x,t) \in Q(\partial\Omega \times \mathbf{R})$ such that $t \leq u_j(x)$.) Then, on the region

$\Omega \cap B_{3\rho/4} \setminus I_j$, $\widetilde{u}_j$ satisfies the equation $\sum_{i=1}^{n} D_i\big((1+|D\widetilde{u}_j|^2)^{-1/2} D_i\widetilde{u}_j\big) = \widetilde{H}(x,\widetilde{u}_j)$, where $\widetilde{H} = H \circ Q^{-1}$. Notice for later reference that

$$(6) \qquad D_{y^{n+1}}\widetilde{H}(y,y^{n+1}) > 0, \quad (y,y^{n+1}) \in B_{\rho/2} \times [-M,M],$$

provided that constant C_0 of Corollary 1 was chosen large enough to begin with (depending now also on κ), and provided $D_z H(x,z) \geq \kappa$ on $(B_\rho \cap \Omega) \times [-M,M]$. Also $\widetilde{u}_j$ satisfies the boundary condition $(1+|D\widetilde{u}_j|^2)^{-1/2}(D\widetilde{u}_j,-1)\cdot(1+|D\psi|^2)^{-1/2}(D\psi,-1) = 1-\epsilon$ on $\partial I_j \cap B_{3\rho/4}$. Now write $\sum_{i=1}^{n} D_i\big((1+|D\psi|^2)^{-1/2} D_i\psi\big) = A(\psi)$, so $A(\psi)$ is the mean curvature of graph $\psi \equiv Q(\partial\Omega \times \mathbf{R}) \cap (B_{3\rho/4} \times \mathbf{R})$ expressed as a function of x corresponding to the point $(x,\psi(x)) \in$ graph ψ.

Thus weakly on $B_{3\rho/4}$ we have the equation

$$\int_{B_{3\rho/4}} (1+|D\widetilde{u}_j|^2)^{-1/2} D_i\widetilde{u}_j D_i\zeta = \int_{B_{3\rho/4}} B_j\zeta + \int_{\partial I_j} \kappa_j\zeta,$$

where

$$B_j = \begin{cases} \widetilde{H}(x,\widetilde{u}_j) & \text{on} \quad B_{3\rho/4} \setminus I_j \\ A(\psi) & \text{on} \quad I_j, \end{cases}$$

and

$$|\kappa_j| \leq C\epsilon_j.$$

By letting $j \uparrow \infty$, we thus have

$$\int_{B_{3\rho/4}} (1+|D\widetilde{u}|^2)^{-1/2} D_i\widetilde{u} D_i\zeta = \int_{B_{3\rho/4}} B\zeta,$$

where

$$B = \begin{cases} \widetilde{H}(x,\widetilde{u}) & \text{on} \quad B_{3\rho/4} \setminus I \\ A(\psi) & \text{on} \quad I, \end{cases}$$

where

$$I = \{x \in B_{3\rho/4} : \widetilde{u}(x) = \psi(x)\}.$$

(Of course I is closed in $B_{3\rho/4}$ because $\widetilde{u}$ is continuous, in fact Lipschitz continuous.) We have thus established the conditions needed to apply Lemma 1 above with $\widetilde{u}$ in place of u and $\widetilde{H}$ in place of H. Hence Lemma 1 implies that

$$\widetilde{u} \quad \text{is of class} \quad C^{1,1} \quad \text{on} \quad B_{3\rho/4}.$$

Of course then $\operatorname{graph} u \cap (B_{\rho/2} \times \mathbf{R})$ is contained in $Q^{-1}\operatorname{graph}\widetilde{u}$, so also has a Lipschitz continuous unit normal which approaches the unit normal of $\partial\Omega \times \mathbf{R}$ uniformly on approach to $\partial\Omega \times \mathbf{R}$; it is then evident that ν, considered as the function $(1+|Du(x)|^2)^{-1/2}(Du(x), -1)$ of the x coordinates in $\Omega \cap B_{\rho/2}$, is continuous up to $\partial\Omega$.

This completes the proof of the first 2 claims of Theorem 2, so now suppose that $k \geq 4$, that $H_z(x,z) \geq \kappa > 0$ for $(x,z) \in (\overline{\Omega} \cap B_\rho) \times [-M, M]$, and suppose that H is of class C^{k-1} and $\partial\Omega \cap B_\rho$ is of class C^k. We can then of course suppose that the domain $\widetilde{\Omega}$ used in the construction of the w_ϵ in the first part of the proof above is of class C^k.

Let $d = \operatorname{dist}(x, \partial\Omega)$ be the distance function of $\partial\Omega$; we know that $d \in C^k(\{x \in \overline{\Omega} \cap B_{3\rho/4} : \operatorname{dist}(x, \partial\Omega) \leq \sigma\})$ for suitable $\sigma > 0$. Now let D' be the tangential gradient operator on $\Omega \cap B_{3\rho/4}$; this $D'f = Df - (Dd \cdot Df)Dd$. By the estimates of [SS], we have

$$\sup_{\Omega \cap B_{3\rho/4}} |D'u_j| \leq C,$$

with C depending only on $\Lambda\rho^2$, M/ρ, $\kappa\rho^2$ and n and hence for each $t \in (0, 3\rho/4)$ we have

$$|u(x) - u(y)| \leq C|x - y|, \quad x, y \in \Omega_t \cap B_{3\rho/4},$$

where $\Omega_t = \{x \in \Omega : \operatorname{dist}(x, \partial\Omega) = t\}$, and where C depending only on $\Lambda\rho^2$, M/ρ, $\kappa\rho^2$ and n.

Letting $j \to \infty$ and using also the Hölder estimate from [SS], we then have the same estimate for u and also that u is Hölder continuous on $\overline{\Omega} \cap B_{3\rho/4}$, and hence in particular the trace Γ of u on $\partial\Omega$ is Lipschitz with Lipschitz constant determined by $\Lambda\rho^2$, M/ρ, $\kappa\rho^2$ and n, and the boundary ∂I of coincidence set I above is actually the Lipschitz manifold $Q(\Gamma)$.

Next note that $\widetilde{u}$, ψ have property (**) of Corollary 1 (because $\operatorname{graph}\widetilde{u}$ is obtained by rotation of $\operatorname{graph} u$ in the x^n–x^{n+1}-plane), and it is easy to check (taking a larger C_0 in Corollary 1 if necessary) that, by virtue of (6), we also have $D_n\big(A(\psi) - \widetilde{H}(x, \psi(x))\big) < 0$ on $B_{\rho/2}$ for small enough ρ. Then by Remark 2.2 we know that $A(\psi) - \widetilde{H}(x, \psi(x)) > 0$ at each boundary point of the coincidence set I. Then free boundary results of Caffarelli [CL] and Kinderlehrer–Nirenberg [KN] imply that Γ is C^{k-1} and that $\widetilde{u} \,|\, (\operatorname{closure} B_{3\rho/4} \setminus I)$ is C^k. Finally, since the trace of u on $\partial\Omega \cap B_{\rho/2}$ is thus of class C^2, we can use the results of [KS] to conclude that u is Hölder continuous with exponent $\frac{1}{2}$.

This completes the proof of Theorem 2.

We conclude this section with an existence theorem for the zero contact angle problem.

For this we need in addition to 0.3 the property that

2.3 $$\pm H(x,z) \geq \beta_{\pm}(z), \quad (x,z) \in \overline{\Omega} \times \mathbf{R},$$

where $\beta_{+}(z) \to \infty$ as $z \to \infty$ and $\beta_{-}(z) \to -\infty$ as $z \to -\infty$.

Theorem 3. *If Ω is a bounded C^2 domain, and if 0.3 and 2.3 hold, then there is a solution $u \in C^2(\Omega)$ of 0.1 such that $\nu \equiv (1+|Du|^2)^{-1/2}(Du,-1) \in C^0(\overline{\Omega})$, such that 0.2 holds with $\theta \equiv 0$ and such that ν is Lipschitz as a function on* graph u. *Furthermore, if $k \geq 4$, $\partial\Omega$ is of class C^k, H of class C^{k-1}, and $H_z > 0$ everywhere in $\Omega \times \mathbf{R}$, then u is Hölder continuous with exponent $\frac{1}{2}$, the trace Γ of u on $\partial\Omega$ is C^{k-1}, and* closure graph u *is a C^k submanifold with boundary Γ.*

Proof. Under the present hypotheses we can apply exactly the same argument that we used in the proof of Theorem 2 for the construction of the w_ϵ, but this time we can work on all of Ω (rather than on the subdomain $\widetilde{\Omega}$) and with the original function H rather than the function H_M. We then have by exactly the same reasoning as in the proof of Theorem 2 that $|w_\epsilon|$ is bounded, the estimates of Theorem 1 hold near any boundary point of Ω, and the $C^{2,\alpha}$ norms of the w_ϵ are locally bounded in Ω independent of ϵ. Thus letting $\epsilon \downarrow 0$ as in the proof of Theorem 2 we see that, in a neighbourhood of each boundary point, all the conclusions of Theorem 2 can be applied to u. Thus the proof is complete.

3. Variational equations of mean curvature type.

Here we want to discuss the analogues of Theorems 1–3 in case we have the non-parametric Euler–Lagrange equation of a general elliptic parametric functional in place of the mean curvature equation.

Recall that a parametric elliptic functional is a functional $\mathcal{F}(M)$ defined on smooth oriented hypersurfaces $M \subset \mathbf{R}^{n+1}$ and having the general form $\mathcal{F}(M) = \int_M F(x,\nu)\, d\mathcal{H}^n$, where ν is the unit normal of the hypersurface and F is a given function on $\mathbf{R}^{n+1} \times \mathbf{R}^{n+1}$ such that $F(x,q)$ is homogeneous of degree one in the variable q and a C^3 function of the variables for $q \neq 0$; the ellipticity condition in a domain $U \subset \mathbf{R}^{n+1}$ is

3.1 $$(\eta - \nu)\cdot(F_p(x,\eta) - F_p(x,\nu)) \geq \mu|\eta-\nu|^2, \quad \eta,\nu \in S^n, \quad x \in U.$$

Corresponding to such a functional on a domain $U = \Omega \times \mathbf{R}$, with $\Omega \subset \mathbf{R}^n$, there is the non-parametric functional obtained by restricting $\mathcal{F}$ to graphs of functions defined and C^1 over Ω. Thus (since the unit normal of the graph of a function $u \in C^2(\Omega)$, $\Omega \subset \mathbf{R}^n$, is $\nu = (1 + |Du|^2)^{-1/2}(Du, -1)$ and since integration over the graph corresponds to using the volume form $(1 + |Du|^2)^{1/2}\, dx$) this non-parametric functional is given by

$$\mathcal{G}(u) = \mathcal{F}(\operatorname{graph} u) = \int_\Omega F(x, u, Du, -1)\, dx,$$

and the Euler–Lagrange operator $\mathcal{M}_\mathcal{G}$ is

$$\mathcal{M}_\mathcal{G}(u) \equiv \sum_{i=1}^n D_i[A_i(x, u, Du)] - F_z(x, u, Du),$$

where $A_i(x, z, p) = F_{p_i}(x, z, p, -1)$, $i = 1, \dots, n$. We here consider the case when $F(x, z, p)$ is independent of the variable z; in this case the Euler–Lagrange operator takes the form

$$\mathcal{M}_\mathcal{G}(u) = \sum_{i=1}^n D_i[A_i(x, Du)].$$

We also assume that F is positive: $F(x, \nu) > 0$ for all $x \in \overline{\Omega}$, $\nu \in S^n$.

Analogous to the mean curvature equation discussed above we consider the equation

$$\sum_{i=1}^n D_i A_i(x, Du) = H(x, u) \quad \text{in} \quad \Omega \tag{3.2}$$

where again $A_i(x, p) = \partial F(x, p, -1)/\partial p_i$, $(x, p) \in \Omega \times \mathbf{R}^n$, and where Ω is a bounded C^2 domain in $\mathbf{R}^n$.

We now want to claim that all of the results of Theorems 1, 2, 3 carry over to the case when we have the equation in 3.2 instead of the mean curvature equation. Specifically:

Theorem 4. *Subject to precisely the same hypotheses, except that we consider the equation* 3.2 *rather than the mean curvature equation* 0.1, *all the regularity and existence conclusions of Theorems* 1–3 *remain valid.*

The proof of this involves rather routine modifications of the arguments already given. We discuss in detail the points where the arguments are at all different:

First, concerning the generalization of Theorem 1 we first need to derive the analogous identities to 1.3, 1.5, 1.6. This we do as follows. Differentiating the equation 3.2 with respect to the variable x^l gives the identity

$$\textbf{3.3} \qquad \sum_{i,j=1}^{n} D_i\big(v^{-1}a^{ij}(x,Du)D_j u_l\big) = H_{x^l}(x,u) + H_z(x,u)u_l,$$

(Cf. 1.3), where $a^{ij}(x,p) = (1+|p|^2)^{1/2}F_{p_ip_j}(x,p,-1)$. Then by multiplying by ν^l (as we did in the derivation of 1.5) we obtain

$$\textbf{3.4} \quad \sum_{i,j=1}^{n} D_i\big(v^{-1}a^{ij}(x,Du)D_j v\big) = \mathcal{C}_a^2 + \sum_l \nu^l H_{x^l}(x,u) + H_z(x,u)|Du|^2,$$

where

$$\textbf{3.5} \qquad \mathcal{C}_a^2 = v^{-2}\sum_{i,j,r,s=1}^{n} a^{ij}g^{rs}u_{ir}u_{js}.$$

Further, using the ellipticity 3.1 of the functional F, it is straightforward to check (see [SL2] for the details) that

$$\textbf{3.6} \qquad C^{-1}\sum_{i,j} g^{ij}\xi_i\xi_j \le \sum_{i,j} a^{ij}\xi_i\xi_j \le C\sum_{i,j} g^{ij}\xi_i\xi_j, \quad \xi\in\mathbf{R}^n$$

with C a fixed constant only depending on the functional $\mathcal{F}$ and not depending on the particular function u. It then easily follows from the identity 3.5 that

$$\textbf{3.7} \qquad C^{-1}\mathcal{C}^2 \le \mathcal{C}_a^2 \le C\mathcal{C}^2,$$

where $\mathcal{C}^2$ is the sum of squares of principal curvatures of graph u, as in §1, and where again C depends only on $\mathcal{F}$ and not on u.

Now using 3.3–3.7 it is very straightforward to modify the argument used in the proof of Theorem 1 in the case when we have equation 3.2 in place of 0.1.

Next we want to establish Theorems 2 and 3 in case we have equation 3.2 in place of 0.1. First we have to discuss the existence and estimates for the functions w_ϵ used in the proof of Theorem 2 in case we have equation 3.2 in place of 0.1. In fact it is necessary to consider the "natural" boundary condition instead of the contact angle boundary condition for w_ϵ in the subdomain $\widetilde{\Omega}$ ($\widetilde{\Omega}$ as in the proof of Theorem 2); thus in the present context we define w_ϵ to be the solution of the problem

$$\sum_{i=1}^{n} D_i A_i(x, Dw) = H(x, w) \quad \text{in} \quad \widetilde{\Omega} \tag{1}$$

$$\widetilde{\eta} \cdot A(x, Dw) = \gamma_\epsilon(x) \quad \text{on} \quad \partial\widetilde{\Omega}, \tag{2}$$

where $\gamma_\epsilon \equiv (1-\epsilon)F(x, v(x))$ on $\partial\widetilde{\Omega}$, where v is a smooth vector field in $\mathbf{R}^{n+1}$ such that $v = (1+|Du|^2)^{-1}(Du, -1)$ at points of $\partial\widetilde{\Omega} \cap \Omega$ which are distance $> \epsilon$ from Σ, and $v = \widetilde{\eta} \cdot F_p(\widetilde{\eta}, 0)$ at points of $\partial\widetilde{\Omega} \cap \partial\Omega$ which are at distance ϵ from Σ (notation as in proof of Theorem 2).

It is important to note here that this boundary condition guarantees that we still get the zero contact angle condition on $\partial\Omega$ as $\epsilon \downarrow 0$, but still implies contact angle bounded away from $0, \pi$ (depending on ϵ). To see this, we recall some inequalities related to the homogeneity and ellipticity of F. First, using this homogeneity and ellipticity, one can check by a simple calculus argument (see [SSA, p. 221]) that

$$C^{-1}|q_1 - q_2|^2 \leq q_1 \cdot (F_p(x, q_1) - F_p(x, q_2)) \leq C|q_1 - q_2|^2 \tag{3}$$

for any unit vectors $q_1, q_2 \in \mathbf{R}^{n+1}$. By virtue of the first inequality in (3) we have

$$\sup_{q \in S^n} q_1 \cdot F_p(x, q) = q_1 \cdot F_p(x, q_1) \equiv F(x, q_1).$$

Hence if $\gamma \in (0, 1)$ then, using also the second inequality in (3) and the fact that $F > 0$, we have for any $q_3 \in S^n$

$$\begin{aligned} q_1 \cdot (\gamma F_p(x, q_3) - F_p(x, q_2)) &\leq q_1 \cdot (\gamma F_p(x, q_1) - F_p(x, q_2)) \\ &= q_1 \cdot (F_p(x, q_1) - F_p(x, q_2)) - (1-\gamma)F(x, q_1) \\ &\leq C|q_1 - q_2|^2 - C^{-1}(1-\gamma) \end{aligned}$$

for suitable C depending only on F, and also (by the same argument with $-q_1$ in place of q_1)

$$(-q_1) \cdot (\gamma F_p(x, q_3) - F_p(x, q_2)) \leq C|q_1 + q_2|^2 - C^{-1}(1-\gamma),$$

so that in particular

$$\begin{aligned} &q_1, q_2, q_3 \in S^n, \gamma \in (0,1) \text{ and } q_1 \cdot (\gamma F_p(x, q_3) - F_p(x, q_2)) = 0 \Rightarrow \\ &\qquad -1 + C^{-1}(1-\gamma) \leq q_1 \cdot q_2 \leq 1 - C^{-1}(1-\gamma), \end{aligned} \tag{4}$$

where C is a constant determined by F alone. Also by using both inequalities in (3) we have

$$
\begin{aligned}
&q_1, q_2 \in S^n \quad \text{and} \quad q_1 \cdot (\gamma F_p(x,q_1) - F_p(x,q_2)) = 0 \Rightarrow \\
&\qquad\qquad C^{-1}(1-\gamma) \le |q_1 - q_2|^2 \le C(1-\gamma).
\end{aligned} \tag{5}
$$

In particular, assuming the boundary condition of (2) for w, we see that by (5)

$$
\left|(\tilde{\eta}, 0) - (1+|Dw|^2)^{-1/2}(Dw, -1)\right|^2 \le C\epsilon \tag{6}
$$

at points of $\partial\Omega \cap \partial\tilde{\Omega}$ at distance $> \epsilon$ from Σ, which says that the contact angle is close to zero as claimed, and by (4) we have

$$
-1 + C^{-1}\epsilon \le (\tilde{\eta}, 0) \cdot (1+|Dw|^2)^{-1/2}(Dw, -1) \le 1 - C^{-1}\epsilon, \tag{7}
$$

everywhere on $\partial\tilde{\Omega}$, which says that the contact angle is bounded away from zero and π (depending on ϵ) as claimed.

Now the existence, regularity and uniqueness properties needed for w_ϵ follow from routine modifications of the arguments described in the proof of Theorem 2. Indeed the proof of the global height bound for w_ϵ (independent of ϵ) is an easy modification of the argument in [CF], and the local interior gradient estimates (and hence $C^{2,\alpha}$ estimates) are established for the equations 3.2 in [SL2]. The proof of existence of w_ϵ for each $\epsilon \in (\frac{1}{2}, 1)$ is proved by the same method as in [UN2] to get a global gradient estimate for w_ϵ which depends on ϵ and consequently global $C^{2,\alpha}$ estimates depending on ϵ. Then as before the Leray–Schauder existence theory can be applied as for example in [UN1, 2]. The remainder of the argument to establish that the solution u has graph with closure which is contained in a $C^{1,1}$ manifold goes as before; indeed in the remainder of the argument only the proof of the tangential gradient estimate requires anything more than very minor modification. The proof of the tangential gradient estimate can also be achieved by fairly routine modification of the arguments used in [SS] for the mean curvature case.

Finally, as in §2, Theorem 3 also follows by specializing the construction above to the case when $\tilde{\Omega} = \Omega$.

4. Concluding Remarks.

Finally we want to mention that the main estimate (i.e. the estimate of Theorem 1) extends to a very large class of divergence-form elliptic equations, ranging from the equations considered above to the uniformly elliptic equations of divergence form.

Indeed the argument described here generalizes without difficulty to any of the very large class of equations considered in [SL2], provided the growth function Λ considered there has no more than polynomial growth (i.e. $\Lambda(p) \leq C(1 + |p|)^k$ for some constants C, k). We do not discuss the details here.

References.

[BDM] E. Bombieri, E. De Giorgi & M. Miranda, *Una maggiorazzione a priori relativa alle ipersuperfici minimali non parametriche*, Arch. Rat. Mech. Anal. **32** (1969), 255–267.

[CF] P. Concus & R. Finn, *On capillary surfaces in a gravitation field*, Acta Math. **132** (1974), 207–223.

[CL] L. Caffarelli, *The regularity of free boundaries in higher dimensions*, Acta Math. **139** (1977), 155–184.

[FG] R. Finn & D. Gilbarg, *Subsonic flows*, Comm. Pure Appl. Math. **10** (1957), 23–63.

[GC] C. Gerhardt, *Regularity of solutions of nonlinear variational inequalities*, Arch. Rational Mech. Anal. **52** (1973), 389–393.

[GT] D. Gilbarg & N. Trudinger, Elliptic Partial Differential Equations of Second Order (2nd Edition), Springer–Verlag, (1983).

[HS] R. Hardt & L. Simon, *Boundary regularity and embedded solutions for the oriented Plateau problem*, Annals of Math. **110** (1979), 439–486.

[KN] D. Kinderlehrer & L. Nirenberg, *Regularity of free boundary problems*, Ann. Scuola Norm. Sup. Pisa **4** (1977), 373–391.

[KS] N. Korevaar & L. Simon, *Continuity estimates for solutions to the prescribed curvature Dirichlet problem*, Math. Zeit. **197** (1989), 457–464.

[MS] J. H. Michael & L. Simon, *Sobolev and Mean-Value inequalities for generalized submanifolds of Riemannian Manifolds*, Comm. Pure and Applied Math. **26** (1973), 361–379.

[SS] L. Simon & J. Spruck, *Existence and regularity of solutions of the capillary surface equation*, Archive for Rat. Mech. and Analysis **61** (1976), 19–34.

[SSA] R. Schoen, L. Simon & F. Almgren, *Regularity and Singularity Estimates for hypersurfaces minimizing parametric elliptic variational integrals*, Acta Math. **139** (1977), 217–265.

[SL1] L. Simon, Lectures on Geometric Measure Theory, Proceedings of the Centre for Mathematical Analysis, Australian National University, **3** (1983).

[SL2] L. Simon, *Interior Gradient Estimates for non-uniformly elliptic equations*, Indiana Univ. Math. Journal **25** (1976), 821–855.

[SL3] L. Simon, *Boundary regularity for solutions of the non-parametric least area problem*, Annals of Math. **103** (1976), 429–455.

[UN1] N. Ural'tseva, *Solvability of the capillarity problem*, Vestnik Leningrad Univ. No. 19, Mat. Meh. Astronom. Vyp. **4** (1973).

[UN2] N. Ural'tseva, *Solvability of the capillarity problem II*, Vestnik Leningrad Univ. No. 1, Mat. Meh. Astronom. Vyp. **1** (1975).

RECEIVED JUNE 2, 1995.

Weak Limits in the Free Boundary Problem for Immersions of the Disk which Minimize a Conformally Invariant Integral

ERNST KUWERT

Introduction.

Let $K \subset \mathbb{R}^n$ be a Lipschitz neighborhood retract and let S be a closed subset of K. In this paper we consider the problem of minimizing a conformally invariant integral $F(\cdot)$ among maps $X : B = \{w = u + iv \in \mathbb{C} : u^2 + v^2 < 1\} \to K$ whose boundary $x = X|_{\partial B}$ belongs to a given homotopy class $\alpha \in \widehat{\pi}_1(S)$ of free loops in S.

In general $F_*(\alpha) := \inf\{F(X) : x \in \alpha\}$ need not be attained; in other words, Dirichlet's principle fails in this setting. We say that $X \in H^{1,2}(B, K)$ is an *interior minimizer* of F if and only if $F(X) = \inf\{F(Y) : Y \in H^{1,2}(B, K), x = y\}$. Let X_k be a minimizing sequence for α, and assume that the sequence is *improved* in the sense that each X_k minimizes F in the interior. In general it may then happen that *the sequence* (X_k) *decomposes into a certain number of disks in the limit*; this is the boundary analogue of the phenomenon of separation of two-spheres studied, for example, by Sacks–Uhlenbeck [16]. In [12] a detailed analysis of the decomposition process is given; here we confine ourselves to the presentation of a simpler result. Although the implied existence statement has been known for a long time [5, Ch.5], we believe that the result is interesting in itself, in particular due to its geometric proof. Under appropriate conditions on F (see p.3), we obtain the following

Theorem. *If the improved minimizing sequence X_k converges weakly to X, then X minimizes F in its own boundary homotopy class $\{x\}$.*

The theorem thus provides evidence for the generally accepted principle that weak limits should minimize "with respect to their own constraint" in problems of this type. The validity of this principle was settled for the volume-constrained Plateau problem by Wente [18] and for the Dirichlet-homotopy problem for harmonic maps from a surface into a two-sphere by Brézis–Coron [3], see also [11]. In these situations the constraint can be written as an analytic expression whose jump in the weak limit is related to the jump of energy by an inequality. Since this is not possible in the present problem, the proof requires a certain geometric construction which is actually a refinement of an idea due to Courant [4]; a beautiful presentation of his argument can be found in [5, Ch.5]. The existence question was first solved—under more restrictive assumptions on K, S, F—by Ye [20], whose approach is partially based on prior work by Tolksdorf [17]. A brief discussion of the problem can also be found in a monograph by Jost [10, pp.147–149]. Béthuel recently showed that weak limits of sequences satisfying a certain Palais–Smale type condition are stationary in the interior [2]. In joint work with F.Duzaar we have obtained a result corresponding to the above for homotopic minimizing sequences of maps from a closed surface into K [6].

1. Preliminaries.

In this paper we assume that $K \subset \mathbb{R}^n$ is closed and admits a retraction $\Pi : U_{\delta_0}(K) = \{p \in \mathbb{R}^n : \mathrm{dist}(p, K) < \delta_0\} \to K$ with Lipschitz constant $\mathrm{Lip}(\Pi) = L < \infty$; here $\mathrm{dist}(\cdot, K)$ is the euclidian distance from K. We also assume that $U_{\delta_0}(K)$ is connected, so that for any two points $p, q \in K$ there is a geodesic in K realizing the intrinsic distance $\mathrm{dist}_K(p, q) < \infty$. Let $S \subset K$ be closed and $U_\delta(S, K) = \{p \in K : \mathrm{dist}_K(p, S) < \delta\}$. In order to obtain a meaningful definition of homotopy classes for loops belonging to the trace space $H^{1/2}(\mathbb{R}/2\pi, S)$ we have to replace the standard notion $\pi_1(S)$, i.e. homotopy classes of maps $x \in C^0(\mathbb{R}/2\pi, S)$, by the inverse limit $\widehat{\pi}_1(S) = \lim_{\delta \to 0} \pi_1(U_\delta(S))$. This is described in detail in [12, Sec.1]; here we outline the procedure. For $x \in C^0(\mathbb{R}/2\pi, \mathbb{R}^n)$ we put $\mathrm{dist}(x, S) = \max\{\mathrm{dist}(x(\vartheta), S) : \vartheta \in \mathbb{R}/2\pi\}$ and introduce the relation

$$x \overset{\delta}{\sim} y \Leftrightarrow \mathrm{dist}(x, S) < \delta,\ \mathrm{dist}(y, S) < \delta, \quad x \text{ is homotopic to } y \text{ in } U_\delta(S). \tag{1}$$

We then consider the sequence space $\Pi_1(S) = \{\mathbf{x} = (x_k) : \text{for any } \delta > 0$ there is a $k_0 \in \mathbb{N}$ such that $x_k \overset{\delta}{\sim} x_l$ for $k, l \geq k_0\}$ and define $\mathbf{x}, \mathbf{y} \in \Pi_1(S)$ to be equivalent if and only if for any $\delta > 0$ we have $x_k \overset{\delta}{\sim} y_l$ for k, l sufficiently large; now we put $\widehat{\pi}_1(S) = \Pi_1(S)/\sim$. The inclusion $S \subset U_\delta(S)$ induces a well-defined map

$$i_\delta^K : \widehat{\pi}_1(S) \to \pi_1(U_\delta(S,K)), \tag{2}$$

and we also have a natural map $i : \pi_1(S) \to \widehat{\pi}_1(S)$. In particular we can define a distance function and a size on $\widehat{\pi}_1(S)$ by letting

$$\begin{aligned} \delta_K(\alpha,\beta) &= \inf\{\delta > 0 : i_\delta^K(\alpha) = i_\delta^K(\beta)\} \in [0,\infty], & (3)\\ \|\alpha\|_K &= \inf\{\delta > 0, : i_\delta^K(\alpha) \text{ contains a constant loop}\} \in [0,\infty]. & (4) \end{aligned}$$

Note that $\delta(\alpha,\beta) \le \delta_K(\alpha,\beta)$, $\|\alpha\| \le \|\alpha\|_K$ (where omitting the subscript K means $K = \mathbb{R}^n$) and conversely $\delta_K(\alpha,\beta) \le (L+2)\delta(\alpha,\beta)$, $\|\alpha\|_K \le (L+2)\|\alpha\|$ if $\delta(\alpha,\beta) \le \delta_0$, $\|\alpha\| \le \delta_0$. To obtain the latter inequalities we use that if $\operatorname{dist}(p,K) \le \frac{\delta}{L+1} \le \delta_0$, then there is a $q \in K$ with $|p-q| \le \frac{\delta}{L+1}$ and thus $|\Pi(p)-p| \le |\Pi(p)-q| + |p-q| = |\Pi(p)-\Pi(q)| + |p-q| \le (L+1)|p-q| \le \delta$. By the same token we can show that if S admits a uniformly continuous retraction $r : U_{\delta_1}(S) \to S$, then $i : \pi_1(S) \to \widehat{\pi}_1(S)$ is bijective and $\delta(\alpha,\beta) \ge \delta_1$ for $\alpha,\beta \in \widehat{\pi}_1(S)$, $\alpha \ne \beta$.

The relevant function spaces $H^{1/2}(\mathbb{R}/2\pi, S)$, $H^1(B,K)$ can be introduced in a similar fashion. For example, to define $H^{1/2}(\mathbb{R}/2\pi, S)$ we consider the sequence space $\mathcal{H}^{1/2}(S) := \{\mathbf{x} = (x_k) : x_k \in (H^{1/2} \cap C^0)(\mathbb{R}/2\pi, \mathbb{R}^n)$, $\operatorname{dist}(x_k, S) \to 0$, for all $\varepsilon > 0$ there is a $k_0 \in \mathbb{N}$ such that $\|x_k - x_l\|_{H^{1/2}} \le \varepsilon$ for $k,l \ge k_0\}$ and divide by the equivalence relation $\mathbf{x} \sim \mathbf{y} \Leftrightarrow \|x_k - y_k\|_{H^{1/2}} \to 0$. Alternatively we could require that the maps take their values in S (K respectively) almost everywhere; the equivalence of both definitions follows from local Poincaré inequalities. Based on an idea of White [19] the author proved in [12, Thm.1] the following

Lemma 1. *$\mathcal{H}^{1/2}(S) \subset \Pi_1(S)$, and this inclusion induces a continuous assignment $\{\cdot\} : H^{1/2}(\mathbb{R}/2\pi, S) \to (\widehat{\pi}_1(S), \delta_K(\cdot,\cdot))$ of a homotopy class $\{x\}$ to any $x \in H^{1/2}(\mathbb{R}/2\pi, S)$. The map $\{\cdot\}$ is constant on the path components of $H^{1/2}(\mathbb{R}/2\pi, S)$.*

Let $F(\cdot,\Omega) : H^1(\Omega,K) \to \mathbb{R}_0^+$ be a functional defined for all bounded domains $\Omega \subset \mathbb{C}$. We shall impose the following conditions on F:

(5) For any biholomorphic equivalence $\Phi : \Omega \xrightarrow{\sim} \Phi(\Omega)$ and any $Y \in H^1(\Phi(\Omega),K)$ we have $F(Y,\Phi(\Omega)) = F(Y \circ \Phi, \Omega)$ (*conformal invariance*).

(6) There are constants $0 < \lambda \leq \Lambda < \infty$ such that $\lambda E(X) \leq F(X) \leq \Lambda E(X)$ for all $X \in H^1(B,K)$; here $E(X) = \frac{1}{2}\int_B |dX|^2\,du\,dv$ is Dirichlet's integral and $F(X) = F(X,B)$ (*coercivity and boundedness*).

(7) If $X_k, X \in H^1(B,K)$ satisfy $X_k|_{\partial B} \equiv X|_{\partial B} = x \in C^0(\mathbb{R}/2\pi, K)$, and if $X_k \to X$ weakly in $H^1(B,\mathbb{R}^n)$, then $F(X) \leq \liminf_{k\to\infty} F(X_k)$ (*weak sequential lower semicontinuity*).

A functional of the type

$$F(X) = \frac{1}{2}\int_B [g(X)(X_u,X_u)+g(X)(X_v,X_v)]\,du\,dv + \int_B \langle \eta(X), X_u \wedge X_v\rangle\,du\,dv$$

satisfies (5), (6) and (7) if $g : K \to \mathrm{Sym}^2(\mathbb{R}^n)$, $\lambda + \|\eta(p)\| \leq \lambda(g(p)) \leq \Lambda(g(p)) \leq \Lambda - \|\eta(p)\|$ hold; here $\|\eta\|$ denotes the comass norm and $\lambda(g), \Lambda(g)$ are the smallest and biggest eigenvalue of g.

Let $\alpha \in \widehat{\pi}_1(S)$ be given and put

$$F_*(\alpha) = \inf\{\liminf_{k\to\infty} F(X_k) : X_k \in H^1(B,K), (X_k|_{\partial B} = x_k) \in \alpha\} \tag{8}$$

We assume without further discussion $F_*(\alpha) < \infty$. A *minimizing sequence for* α is a sequence X_k with $(x_k) \in \alpha$ and $F(X_k) \to F_*(\alpha)$; the sequence will be called *improved* if X_k minimizes F in the interior. Using (6), (7) and a diagonal argument [12, Sec.1, Lemma 9] we can show that an improved minimizing sequence can always be found. For $X : \Omega \to K$ and $0 < \mu < 1$ the intrinsic Hölder seminorm is given by

$$\text{höl}_\mu(X,\Omega) = \sup\left\{\frac{\mathrm{dist}_K(X(w),X(w'))}{|w-w'|^\mu} : w, w' \in \Omega, w \neq w'\right\}. \tag{9}$$

The following statements are proved in [12, Sec. 1, Sec. 6] and are essentially due to Morrey [14, pp.380–381].

Lemma 2. *Suppose that* $X \in H^1(B,K)$ *minimizes* F *in the interior and that* $F(X) \leq D < \infty$. *Then the following hold:*

(10) *For* $0 < d < 1$ *we have* $\text{höl}_\mu(X, B_{1-d}) \leq \frac{c_0\sqrt{D}}{d\mu}$, *where* $\mu = \mu_0 \min\{\frac{1}{D}, 1\}$, $\mu_0 = \min\{\frac{\lambda\pi\delta_0^2}{2}, \frac{\lambda}{2\Lambda L^2}\} \in (0, \frac{1}{2}]$ *and* $c_0 = 16\sqrt{\frac{2}{\pi\lambda}}$.

(11) $\mathrm{dist}_K(X,S) \leq \frac{c_0}{\mu_0}\max\{\sqrt{D}, \sqrt{D}^3\}$.

(12) $\lim_{|w|\to 1} \operatorname{dist}_K(X(w), S) = 0$, *and for any sequence* $r_k < 1$ *with* $r_k \to 1$ *the sequence* $(x_k(\vartheta) = X(r_k e^{i\vartheta}))$ *is admissible in the definition of* $\{x\} \in \widehat{\pi}_1(S)$ *as given in lemma 1.*

Corollary 1. *For* $\alpha \in \widehat{\pi}_1(S)$ *and* $a_1 := \min\{(\frac{\mu_0}{c_0})^2, (\frac{\mu_0}{c_0})^{2/3}\}$ *we have*

$$F_*(\alpha) \geq a_1 \min\{\|\alpha\|_K^2, \|\alpha\|_K^{2/3}\}. \tag{13}$$

If $K = \mathbb{R}^n$ *and if* $F(X) = E(X)$, *then the estimates* (11), (13) *improve to* $\operatorname{dist}(X, S) \leq \sqrt{\frac{E(X)}{\pi}}$ *(for harmonic* X*) and* $E_*(\alpha) \geq \pi\|\alpha\|^2$. *We do not know whether one can improve* (13) *to an estimate with linear growth for* $\|\alpha\|_K$ *large in the considered generality.*

We shall also need the following special case of a result due to Luckhaus [13]; see also [12, Sec.6, Lemma 4].

Lemma 3. *Let* $X_k \in H^1(B, K)$ *be a sequence of interior minimizers for* F *which converges weakly to* $X \in H^1(B, K)$. *Then for any* $R < 1$

$$F(X, B_R(0)) = \lim_{k\to\infty} F(X_k, B_R(0)). \tag{14}$$

2. Two cut-and-paste lemmas.

The first lemma is a standard consequence of the coarea formula and is proved for example in [5, Sec.5.1]; see also [12, Sec.2, Lemma 2].

Lemma 4. *Let* $X_k \in (H^1 \cap C^0)(B, \mathbb{R}^n)$ *be a sequence with* $E(X_k) \leq \frac{1}{2}E < \infty$ *for all* k. *Then there is a set* $I \subset \mathbb{R}/2\pi$ *of measure at least* π *with the following property: for any* $\theta \in I$ *there exists a subsequence* $\mathcal{K} = (k(l, \theta))$ *such that for all* $k \in \mathcal{K}$ *we have* $X_k(\cdot, \theta) \in H^1((\frac{1}{2}, 1), \mathbb{R}^n)$ *in polar coordinates, and in fact*

$$\int_{1/2}^1 \left|\frac{\partial X_k}{\partial r}\right|^2 (r, \theta)\, dr \leq \frac{2E}{\pi}; \text{ furthermore} \tag{15}$$

$$|X_k(r, \theta) - X_k(r', \theta)|^2 \leq \frac{2E}{\pi}|r - r'| \text{ for all } r, r' \in \left(\tfrac{1}{2}, 1\right). \tag{16}$$

Lemma 5. *Let $u \in C^0([-\varepsilon,\varepsilon],\mathbb{R}^n)$ and define $U : [0,\varepsilon] \times [-\varepsilon,\varepsilon] \to \mathbb{R}^n$ by $U(s,t) = u(\frac{\varepsilon-s}{\varepsilon}t)$. Then the curves $U(s,\cdot)$ are a continuous deformation of $U(0,\cdot) = u$ into $U(\varepsilon,\cdot) \equiv u(0)$ with $\mathrm{im}(U) = \mathrm{im}(u)$, $U(s,\varepsilon) = u(\varepsilon - s)$ and $U(s,-\varepsilon) = u(s-\varepsilon)$. In addition we have the estimate*

$$E(U) \leq \varepsilon \int_{-\varepsilon}^{\varepsilon} |\dot{u}(t)|^2\, dt. \tag{17}$$

Proof. We have $\frac{\partial U}{\partial s}(s,t) = -\frac{t}{\varepsilon}\dot{u}(\frac{\varepsilon-s}{\varepsilon}t)$, $\frac{\partial U}{\partial t}(s,t) = \frac{\varepsilon-s}{\varepsilon}\dot{u}(\frac{\varepsilon-s}{\varepsilon}t)$. Therefore

$$\begin{aligned} E(U) &= \frac{1}{2}\int_0^{\varepsilon}\int_{-\varepsilon}^{\varepsilon}\left\{\left(\frac{t}{\varepsilon}\right)^2 + \left(\frac{\varepsilon-s}{\varepsilon}\right)^2\right\}\left|\dot{u}\left(\frac{\varepsilon-s}{\varepsilon}t\right)\right|^2 dt\, ds \\ &=: \frac{1}{2\varepsilon^2}\{I_1 + I_2\}, \end{aligned}$$

where

$$I_1 = \int_{-\varepsilon}^{\varepsilon}\int_0^{\varepsilon} t^2 \left|\dot{u}\left(\frac{\varepsilon-s}{\varepsilon}t\right)\right|^2 ds\, dt,\ I_2 = \int_0^{\varepsilon}\int_{-\varepsilon}^{\varepsilon}(\varepsilon-s)^2\left|\dot{u}\left(\frac{\varepsilon-s}{\varepsilon}t\right)\right|^2 dt\, ds.$$

We treat the two integrals I_1, I_2 separately. In I_1 we use $\sigma := t\frac{\varepsilon-s}{\varepsilon}$ instead of s so that $\int_0^{\varepsilon}|\dot{u}(\frac{\varepsilon-s}{\varepsilon}t)|^2\, ds = -\frac{\varepsilon}{t}\int_t^0 |\dot{u}(\sigma)|^2\, d\sigma \leq \frac{\varepsilon}{|t|}\int_{-\varepsilon}^{\varepsilon}|\dot{u}(\sigma)|^2\, d\sigma$, whence

$$I_1 \leq \int_{-\varepsilon}^{\varepsilon} t^2 \frac{\varepsilon}{|t|}\int_{-\varepsilon}^{\varepsilon}|\dot{u}(\sigma)|^2\, d\sigma\, dt = \varepsilon^3 \int_{-\varepsilon}^{\varepsilon}|\dot{u}(\sigma)|^2\, d\sigma.$$

In I_2 we substitute $\frac{\varepsilon-s}{\varepsilon}t =: \tau$ for t and have

$$\int_{-\varepsilon}^{\varepsilon}|\dot{u}(\frac{\varepsilon-s}{\varepsilon}t)|^2\, dt = \frac{\varepsilon}{\varepsilon-s}\int_{s-\varepsilon}^{\varepsilon-s}|\dot{u}(\tau)|^2\, d\tau \leq \frac{\varepsilon}{\varepsilon-s}\int_{-\varepsilon}^{\varepsilon}|\dot{u}(\tau)|^2\, d\tau.$$

Thus

$$I_2 \leq \int_0^{\varepsilon}(\varepsilon-s)^2\frac{\varepsilon}{\varepsilon-s}\int_{-\varepsilon}^{\varepsilon}|\dot{u}(\tau)|^2\, d\tau\, ds = \frac{\varepsilon^3}{2}\int_{-\varepsilon}^{\varepsilon}|\dot{u}(\tau)|^2\, d\tau,$$

and statement (17) follows. □

3. Proof of the theorem.

Theorem. *Let $K \subset \mathbb{R}^n$ be a Lipschitz neighborhood retract with retraction $\Pi : U_{\delta_0}(K) \to K$, $\mathrm{Lip}(\Pi) = L < \infty$, and suppose $U_{\delta_0}(K)$ is connected. Let $S \subset K$ be closed and let F be a functional satisfying* (5), (6), *and* (7). *Suppose that $X_k \in H^1(B, K)$ is an improved minimizing sequence for $\alpha \in \widehat{\pi}_1(S)$. If X_k converges weakly to X, then X minimizes F in its own boundary homotopy class:*

$$F(X) = F_*(\{x\}). \tag{18}$$

Proof. We start with the following remark: if $X \in H^1(\Omega, K)$ and if $\Omega = \Omega_1 \cup \Omega_2 \cup \Sigma$, where Ω_1, Ω_2 are open and disjoint and $\mathcal{L}^2(\Sigma) = 0$, then $F(X, \Omega) = F(X, \Omega_1) + F(X, \Omega_2)$. This follows from the fact that using (6) we can write $F(X, \Omega) = \int_\Omega \theta_X \, du\, dv$, where the density function θ_X belongs to $L^1(\Omega)$. By (12) we can find a sequence of radii (r_l), $0 < 1 - r_l \le \frac{1}{l}$, such that $F(X, B \setminus B_{r_l}) \le \frac{1}{2l}$ and $\max\{\mathrm{dist}_K(X(w), S) : r_l \le |w| < 1\} \le \frac{1}{2l}$. By (10) X_k converges to X locally uniformly on B, and by (14) the convergence is also in F locally in B (we do not use locally strong convergence in $H^1(B, \mathbb{R}^n)$). We can select a subsequence $\mathcal{K} = (k(l))$ such that

$$|F(X_k, B_{r_l}) - F(X)| \le \frac{1}{l} \text{ for } k = k(l), \text{ and} \tag{19}$$

$$\max\{\mathrm{dist}_K(X_k(r_l e^{i\vartheta}), X(r_l e^{i\vartheta})) : \vartheta \in \mathbb{R}/2\pi\} \le \frac{1}{2l}, \tag{20}$$

again for $k = k(l)$. Renumbering, we can assume $k(l) = l$, i.e. (19), (20) hold with l replaced by k. Now let $\xi_k(\vartheta) = X_k(r_k e^{i\vartheta})$; then $\mathrm{dist}_K(\xi_k(\vartheta), X(r_k e^{i\vartheta})) \le \frac{1}{2k}$. Let $\widetilde{H}_k$ be the affine homotopy between ξ_k and $X(r_k, \cdot)$. Then $\mathrm{dist}(\widetilde{H}_k, K) \le \frac{1}{4k}$ so that for $k > \frac{1}{4\delta_0}$ we can consider the homotopy $H_k := \Pi \circ \widetilde{H}_k$ in K. We have $\mathrm{dist}_K(H_k, S) \le \max\{L|\xi_k(\vartheta) - X(r_k e^{i\vartheta})| + \mathrm{dist}_K(X(r_k e^{i\vartheta}), S) : \vartheta \in \mathbb{R}/2\pi\} \le \frac{L+1}{2k}$. Recalling (12) we obtain

$$\begin{aligned} &(\xi_k(\vartheta) = X_k(r_k e^{i\vartheta})) \text{ is a representative of } \{x\} \in \widehat{\pi}_1(S) \text{ with} \\ &\mathrm{dist}_K(\xi_k, S) \le \frac{1}{k} \text{ and } \xi_k \sim \xi_l \text{ in} \\ &U_\delta(S, K) \text{ for } k, l > \max\left\{\tfrac{1}{4\delta_0}, \tfrac{L+1}{2\delta}\right\}. \end{aligned} \tag{21}$$

By (6) we have $E(X_k) \le \frac{1}{2}E$ for some $E < \infty$. According to lemma 4 there exists an angle $\theta \in \mathbb{R}/2\pi$ such that after passing to a further subsequence we additionally have

$$\int_{1/2}^{1} \left|\frac{\partial X}{\partial r}\right|^2 (r,\theta)\, dr \leq \frac{2E}{\pi}, \tag{22}$$

$$\max\{\mathrm{dist}_K(X_k(re^{i\vartheta}), S) : r_k \leq r \leq 1\} \leq \frac{1}{k}, \text{ and} \tag{23}$$

$$q_k := X_k(\sqrt{r_k}e^{i\vartheta}) \to q \in S. \tag{24}$$

We introduce $D_k := \{re^{i\vartheta} : r_k < r < 1, \vartheta \neq \theta\}$ and put

$$F_k := F(X_k, D_k). \tag{25}$$

Let $\varepsilon_k \in (0, \log \frac{1}{\sqrt{r_k}})$ be a parameter to be chosen later and consider the rectangle $Q_k := \{\zeta = s + it \in \mathbb{C} : \varepsilon_k < s < 2\pi + \varepsilon_k, \log \sqrt{r_k} < t < \log \frac{1}{\sqrt{r_k}}\}$. The map $\Phi(\zeta) := \exp\left[-\log \frac{1}{\sqrt{r_k}} + i(\zeta + \theta - \varepsilon_k)\right]$ is the biholomorphic equivalence between Q_k and D_k; note that

$$\Phi_k(\varepsilon_k + it) = \Phi_k(2\pi + \varepsilon_k + it) \text{ for all } t. \tag{26}$$

The four corners $\mathbf{1}_k = 2\pi + \varepsilon_k - i\log\frac{1}{\sqrt{r_k}}$, $\mathbf{2}_k = 2\pi + \varepsilon_k + i\log\frac{1}{\sqrt{r_k}}$, $\mathbf{3}_k = \varepsilon_k + i\log\frac{1}{\sqrt{r_k}}$, $\mathbf{4}_k = \varepsilon_k - i\log\frac{1}{\sqrt{r_k}}$ of Q_k correspond to the four corners of D_k under Φ_k. On Q_k we obtain the map $Z_k := X_k \circ \Phi_k$.

We now turn to comparison maps for X. We can assume that we have an improved minimizing sequence $\widetilde{Y}_k \in H^1(B, K)$ such that

$$F(\widetilde{Y}_k) \leq F_*(\{x\}) + \frac{1}{k}, \text{ and} \tag{27}$$

$$\widetilde{y}_k \sim \widetilde{y}_l \text{ in } U_\delta(S, K) \text{ for } k, l \geq \tfrac{1}{\delta}. \tag{28}$$

Replacing $\widetilde{Y}_k(w)$ by $\widetilde{Y}_k(\rho_k w)$ with appropriate $\rho_k \in (0,1)$, $\rho_k \to 1$, and using (12) we can also assume that

$$\widetilde{L}_k := \int_0^{2\pi} |\widetilde{y}_k'(\vartheta)|^2 \, d\vartheta < \infty.$$

Let $P_k := \{\zeta = s + it : -2 - \varepsilon_k < s < -1 - \varepsilon_k, -1/2 < t < 1/2\}$ and denote by $\wp_k : P_k \to B$ a biholomorphic equivalence; note that $\wp_k$ is an appropriately scaled Weierstraß $\wp$-function. Let $Y_k := \widetilde{Y}_k \circ \wp_k$ and observe that we still have

$$L_k := \int_{-1/2}^{1/2} \left|\frac{\partial Y_k}{\partial t}\right|^2 (-1 - \varepsilon_k + it)\, dt < \infty, \tag{29}$$

with L_k *independent of* ε_k.

We extend Y_k, Z_k to $P_{\varepsilon_k} := [-1-\varepsilon_k, -1] \times [-\varepsilon_k, \varepsilon_k]$ and $Q_{\varepsilon_k} := [0, \varepsilon_k] \times [-\varepsilon_k, \varepsilon_k]$ (respectively) by means of the construction of lemma 5. Then from (17) we have the estimate

$$E(Y_k, P_{\varepsilon_k}) \leq \varepsilon_k L_k, \tag{30}$$

Furthermore

$$\begin{aligned} E(Z_k, Q_{\varepsilon_k}) &\leq \varepsilon_k \int_{\log\sqrt{r_k}}^{\log\frac{1}{\sqrt{r_k}}} \left|\frac{\partial Z_k}{\partial t}\right|^2 (\varepsilon_k + it)\, dt \\ &= \varepsilon_k \int_{r_k}^{1} \left|\frac{\partial X_k}{\partial r}\right|^2 (re^{i\theta}) r\, dr, \end{aligned}$$

and thus from (22)

$$E(Z_k, Q_{\varepsilon_k}) \leq \frac{2E}{\pi}\varepsilon_k. \tag{31}$$

We are now aiming to connect Y_k and Z_k, thereby reconstructing the original homotopy class α. Introducing $\mathbf{5}_k := \varepsilon_k$, $\mathbf{6}_k := 2\pi + \varepsilon_k$ and restricting Z_k to the part of ∂Q_k corresponding to $\mathbf{5}_k - \mathbf{3}_k - \mathbf{2}_k - \mathbf{6}_k$, we obtain a *loop* z_k which is freely homotopic to ξ_k in $U_{1/k}(S, K)$. Namely, consider the trapezoid $T_k = \{s + it : 0 \leq t \leq \log\frac{1}{\sqrt{r_k}}, t + \varepsilon_k - \log\frac{1}{\sqrt{r_k}} \leq s \leq 2\pi + \varepsilon_k + \log\frac{1}{\sqrt{r_k}} - t\}$ and define a map $H_k : T_k \to K$ by putting

$$H_k(s+it) = \begin{cases} Z_k(\varepsilon_k + i(s + \log\frac{1}{\sqrt{r_k}} - \varepsilon_k)) & \text{for } s \leq \varepsilon_k \\ Z_k(s + i\log\frac{1}{\sqrt{r_k}}) & \text{for } \varepsilon_k \leq s \leq 2\pi + \varepsilon_k \\ Z_k(2\pi + \varepsilon_k + i(2\pi + \varepsilon_k + \log\frac{1}{\sqrt{r_k}} - s)) & \text{for } s \geq 2\pi + \varepsilon_k. \end{cases}$$

By (21), (23) and (26) restriction of H_k to horizontal lines of T_k gives the desired homotopy with deformation parameter t.

Let $A = \{re^{i\vartheta} : \frac{1}{2} < r < 1\}$ and let $\widetilde{z}_k : \{\frac{1}{2}e^{i\vartheta} : \vartheta \in \mathbb{R}/2\pi\} \to K$, $\widetilde{y}_k : \{e^{i\vartheta} : \vartheta \in \mathbb{R}/2\pi\} \to K$ be reparametrizations of z_k, y_k with $\widetilde{z}_k(1/2) = z_k(\mathbf{5}_k) = q_k$, $\widetilde{y}_k(1) = Y_k(-1-\varepsilon_k) =: p_k$; here we use the counterclockwise orientation on both circles. By (21), (28) and the above deformation argument we have an extension $H_k \in C^0(\bar{A}, U_{\frac{L+1}{k}}(S, K))$ for $k > \frac{1}{4\delta_0}$. Let $\widetilde{c}_k := H_k(\frac{1}{2}(1-s))$ for $-1 \leq s \leq 0$ so that $\widetilde{c}_k(-1) = p_k$ and $\widetilde{c}_k(0) = q_k$. Finally let $c_k : [-1, 0] \to U_{\frac{L+1}{k}}(S, K)$ be a curve parametrized proportionally to its arclength with total length $l_k < \infty$, which is homotopic to $\widetilde{c}_k$ in

$U_{\frac{L+1}{k}}(S,K)$ *with endpoints fixed.* For instance we can take c_k to be a piecewise geodesic in K suitably inscribed to $\widetilde{c}_k$; note that l_k is independent of ε_k by construction. Let $R_k := (-1,0)\times(-\varepsilon_k,\varepsilon_k)$, $\Omega_k := P_k \cup P_{\varepsilon_k} \cup R_k \cup Q_{\varepsilon_k} \cup Q_k$, and obtain a map $W_k : \Omega_k \to K$ by putting $W_k = Y_k$ on $P_k \cup P_{\varepsilon_k}$, $W_k = Z_k$ on $Q_{\varepsilon_k} \cup Q_k$ and $W_k(s+it) = c_k(s)$ for $s+it \in R_k$. We have

$$E(W_k, R_k) = l_k^2 \varepsilon_k. \tag{32}$$

We now claim that the sequence $(w_k := W_k|_{\partial\Omega_k})$ belongs to α. Let $\delta > 0$ be given and let $k > \max\{\frac{1}{4\delta_0}, \frac{L+1}{\delta}\}$. We can then first deform $w_k|_{[-1,0]\times\{\pm\varepsilon_k\}} = c_k$ to $\widetilde{c}_k$ in $U_\delta(S,K)$ while leaving the endpoints fixed, thus obtaining a new loop $\widetilde{w}_k$. The restriction of $\widetilde{w}_k$ to $\mathbf{6}_k - \mathbf{2}_k - \mathbf{3}_k - \mathbf{7}_k$, where $\mathbf{7}_k = -i\varepsilon_k$, is a reparametrization of $c_k * \widetilde{y}_k * c_k^{-1} * \widetilde{z}_k^{-1}$, with basepoint q_k. Since this loop bounds a slit annulus—a disk—in $U_\delta(S,K)$, it can be deformed to the constant q_k in $U_\delta(S,K)$. After performing this homotopy we are left with the restriction of Z_k to $\mathbf{5}_k - \mathbf{4}_k - \mathbf{1}_k - \mathbf{6}_k$. By a trapezoid homotopy analogous to the above we can deform this further to the restriction of Z_k to $\mathbf{4}_k - \mathbf{1}_k$, again in $U_\delta(S,K)$. But this is just a reparametrization of $x_k = X_k|_{\partial B}$, and thus we have shown $w_k \sim x_k$ in $U_\delta(S,K)$ for $k > \max\{\frac{1}{4\delta_0}, \frac{L+1}{\delta}\}$. This implies that (w_k) belongs to $\Pi_1(S)$ and that it represents the class $\alpha \in \widehat{\pi}_1(S)$. Since the domain Ω_k is biholomorphically equivalent to the disk, the sequence (W_k) is admissible for comparison for $F_*(\alpha)$.

Let us denote by μ_k a quantity which goes to zero as $k \to \infty$. As X_k is a minimizing sequence, we have $F(W_k) \geq F_*(\alpha) - \mu_k \geq F(X_k) - 2\mu_k = F(X_k, B_{r_k}(0)) + F(X_k, D_k) - 2\mu_k \geq F(X) - \frac{1}{k} + F_k - 2\mu_k$; here we have used (19) and (25). On the other hand (6), (27), (30), (31), (32) and (25)—of course always using the conformal invariance (5)—imply

$$\begin{aligned}
F(W_k) &= F(Y_k, P_k) + F(Y_k, P_{\varepsilon_k}) + F(W_k, R_k) + F(Z_k, Q_{\varepsilon_k}) + F(Z_k, Q_k) \\
&\leq F(Y_k, P_k) + \Lambda E(Y_k, P_{\varepsilon_k}) + \Lambda E(W_k, R_k) \\
&\quad + \Lambda E(Z_k, Q_{\varepsilon_k}) + F(Z_k, Q_k) \\
&\leq F_*(\{x\}) + \frac{1}{k} + F_k + \Lambda\left(L_k\varepsilon_k + \frac{2E}{\pi}\varepsilon_k + l_k^2\varepsilon_k\right).
\end{aligned}$$

Combining the two estimates and cancelling F_k we arrive at

$$F(X) \leq F_*(\{x\}) + \frac{2}{k} + 2\mu_k + \Lambda\left(L_k + \frac{2E}{\pi} + l_k^2\right)\varepsilon_k. \tag{33}$$

Since none of the constants depend on ε_k, we can first let $\varepsilon_k \to 0$ and then $k \to \infty$ to obtain $F(X) \leq F_*(\{x\})$; thus X minimizes in its own boundary homotopy class. □

In [12] the idea of the proof above is used in an inductive framework to construct all the minimizers into which the sequence decomposes; these can be countably many due to the generality of S. Note that in the so-called *thread problem* a decomposition into infinitely many parts can occur even if the data are smooth [1], [5, Ch.10], [12, Sec.4]. If one is only interested in the existence of *some* F-minimizer in K with free boundary on S, then the result above already provides a satisfactory answer. To see this, let us define

(34) $\varepsilon_\infty(S,K) = \inf\{\varepsilon > 0 :$ There is a sequence X_k of interior minimizers for F with the following properties:

- $x_k = X_k|_{\partial B} \in C^0(\mathbb{R}/2\pi, K)$, $\operatorname{dist}(x_k, S) \to 0$, there is a $\delta > 0$ such that infinitely many of the x_k are not contractible in $U_\delta(S,K)$,
- $\limsup_{k\to\infty} F(X_k) \le \varepsilon$,
- the set of accumulation points of the sequence $X_k(B) \subset K$ is contained in S $\}$.

Geometrically speaking, $\varepsilon_\infty(S,K)$ is the minimal F-energy of a homotopically nontrivial sequence which escapes to infinity. If there exists a closed set L with $S \cap L = \emptyset$ such that for any homotopically nontrivial sequence we have $X_k(B) \cap L \neq \emptyset$ for k sufficiently large ("a link"), then from (11), (12) and (13) we have the lower bound

$$\varepsilon_\infty(S,K) \ge a_1 \min\{\operatorname{dist}_\infty(S,L)^2, \operatorname{dist}_\infty(S,L)^{2/3}\}, \tag{35}$$

where a_1 is as in corollary 1 and $\operatorname{dist}_\infty(S,L) = \inf\{\liminf_{k\to\infty} \operatorname{dist}_K(p_k, S) : p_k \in L, p_k \to \infty\}$. In particular if S is compact or if we can find a set L as above with $\operatorname{dist}_\infty(S,L) = +\infty$ (e.g. L compact), then we have $\varepsilon_\infty(S,K) = +\infty$.

Corollary 2. *Let K, S, F be as in the theorem and suppose that there is a nontrivial homotopy class α with $F_*(\alpha) < \varepsilon_\infty(S,K)$. Then there is a nontrivial homotopy class $\alpha_0 \in \widehat{\pi}_1(S)$ for which $F_*(\alpha_0) \in (0, F_*(\alpha)]$ is attained.*

Proof. Let $\widetilde{X}_k$ be an improved minimizing sequence for α. By assumption there exist points $p \in K \setminus S$, $w_k \in B$, such that $\widetilde{X}_k(w_k) \to p$. Let $h_k(w) = \frac{w+w_k}{1+\bar{w}_k w} \in \operatorname{Aut}(B)$ and pass to $X_k := \widetilde{X}_k \circ h_k$, so that $X_k(0) \to p$. As $\operatorname{dist}(x_k, S) \to 0$ we may use (10) to select a subsequence which converges weakly to a *nonconstant* map $X \in H^1(B,K)$ with $X(0) = p$. The theorem implies that X minimizes with respect to the homotopy class $\alpha_0 := \{x\}$. $\square$

In general the number $\varepsilon_0(S,K) := \inf\{F_*(\alpha) : \alpha \in \widehat{\pi}_1(S), \|\alpha\|_K > 0\}$ can be zero and even if it is positive, it need not be attained by a homotopy class. On the other hand, if $\varepsilon_0(S,K) > 0$ is attained by a homotopy class, then the proof of the theorem implies that it is in fact attained by a minimizer in that class; see also [5, Ch.5].

The regularity question for F-minimizers was solved by Nitsche [15], Goldhorn–Hildebrandt [7], Hildebrandt [8] and Hildebrandt–Kaul [9]. Suppose that S satisfies a (δ_1,κ)–chord–arc condition in K, i.e. any two points $p,q \in S$ with $\text{dist}_K(p,q) < \delta_1$ can be connected *in* S by an arc γ of length $L(\gamma) \leq \kappa \, \text{dist}_K(p,q)$. Suppose also that there is a uniformly continuous retraction $r : U_{\delta_1}(S,K) \to S$. Then any minimizer will be of class $C^{0,\mu}(\bar{B})$ where $\mu = \frac{\lambda}{\Lambda L^2(1+\kappa^2)} \in (0,1/2]$, see [12, Sec.6, Cor.1] or [5, Ch.7] for a proof. For higher order regularity theory we refer to [5, Vol.2].

References.

[1] Alt, H. W., Die Existenz einer Minimalfläche mit freiem Rand vorgeschriebener Länge. *Arch. Ration. Mech. Anal.* **51** (1973), 304–320.

[2] Béthuel, F., Weak limits of Palais–Smale sequences for a class of critical functionals. *Calc. Var.* **1** (1993), 267–310.

[3] Brézis, H., and Coron, J.-M., Large solutions for harmonic maps in two dimensions. *Commun. Math. Phys.* **92** (1983), 203–215.

[4] Courant, R., The existence of minimal surfaces of given topological structure under prescribed boundary conditions. *Acta Math.* **72** (1940), 51–98.

[5] Dierkes, U., Hildebrandt S., Küster, A., and Wohlrab, O., Minimal surfaces, Vol.1: Boundary value problems, Vol.2: Boundary regularity. Grundlehren math. Wiss. **295**, **296**, Springer 1992.

[6] Duzaar, F., and Kuwert, E., Weak limits of minimizing sequences for the m-energy of homotopic maps $u : M^m \to K$ are minimizers. In preparation.

[7] Goldhorn, K., and Hildebrandt, S., Zum Randverhalten der Lösungen gewisser zweidimensionaler Variationsprobleme unter freien Randbedingungen. *Math. Z.* **118** (1973), 241–253.

[8] Hildebrandt, S., Ein einfacher Beweis für die Regularität der Lösungen gewisser zweidimensionaler Variationsprobleme unter freien Randbedingungen. *Math. Ann.* **194** (1972), 316–331.

[9] Hildebrandt, S., and Kaul, H., Two-dimensional variational problems with obstructions, and Plateau's problem for H-surfaces in a Riemannian manifold. *Commun. Pure Appl. Math.* **25** (1972), 187–223.

[10] Jost, J., Two-dimensional geometric variational problems. Wiley Interscience, Chichester, New York 1991.

[11] Kuwert, E., Minimizing the energy of maps from a surface into a 2-sphere with prescribed degree and boundary values. *Manuscr. Math.* **83** (1994), 31–38.

[12] Kuwert, E., Area-minimizing immersions of the disk with boundary in a given homotopy class: A general existence theory. Submitted as Habilitation thesis at the University of Bonn, 1995.

[13] Luckhaus, S., Partial Hölder continuity for minima of certain energies among maps into a Riemannian manifold. *Indiana Univ. Math. J.* **37** (1988), 349–367.

[14] Morrey, C. B., Multiple integrals in the Calculus of Variations. Grundlehren math. Wiss. **130**, Springer 1966.

[15] Nitsche, J. C. C., Minimal surfaces with partially free boundary. Least area property and Hölder continuity for boundaries satisfying a chord–arc condition. *Arch. Ration. Mech. Anal.* **39** (1970), 131–145.

[16] Sacks, J., and Uhlenbeck, K., The existence of minimal immersions of two-spheres. *Ann. Math.* **113** (1981), 1–24.

[17] Tolksdorf, P., On minimal surfaces with free boundaries in given homotopy classes. *Ann. Inst. H. Poincaré* **2** (1985), 157–165.

[18] Wente, H., A general existence theorem for surfaces of constant mean curvature. *Math. Z.* **120** (1971), 277–288.

[19] White, B., Infima of energy functionals in homotopy classes of mappings. *J. Differ. Geom.* **23** (1986), 127–142.

[20] Ye, R., On the existence of area-minimizing surfaces with free boundary. *Math. Z.* **206** (1991), 321–331.

RECEIVED APRIL 24, 1995.

Asympotitically Conic Elliptic Operators and Liouville Type Theorems

FANG HUA LIN [1]

1. Introduction.

It is well known that a harmonic function on the Euclidian space $\mathbb{R}^n$ having polynomial growth is necessarily a polynomial. In particular the space of harmonic functions on $\mathbb{R}^n$ with a given polynomial growth rate at the infinity is always finite dimensional. Moreover, any nonconstant harmonic function has to grow at least linearly.

Let us consider an elliptic operator L on $\mathbb{R}^n$ of the form

$$L = \frac{\partial}{\partial x^i}\left[A^{ij}(x)\frac{\partial}{\partial x^j}\cdot\right] \tag{1.1}$$

with $A^{ij}(x)$ are bounded, measurable functions on $\mathbb{R}^n$ satisfying

$$\lambda I \leq (A^{ij}(x)) \leq \Lambda I, \tag{1.2}$$

for all $x \in \mathbb{R}^n$ and for some positive constants λ, Λ.

A solution u of $Lu = 0$ in $\mathbb{R}^n$ is called a L-harmonic function. Then, in general, nothing similar for polynomial growth L-harmonic functions is known as that for harmonic functions, except the case that L is the Laplace-Beltrami operator for an uniformly elliptic metric defined on $\mathbb{R}^2$, (see [LN]). For the latter case, one can even characterize the optimal minimal growth rate in terms of ellipticity. When the dimension $n \geq 3$, the DeGiorgi-Moser Harnack type inequality (see [GT]) implies that any non-constant L-harmonic function on $\mathbb{R}^n$ has to grow like a power of $|x|$. It is however not known what is the minimal possible growth rate with the given ellipticity and the dimension n. On the other hand, even with the additional assumption

[1]The authors research is supported by NSF Grants DMS-9149555 and DMS-9401546. This work was done while the author was a visiting member of the Institute for Advanced Study. The author wishes to acknowledge the support of IAS through the NSF Grant, DMS-9304580.

that $A^{ij}(x)$ are smooth functions on $\mathbb{R}^n, n \geq 3$, it is not known in general that if there is a polynomial growth, nonconstant L-harmonic function on $\mathbb{R}^n$. By applying a theorem of P. Lax [L], it is not hard to show there are many nonconstant L-harmonic functions on $\mathbb{R}^n$ whenever $A^{ij}(x)$ are local lipschitz continuous, cf. [GW]. Here the assumption on the uniform ellipticity (1.2) is not even required, though one certainly does not know the growth rate of such L-harmonic functions.

In [AL], M. Avellaneda and the author studied a special class of elliptic operators L. Under the assumptions (1.1), (1.2) and, in addition, that $A^{ij}(x)$'s are periodic functions in each variable $x_k, k = 1, 2, \ldots n$. We showed that polynomial growth L-harmonic functions are necessarily polynomials with periodic coefficients. Moreover, there is a linear isomorphism between harmonic polynomials and polynomial growth L-harmonic functions on $\mathbb{R}^n$. We should point out that our original proof in [AL] used the theory of homogenization, which enabled us to bring the information at the infinity to a finite region, then we applied the unique continuation property of solutions of $Lu \equiv 0$. For the latter reason we made the additional assumption that $A^{ij}(x)$'s are also lipschitz continuous. However, if one examines the proof presented in [AL], one will notice that we had actually proved: if u is a polynomial growth L-harmonic function on $\mathbb{R}^n$ with L satisfying (1.1) - (1.2) and with the periodicity on the coefficients, then for any $R \geq R_0$, for some positive R_0 which may also depend on u, there is a polynomial $Q_R(x)$ with periodic coefficients and bounded degree such that $u \equiv Q_R$ on B_R. It is then clear that u has to be a polynomial with periodic coefficients globally. In other words, the unique continuation property is not needed. This latter fact was also observed in [MS]. Where they gave an elegant proof of the theorem in [AL]. Moreover, the proof also works for a class of nonlinear problems. However, the DeGiorgi-Moser estimates for such equations are required for [MS]. The proof of [AL] works equally well for elliptic systems.

The purpose of this article is to generalize the result of [AL] to a larger class of elliptic operators, which we shall call them asymptotically conic. Roughly speaking, this is a class of elliptic operators that in certain weak sense close, at the infinity, to some Laplace-Beltrami operators on Riemannian cones. Note that the operators considered in [AL] are just those that are weakly close to the ordinary Laplacian at the infinity. We shall establish the finite dimensionality of the spaces of L-harmonic functions with given polynomial growth rates. In fact, we have a precise dimension estimate on such spaces in terms of spectral informations of certain limiting operators.

To study these asymptotically conic elliptic operators is also motivated by problems concerning polynomial growth harmonic functions on certain complete Riemannian manifolds, such as area-minimizing hypersurfaces in R^{n+1} or complete manifolds of nonnegative Ricci curvature. Indeed, it is well-known that there are cone structures for complete area-minimizing hypersurfaces in R^{n+1} at the infinity (cf. [F, chapter 5]). The same is true for complete Riemannian manifolds with nonnegative Ricci curvature, quadratic curvature decay and Euclidian volume growth, see [BKN], [CC]. In the case that such cone structure is unique (cf. [SL1, 2] and [CT]), then the Laplace-Beltrami operators on such manifolds will tend to the corresponding Laplace-Beltrami operators on the cones. In particular, they are asymptotically conic. Since the cone structures (such as tangent cones for area-minimizing submanifolds at the infinity) may not be unique in general, that suggests us to define a class of weakly asymptotically conic elliptic operators. One consequence of our results concerning these asymptotically conic elliptic operators is that polynomial growth harmonic functions on complete area-minimizing hypersurfaces are of finite dimension (depending also on the growth rate) provided that the tangent cones at the infinity are unique. A similar statement is also true for complete, nonnegative Ricci curvature manifolds.

The author wishes to dedicate this paper to Professor Stefan Hildebrandt on the occasion of his sixtieth birthday.

2. Asymptotically Conic Operators.

Let $\mathcal{L}(\lambda, \Lambda)$ denote all elliptic operators L of form (1.1) and satisfying (1.2). Let Ω be a bounded, Lipschitz domain in $\mathbb{R}^n$, then any $L \in \mathcal{L}(\lambda, \Lambda)$ defines a map $L : H_0^1(\Omega) \to H^{-1}(\Omega)$, by $Lu = \frac{\partial}{\partial x^i}\left[A^{ij}(x)\frac{\partial}{\partial x^j}u\right]$. Moreover, L is invertible, in sense that for any $f \in H^{-1}(\Omega)$, there is an unique $u \in H_0^1(\Omega)$ satisfying $Lu = f$, and $\|u\|_{H_0^1(\Omega)} \leq C(L,\Omega)\|f\|_{H^{-1}(\Omega)}$.

Definition 2.1. We say that a sequence of operators

$$L_k (k = 1, 2, \dots ,) : H_0^1(\Omega) \to H^{-1}(\Omega)$$

is G-convergent, as $k \to \infty$, to $\hat{L}$ (and we write $L_k \overset{G}{\Longrightarrow} \hat{L}$) if for any $f, g \in H^{-1}(\Omega)$, $\lim_{k\to\infty} \langle g, L_k^{-1} f\rangle = \langle g, \hat{L}^{-1} f\rangle$.

A theorem of DeGiorgi and Spagnolo [DGS] says that, for any bounded Lipschitz domain $\Omega \subset \mathbb{R}^n$, $\mathcal{L}(\lambda, \Lambda)$ is G-compact in the sense that for any sequence $\{L_k\}$ in $\mathcal{L}(\lambda, \Lambda)$, there are an operator $\hat{L} \in \mathcal{L}(\lambda, \Lambda)$ and a subsequence

$\{L_{k'}\}$ of $\{L_k\}$ such that $L_{k'} \overset{G}{\Longrightarrow} \hat{L}$. Here we view $\mathcal{L}(\lambda,\Lambda)$ as operators from $H_0^1(\Omega)$ to $H^{-1}(\Omega)$.

Many properties of G-convergence and their applications can be found in the survey paper [ZKOK]. For the readers convenience, we list here a few basic properties of G-convergence of operators in $\mathcal{L}(\lambda,\Lambda)$.

(i) If $L_k \overset{G}{\Longrightarrow} L$ and if $f_k \Longrightarrow f$ in $H^{-1}(\Omega)$, as $k\to\infty$, and $L_k u_k = f_k$, then $u_k \rightharpoonup u$ weakly in $H_0^1(\Omega)$, as $k\to\infty$, and $Lu=f$.

(ii) If $L_k \overset{G}{\Longrightarrow} L$ as $k\to\infty$, then $L_k-\lambda I \overset{G}{\Longrightarrow} L-\lambda I$ for any $\lambda\geq 0$. Let $\{\lambda_i^k\}_{i=1}^\infty$ be eigenvalues of $-L_k$, such that $0<\lambda_1^k\leq\lambda_2^k\leq\ldots$, and let $\{\lambda_i\}_{i=1}^\infty$ be eigenvalues of L. Then, for any i, $\lambda_i^k\to\lambda_i$ as $k\to\infty$.

(iii) If $L_k \overset{G}{\Longrightarrow} L$, as $k\to\infty$, and $L_kU_k = f\in H^{-1}(\Omega), U_k\in H_0^1(\Omega)$, for all $k=1,2,\ldots$, then $U_k\rightharpoonup U$ weakly in $H_0^1(\Omega)$, and $\langle -L_kU_k,U_k\rangle \longrightarrow \langle -LU,U\rangle$.

In the above we have fixed a Lipschitz domain Ω, and have regarded $\{L_k\}$ and L as operators from $H_0^1(\Omega)$ to $H^1(\Omega)$. The following properties will be also useful to us (again we assume $L_k, k=1,2,\ldots$, and L are in $\mathcal{L}(\lambda,\Lambda)$).

(iv) Let $L_kU_k=0$ in B_2, for $k=1,2,\ldots$, and $L_k \overset{G}{\Longrightarrow} L$ as operators from $H_0^1(B_2)$ to $H^{-1}(B_2)$. Suppose $\|U_k\|_{L^2(B_2)}\leq 1$ and $U_k\rightharpoonup U$ weakly in $L^2(B_2)$, then $LU=0$ in B_2. Moreover $U_k\rightharpoonup U$ weakly in $H^1_{\mathrm{loc}}(B_2)$. In addition, one also has

$$A_k^{ij}(x)\frac{\partial}{\partial x_j}U_k \rightharpoonup A^{ij}(x)\frac{\partial}{\partial x_j}U \text{ weakly in } L^2_{\mathrm{loc}}(B_2).$$

Here

$$L_k=\frac{\partial}{\partial x_i}[A_k^{ij}(x)\frac{\partial}{\partial x_j}], L=\frac{\partial}{\partial x_i}[A^{ij}(x)\frac{\partial}{\partial x_j}].$$

(v) A sequence of operators $L_k, k=1,2,\ldots$, in $\mathcal{L}(\lambda,\Lambda)$ defined on $\mathbb{R}^n$ is called G-convergent to $L\in\mathcal{L}(\lambda,\Lambda)$ if $L_k \overset{G}{\Longrightarrow} L$ for any bounded Lipschitz domain $\Omega\subseteq\mathbb{R}^n$. Suppose $L_k \overset{G}{\Longrightarrow} L$ on $\mathbb{R}^n$, then the fundamental solutions $G_k(x,y)$ of L_k converges to the fundamental solution $G(x,y)$ of L on $\mathbb{R}^n\times\mathbb{R}^n$. Moreover, the convergence is locally uniform on the set $\mathbb{R}^n\times\mathbb{R}^n\backslash\{(x,y):x=y\}$.

The proofs of (i), (ii), (iii) and (v) can be found in [ZKOK] and references therein. To prove (iv), we first find a $\rho\in(1,2)$ and ρ close to 2, such that $u_k \rightharpoonup u$ in $H^1(\partial B_\rho)$ weakly (by taking a subsequence of k's if necessary).

Let $w_k = u_k - u + h_k$, here h_k is the harmonic extension of $u - u_k$ on ∂B_ρ into B_ρ. Thus $w_k = 0$ on ∂B_ρ. Then

$$w_k(x) = \int_{B_\rho} G_k(x,y) \frac{\partial}{\partial y_i} \left[a_k^{ij}(y) \frac{\partial}{\partial y_i} (h_k - u) \right] dy.$$

For any f defined on B_ρ, and $x \in B_{\rho-0}$, one defines $f^\sigma(x) = ⨍_{B_\sigma(x)} f$. We have

$$\begin{aligned} w_k^\sigma(x) &= \int_{B_\rho} G_k^\sigma(x,y) \frac{\partial}{\partial y_i} \left[a_k^{ij}(y) \frac{\partial}{\partial y_j} (h_k - u) \right] dy \\ &= \int_{B_\rho} \frac{\partial}{\partial y_i} G_k^\sigma(x,y)\, a_k^{ij}(y) \cdot \frac{\partial}{\partial y_j} (u - h_k)\, dy , \end{aligned}$$

for all $x \in B_{\rho-\sigma}$.

Since $h_k \to 0$ in $H^1(B_\rho)$ and since

$$a_k^{ij}(y) \frac{\partial}{\partial y_i} G_k^\sigma(x,y) \rightharpoonup a^{ij}(y) \frac{\partial}{\partial y_i} G^\sigma(x,y)$$

weakly in $L^2(B_\rho)$, as $k \to \infty$, (This follows from a theorem of L. Tartar, see reference in [ZKOK]), here a^{ij}, G are coefficients and green's function of L, we obtain that

$$w_k^\sigma(x) \to w^\sigma(x) = \int_{B_\rho} \frac{\partial}{\partial y_i} G^\sigma(x,y)\, a^{ij}(y) \frac{\partial}{\partial y_j} u(y)\, dy$$

As $w(x) \equiv 0$, $w^\sigma(x) = 0$. By taking $\sigma \to 0$, we obtain $L\,u = 0$ in B_ρ. As ρ can be chosen arbitrary close to 2, we obtain the conclusion (iv).

Examples 2.2.

(a) Let $L \in \mathcal{L}(\lambda, \Lambda)$, and $A^{ij}(x)$ are periodic functions of $x_1, \ldots, x_n$. Then $L_\epsilon = \frac{\partial}{\partial x_i}[A^{ij}(x/\epsilon)\frac{\partial}{\partial x_j}] \stackrel{G}{\Longrightarrow} \Delta$, as $\epsilon \to 0^+$, on $\mathbb{R}^n$.

(b) Suppose $L_k = \frac{\partial}{\partial x^i}[A_k^{ij}(x)\frac{\partial}{\partial x_j}] \in \mathcal{L}(\lambda, \Lambda)$, and $A_k^{ij}(x) \Longrightarrow A^{ij}(x)$ in $L^1(\Omega)$ or $A_k^{ij}(x) \to A^{ij}(x)$ a.e. in Ω, for $i, j = 1, \ldots, n$, then $L_k \stackrel{G}{\Rightarrow} L \equiv \frac{\partial}{\partial x^i}[A^{ij}(x)\frac{\partial}{\partial x^j}]$ as operators from $H_0^1(\Omega)$ to $H^{-1}(\Omega)$. As a consequence, if $L = \frac{\partial}{\partial x^i}[A^{ij}(x)\frac{\partial}{\partial x^j}]$ has the property that $A^{ij}(x) \to \delta_{ij}$, as $|x| \to \infty$, uniformly, for $i, j = 1, \ldots, n$, then $L_\epsilon = \frac{\partial}{\partial x^i}\left[A^{ij}(x/\epsilon)\frac{\partial}{\partial x^j}\right] \stackrel{G}{\Rightarrow} \Delta$, as $\epsilon \to 0^+$, on $\mathbb{R}^n$.

Remark 2.3. The statement,

$$\lim_{k\to\infty}\langle L_k U, V\rangle = \langle U, V\rangle,$$

for all $U, V \in H_0^1(\Omega)$, does not imply $L_k \overset{G}{\Rightarrow} L$. In fact, it is equivalent to

$$\lim_{k\to\infty}\int_\Omega \left(A_k^{ij}(x) - A^{ij}(x)\right)\phi(x)dx = 0,$$

for all $\phi \in C_0^\infty(\Omega)$.

On the other hand we note that $\|(L_k - L)\phi\|_{H^{-1}(\Omega)} \to 0$, as $k \to \infty$, uniformly for ϕ on the unit sphere of $H_0^1(\Omega)$, or even $\|(L_k - L)\phi\|_{H^{-1}(\Omega)} \to 0$, as $k \to \infty$, for each $\phi \in H_0^1(\Omega)$, implies that $L_k \overset{G}{\Rightarrow} L$.

Indeed, the former convergence is equivalent to $\|A_k^{ij} - A^{ij}\|_{L^\infty(\Omega)} \to 0$, as $k \to \infty$, and the latter is equivalent to

$$\|(-L_k + \lambda I)^{-1} f - (-L + \lambda I)^{-1} f\|_{H_0^1(\Omega)} \to 0, \text{ as } k \to \infty,$$

for each $f \in H^{-1}(\Omega)$, and $\lambda \geq 0$. (see [ZKOK]).

Definition 2.4. Let $L \in \mathcal{L}(\lambda, \Lambda)$, then L is called asymptotically conic if there is an operator $L_{\mathbb{C}} \in \mathcal{L}(\lambda, \Lambda)$ of the form $L_{\mathbb{C}} = \text{div}_M[\mu(\theta)\nabla_M]$ such that $L_\epsilon = \frac{\partial}{\partial x^i}[A^{ij}(x/\epsilon)\frac{\partial}{\partial x^j}] \overset{G}{\Rightarrow} L_{\mathbb{C}}$, as $\epsilon \to 0^+$. Here $L = L_1$, and div_M, ∇_M are divergent and gradient operators on $M = (R^n, g)$. Where the metric g is given by $dr^2 + r^2 g_{ij}(\theta)d\theta^i d\theta^j$, (r,θ) is the polar-coordinate system on R^n and μ is a strictly positive and bounded function of $\theta \in \mathbb{S}^{n-1}$.

Let $\mathcal{L}_{\mathbb{C}}(\lambda, \Lambda)$ be all such operators $L_{\mathbb{C}} \in \mathcal{L}(\lambda, \Lambda)$. Then an operator $L \in \mathcal{L}(\lambda, \Lambda)$ is called weakly asymptotically conic if for any sequence $\epsilon_k \downarrow 0, k = 1, 2, \ldots$, then are a subsequence $\{\epsilon_{k'}\}$ of $\{\epsilon_k\}$ and, an operator $\hat{L} \in \mathcal{L}_{\mathbb{C}}(\lambda, \Lambda)$ such that

$$L_{\epsilon_{k'}} \overset{G}{\Longrightarrow} \widehat{L} \text{ on } \mathbb{R}^n.$$

Note that if $L_{\mathbb{C}} \in \mathcal{L}(\lambda, \Lambda)$, then

$$C_1(n,\lambda,\Lambda) \leq \mu(\theta) \leq C_2(n,\lambda,\Lambda) \tag{2.5}$$

and

$$C_1 I_{n-1} \leq (g_{ij}(\theta)) \leq C_2 I_{n-1}, \theta \in \mathbb{S}^{n-1}, \tag{2.6}$$

for some positive constants C_1, C_2 depending only on n, λ and Λ.

Next we introduce the following definition for a L-harmonic function u to be of polynomial growth. It is obvious there are many ways to define polynomial growth L-harmonic functions we found the following one is easiest to work with.

Definition 2.7. A L-harmonic function u on $\mathbb{R}^n, L \in \mathcal{L}(\lambda, \Lambda)$ is called of polynomial growth if there is a positive number β such that

$$\underline{\lim}_{R\to\infty} S(R)/R^\beta < \infty.$$

Where $S(R) = \fint_{B_R(0)} u^2(x)dx$.

Remark 2.8. By the DeGiorgi-Moser estimate, one has

$$\sup_{B_{R/2}(0)} |u|^2(x) \leq C(n, \lambda, \Lambda) \fint_{B_R(0)} u^2(x)dx.$$

Therefore, if u is of polynomial growth, then

$$\underline{\lim}_{R\to\infty} \frac{M(R)}{R^\beta} < \infty. \tag{2.9}$$

Here $M(R) = \sup_{B_R(0)} |u(x)|^2$. Thus our definition of polynomial growth coincides with the classical one. One also attempts to replace $S(R)$ by $H(R) = \fint_{\partial B_R} |u|^2$. It is, however, not known to the author if is true in general.

$$M(R/2) \leq C(n, \lambda, \Lambda)H(R)$$

In the case L is weakly asymptotically conic, we also consider L-harmonic functions of strict-polynomial growth in the following sense

$$S(2R) \leq 2^\beta S(R) \tag{2.10}$$

for some $\beta > 0$ and for all $R \geq R_0$ (which may depend on u).

Let $L \in \mathcal{L}(\lambda, \Lambda)$ be an asymptotically conic operator. Thus

$$L_\epsilon = \frac{\partial}{\partial x_i}[A^{ij}(x/\epsilon)\frac{\partial}{\partial x_j}] \overset{G}{\Longrightarrow} L_{\mathbb{C}}$$

for some $L_{\mathbb{C}}$ as in the definition 2.4.

Let

$$L_\Sigma = \frac{1}{\mu(\theta)\sqrt{g(\theta)}}\frac{\partial}{\partial\theta_i}[g^{ij}(\theta)\sqrt{g(\theta)}\mu(\theta)\cdot\frac{\partial}{\partial\theta_j]}$$

be the angular part of $L_\mathbb{C}$, which is defined on $\mathbb{S}^{n-1}$ where $g(\theta) = \det(g_{ij}(\theta))$, and $(g^{ij}(\theta))$ is the inverse of $(g_{ij}(\theta))$.

Let $0 = \mu_0 < \mu_1 \leq \mu_2 \leq \cdots \leq \mu_k \leq \ldots$ be eigenvalues of $-L_\Sigma$, and let $\{\phi_j\}_{j=0}^\infty$ be the corresponding eigenfunctions such that

$$L_\epsilon\phi_j + \mu_j\phi_j = 0 \text{ on } \mathbb{S}^{n-1}, \text{ for } j = 0, 1, 2, \ldots$$

and $\int\limits_{\mathbb{S}^{n-1}} \phi_i\phi_j\mu(\theta)\sqrt{g(\theta)}d\theta = \delta_{ij}$, for $i, j = 0, 1, 2 \ldots.$ Note that we may simply take $\phi_0 = 1/\int_{\mathbb{S}^{n-1}}\mu\sqrt{g}d\theta$.

Let v be a $L_\mathbb{C}$-harmonic function defined on the ball $B_{R_0} = \{x \in \mathbb{R}^n, |x| \leq R_0\}$, then

$$v(x) = \sum_{j=0}^\infty a_j r^{\lambda_j}\phi_j(\theta), x = (r, \theta). \tag{2.11}$$

Here $\lambda_j = \frac{\sqrt{(n-2)^2+4\mu_j}-(n-2)}{2}$, for $j = 0, 1, 2, \ldots$, and a_j's are some constants depending on v.

Lemma 2.12. *Let v be a $L_\mathbb{C}$-harmonic function defined on the ball $B_{R_0}, 0 < R_0 \leq \infty$. Set*

$$H(r) = \fint_{\partial B_r} \mu v^2 = \int_{\mathbb{S}^{n-1}} \mu(\theta)v^2(r,\theta)\sqrt{g(\theta)}d\theta,$$

$$D(r) = \fint_{B_r} \mu|\nabla_M v|^2 = \frac{1}{r^{n-2}}\int_0^r\int_{\mathbb{S}^{n-1}} \mu(\theta)[|\frac{\partial v}{\partial\rho}|^2 + \frac{1}{\rho^2}g^{ijh}(\theta)\frac{\partial v}{\partial\theta^i}\frac{\partial v}{\partial\theta^j}]\cdot$$

$$\cdot\rho^{n-1}\sqrt{g(\theta)}d\theta d\rho,$$

$$N(r) = \frac{D(r)}{H(r)}, \text{ and}$$

$$S(r) = \fint_{B_R} \mu v^2 = \frac{1}{r^n}\int_{B_r} \mu(\theta)v^2(r,\theta)\sqrt{g(\theta)}dx.$$

Then a) $N(r)$ is a monotone nondecreasing function on $(0, R_0)$;

b) $\log H(e^t)$ is a convex function of $t \in (-\infty, \log R_0)$;

c) $\log S(e^t)$ is a convex function of $t \in (-\infty, \log R_0)$.

Proof. a) is proven in [GL].

b) is equivalent to a) by the calculations in [GL]. In fact, b) also follows from the proof of c) below. To show c) we use the representation (2.11). Thus

$$S(e^t) = \sum_{j=0}^{\infty} a_j^2 e^{2\lambda_j t} \cdot \frac{1}{2\lambda_j + n}.$$

Then the convexity of $\log S(e^t)$ follows from the Cauthy-Schwartz inequality:

$$\Big(\sum_{j=0}^{\infty} \frac{a_j^2}{2\lambda_j + n} e^{2\lambda_j t} 4\lambda_j^2\Big) \cdot \Big(\sum_{j=0}^{\infty} \frac{a_j^2}{2\lambda_j + n} e^{2\lambda_j t}\Big)$$
$$\geq \Big(\sum_{j=0}^{\infty} \frac{a_j^2}{2\lambda_j + n} e^{2\lambda_j t} 2\lambda_j\Big)^2.$$

□

Let $L \in \mathcal{L}(\lambda, \Lambda)$ be asymptotically conic. Given a $\beta \geq 0$. Let H_β be the space of L-harmonic functions on $\mathbb{R}^n$ such that $\frac{\lim}{R\to\infty} \frac{S(R)}{R^{2\beta+\epsilon}} = 0$, for any $\epsilon > 0$. Denoted by $Q(\beta)$ the number of $\{j : \lambda_j \leq \beta\}$. Thus $\lambda_Q \leq \beta < \lambda_{Q+1}$. Let $\delta(\beta) = \lambda_{Q+1} - \beta$, and $N = \beta + \frac{1}{2}\delta(\beta), \beta \leq N' < N$. It is not clear that H_β is a linear space. This is also a consequence of the following:

Theorem 2.13. *The space $H_{N'}$ is finite dimensional. Moreover, $\dim H_{N'} = Q(\beta)$ In particular, $H_{N'} = H_{N''}$ whenever $\lambda_Q \leq N' < N'' < \lambda_{Q+1}$.*

Proof. By definition, for any $u \in H_{N'}$, one has

$$\frac{\lim}{R\to\infty} R^{-2N} S(R) = \frac{\lim}{R\to\infty} R^{-2N} \int_{B_R} u^2 = 0.$$

Here $S(R) = R^{-n} \int_{B_R} \mu(\theta)|u|^2 \sqrt{g(\theta)}dx$.

In particular, there is a sequence of $R_i \to \infty$ such that

$$S(2R_i) \leq 2^{2N} S(R_i), \text{ for } i = 1, 2, \ldots. \tag{2.14}$$

Next, we want to show there is a $R_0 > 0$ depending only on $N, L, L_{\mathbb{C}}$, (but it is independent of $u \in H_{N'}$) such that

$$S(2R) \leq 2^{2N} S(R) \Longrightarrow S(R) \leq 2^{2N} S(R/2) \tag{2.15}$$

whenever $R \geq R_0$.

Combining (2.14) and (2.15), one obtains the following crucial growth estimate:

$$S(R) \leq C(N) S(R_0) \left(\frac{R}{R_0} \right)^{2N}, \text{ for } R \geq R_0, \tag{2.16}$$

and for $u \in \mathcal{H}_{N'}$.

To show (2.15), we argue by a contradiction. Suppose there is a sequence of $R_j \to \infty$, and a sequence of L-harmonic functions u_j defined on $B_{2R_j}(0)$ such that

$$S(2R_j) = \fint_{B_{2R_j}} |u_j|^2 \leq 2^{2N} \fint_{B_{R_j}} |u_j|^2 = 2^{2N} S(R_j),$$

and that $S(R_{j/2}) \leq 2^{-2N} S(R_j)$ for $j = 1, 2 \ldots$.

Let $V_j = u_j(R_j x)/\sqrt{S(R_j)}$, and $L_j = \frac{\partial}{\partial x^k}[A^{k\ell}(R_j x)\frac{\partial}{\partial x^\ell}]$. Then $L_j V_j = 0$ in B_2 with

$$\fint_{B_2} |V_j|^2 \leq 2^{2N}, \fint_{B_1} |V_j|^2 = 1 \text{ and}$$

$$\fint_{B_{1/2}} |V_j|^2 \leq 2^{-2N}, \text{ for } j = 1, 2, \ldots .$$

By the assumption, $L_j \overset{G}{\Longrightarrow} L_{\mathbb{C}}$, and, by taking a subsequence if necessary, we may assume $V_j \rightharpoonup V$ weakly in $L^2(B_2)$. By property (iv) of G-convergence, one has $L_{\mathbb{C}} V = 0$ in B_2 and $V_j \to V$ weakly in $H^1_{\text{loc}}(B_2)$ and strongly in $L^2_{\text{loc}}(B_2)$. For V, we have

$$S(1) = \fint_{B_1} |V|^2 = \fint_{B_1} \mu(\theta)|V|^2 \sqrt{g(\theta)} dx = 1,$$

$$S(2) = \fint_{B_2} |V|^2 \leq 2^{2N},$$

$$S(\tfrac{1}{2}) = \fint_{B_{1/2}} |V|^2 \leq 2^{-2N}.$$

Since, by Lemma 2.12, $\log S(e^t)$ is a convex function of t, we see that $S(r) = r^{2N}$, for all $r \in (0,2)$. The latter is impossible as $N \neq \lambda_j$ for $j = 0, 1, 2 \ldots$. Therefore we proved (2.15), and hence (2.16).

Note that, in the above proof, we may replace N by any N'' such that $N' < N'' < \lambda_{Q+1}$. Thus the set of L-harmonic functions in $H_{N'}$ form a linear space by (2.16). To show $\dim H_{N'} \leq Q$, we assume, to the contrary, that there are at least $Q+1$ linearly independent functions in $H_{N'}$. Since any $U \in H_{N'}$ satisfies (2.16), these $Q+1$ functions must be also linearly independent on any $B_R, R \geq R_0$.

Now we fix a $R \geq R_0$, by Grame-Schmidt process we may find $Q+1$ functions $U_1, U_2, \ldots, U_{Q+1}$ in $H_{N'}$ such that

$$\fint_{B_R} U_k U_\ell = R^{-n} \int_{B_R} \mu(\theta)\sqrt{g(\theta)} U_k U_\ell dx = \delta_{k\ell},$$

for $k, \ell = 1, 2 \ldots, Q+1$.

Let $L_\epsilon = \frac{\partial}{\partial x_i}[a^{ij}(x/\epsilon)\frac{\partial}{\partial x_j}], \epsilon = \frac{1}{R}$, and let $V_k^\epsilon(x) = U_k(Rx)$, for $k = 1, 2, \ldots, Q+1$. Then by the fact that $L_\epsilon \overset{G}{\Longrightarrow} L_{\mathbb{C}}$ on $\mathbb{R}^n$ and

$$\fint_{B_r} |V_k^\epsilon|^2 = \frac{1}{r^n} \fint_{B_r} \mu(\theta)\sqrt{g(\theta)}|V_k^\epsilon|^2 dx \leq C(N) r^{2N},$$

for all $1 \leq r < \infty, k = 1, 2, \ldots, Q+1$, and $0 < \epsilon \leq 1/R_0$. (cf. (2.16)).

We, therefore, may assume, by taking subsequences if necessary, that

$$V_k^\epsilon(x) \to V_k(x) \text{ in } L^2_{\text{loc}}(\mathbb{R}^n)$$

and weakly in $H^1_{\text{loc}}(\mathbb{R}^n)$, as $\epsilon \to 0^+$, for $k = 1, 2, \ldots, Q+1$. $L_{\mathbb{C}} V_k = 0$ in $\mathbb{R}^n$, and

$$\fint_{B_1} V_k V_\ell = \int_{B_1} \mu(\theta)\sqrt{g(\theta)} V_k V_\ell dx = \delta_{,\ell},$$

for $k, \ell = 1 \ldots, Q+1$.

Moreover, each V_k satisfies the growth estimate

$$\frac{1}{r^n} \int_{B_r} \mu(\theta)\sqrt{g(\theta)}|V_k|^2 dx \leq C(N) r^{2N}, \text{ for } r \geq 1.$$

Thus $V_k(x) = \sum_{\lambda_j \leq N} a_{j,k} r^{\lambda_j} \phi_j(\theta)$, for $k = 1, 2, \ldots, Q+1$, by (2.11), (2.12).

Since $Q =$ number of $\{j : \lambda_j \leq \beta < N\}$, we see it is impossible to find $Q+1$ functions $V_1, V_2, \dots, V_{Q+1}$ which also satisfy

$$\fint_{B_1} V_k V_\ell = \delta_{k,\ell}, k, \ell = 1, \dots, Q+1.$$

This contradiction shows that $\dim H_{N'} \leq Q$.

Next we want to show that $\dim H_{N'} \geq Q$. Let $\phi_0(\theta), \phi_1(\theta), \dots, \phi_{Q-1}(\theta)$ be the eigenfunctions of L_Σ on $\mathbb{S}^{n-1}$ (as before), so that

$$\int_{\mathbb{S}^{n-1}} \mu(\theta)\sqrt{g(\theta)}\phi_k\phi_\ell d\theta = \delta_{k,\ell}, k, \ell = 0, 1, \dots, Q-1.$$

We want to construct Q linearly independent L-harmonic functions on $\mathbb{R}^n, V_1, \dots, V_Q$ such that

$$\int_{B_R} |V_k k|^2 = R^{-n} \int_{B_R} \mu(\theta)\sqrt{g(\theta)}|V_k|^2 dx \leq C(N)\left(\frac{R}{R_0}\right)^{2N}, \int_{B_{R_0}} |V_k|^2,$$

for $k = 1, \dots, Q$. Then, from the first part of our proof, we have $V_k \in H_{N'}$, and, thus $\dim H_{N'} \geq Q$.

We will inductively construct these $V_1, \dots, V_Q$. To start with we let $V_1 \equiv 1$. Suppose, we have constructed $V_1 \dots, V_k, k$ linearly independent L-harmonic functions in $H_{N'}$. We consider following problems, for a sequence of $\{R_i\}, R_i \to \infty$ as $i \to \infty$,

$$\begin{cases} LU_i &= 0 \text{ in } B_{R_i}(0), \\ U_i &= \sum_{j=0}^{k} a_j R_i^{\lambda_j}\phi_j(\theta) \text{ on } \partial B_{R_i}(0). \end{cases}$$

Here $a_0, \dots, a_k$ are some constant, to be chosen later.

Let

$$L_i = \frac{\partial}{\partial x_k}[a^{k\ell}(R_i x)\frac{\partial}{\partial x_\ell}], V_i(x) = \frac{U_i(R_i x)}{\sqrt{S(R_i)}}, S(R_i)$$
$$= R_i^{-n} \int_{B_{R_i}} \mu(\theta)\sqrt{g(\theta)}|U_i|^2 dx.$$

Then

$$\begin{cases} L_i V_i &= 0 \text{ in } B_1(0) \\ V_i &= \sum_{j=0}^{k} c_j\phi_j(\theta) \text{ on } \partial B_1(0) \end{cases}$$

for some constant $c_1, \dots c_k$ (which may also depend on i). Since

$$\int_{B_1} \mu(\theta)\sqrt{g(\theta)}|V_i|^2 dx = 1,$$

we conclude that all c_j's are uniformly bounded (independent of i). For otherwise, we may normalize the boundary data so that $\sum_{j=0}^{k} c_j^2 = 1$ and that $V_i \to 0$ in $L^2(B_1)$.

Then since $L_i \overset{G}{\Longrightarrow} L_{\mathbb{C}}$, we would find a solution

$$\begin{cases} L_{\mathbb{C}}V & = 0 \text{ in } B_1 \\ V & = \sum_{j=0}^{k} c_j\phi_j(\theta) \text{ on } \partial B_1, \sum_{j=0}^{k} c_j^2 = 1 \end{cases}$$

and $V \equiv 0$. It is impossible. This shows c_j's are uniformly bounded. Then apply the G-convergence property (iv) again to conclude (by taking subsequences, if necessary), $V_k \to V$ in $L^2(B_1)$ and $V_i \to V$ weakly in $H^1(B_1)$.

By (2.11), $V(x) = \sum_{j=0}^{k} c_j r^{\lambda_j}\phi_j(\theta)$ in B_1.

In particular,

$$\int_{B_1} \mu(\theta)\sqrt{g(\theta)}|V|^2 dx \le 2^{2\lambda_k} 2^n \int_{B_{1/2}} \mu(\theta)\sqrt{g(\theta)}|V|^2 dx$$

$$< 2^{2N}2^n \int_{B_{1/2}} \mu(\theta)\sqrt{g(\theta)}|V|^2 dx, \text{ as } N > \lambda_k, k < Q.$$

Hence

$$\fint_{B_1} |V_i|^2 \le 2^{2N} \fint_{B_{1/2}} |V_i|^2,$$

for all sufficiently large i. That is

$$\fint_{B_{R_i}} |U_i|^2 \le 2^{2N} \fint_{B_{R_{i/2}}} |U_i|^2, \text{ for large } i\text{'s}.$$

¿From the proof of (2.15) - (2.16), one has

$$\fint_{B_R} |U_i|^2 \le C(N)\left(\frac{R}{R_0}\right)^{2N} \fint_{B_{R_0}} |U_i|^2,$$

for all large i and all $R_0 \leq R \leq R_i$.

Note that the validity of the last estimate is independent of the choices of $a_0, a_1, , \dots , a_k$. Now we choose $a_0, a_1, \dots , a_k$ so that the followings are true:

$$\int_{B_{R_0}} \mu(\theta)\sqrt{g(\theta)}U_i \cdot V_j dx = 0, \text{ for } j = 1, 2, \dots , k,$$

and $\int_{B_{R_0}} \mu(\theta)\sqrt{g(\theta)}|U_i|^2 dx = 1$.

Finally we let $i \to +\infty$. Then since $\int B_R |U_i|^2 \leq C(N)\left(\frac{R}{R_0}\right)^{2N}$, for all large i and $R_i \geq R \geq R_0$, we see that

$$U_i \to U \text{ in } L^2_{\text{loc}}(\mathbb{R}^n) \text{ and weakly in } H^1_{\text{loc}}(R^n).$$

Moreover $LU = 0$ in $\mathbb{R}^n$, and $U \in H_N$. By the above argument again, $U \in H_{N'}$. Let $V_{k+1} = U$, then $V_1, V_2, \dots , V_k, V_{k+1}$ are clearly $k+1$ linearly independent L-harmonic functions in $H_{N'}$. □

Remark 2.17. It is an open problem if Theorem 2.13 remains true for weakly asymptotically conic operators. One of the key factors we used is $N \neq \lambda_j$, for $j = 0, 1, 2, \dots$. From this choice of N we then deduce (2.15) and (2.16). In general (2.15) fails in fact (see example (2.18) below). Nevertheless, we believe that $H_{N'}$ is always finite dimensional. It involves some difficulties with changes of eigenvalues and eigenfunctions as asymptotic behavior of L changes for different scales.

Example 2.18. Consider a Riemannian metric on $\mathbb{R}^n$ given by $g = dr^2 + r^2 f^2(r) d\theta^2$ in the polar coordinates of $\mathbb{R}^n$ where $f(r)$ is *either*

$$f(r) = 1 + a^2 \sin^2(\log\log(e + r)), \text{ for some } 0 < a < \infty,$$

or

$$f(r) = 1 - a^2 \sin(\log\log(e + r)), \text{ for some } 0 < a < 1.$$

Then it is not hard to see the corresponding Laplace-Beltrami operator Δg on $\mathbb{R}^n$ is uniformly elliptic and weakly asymptotically conic. But Δg is not asymptotically conic. It is also not hard to show that any polynomial growth Δg-harmonic functions on $\mathbb{R}^n$ has its growth rates oscilating between two numbers. In particular the estimate (2.15) can't be true. Though it is obvious in this case that Theorem 2.13 remains valid.

On the other hand, it follows from the proof of Theorem 2.13 that

Theorem 2.19. *Let L be weakly asymptotically conic operator in $\mathcal{L}(\lambda, \Lambda)$. Then the space P_β of strictly polynomial grow L-harmonic functions, for any $\beta > 0$, is finite dimensional. Here P_β denotes all those L-harmonic function U on $\mathbb{R}^n$ such that the corresponding function $S(R)$ satisfying $S(2R) \leq 2^{2\beta} S(R)$ for all $R \geq R_0(U, L)$.*

Proof. Since (2.5)-(2.6) and Wyle's asymptotic formula for large eigenvalues, we see that $\lambda_k(L_{\mathbb{C}}) > \beta$ for all $L_{\mathbb{C}} \in \mathcal{L}(\lambda, \Lambda)$ and all $k \geq k_0(\beta, n, \lambda, \Lambda)$.

Then we can show as in the proof of theorem 2.13 that P_β contains at most k_0 linearly independent functions $u_j, j = 1, 2, \ldots, k$, that satisfying the estimate $S(2R) \leq 2^{2\beta} S(R)$ for all $R \geq R_0(L, U_1, \ldots, U_k)$. That is $\dim P_\beta \leq k_0$. □

3. Application.

Let M be a complete, area-minimizing hypersurface in R^{n+1}. For any $\epsilon > 0$ let $\eta_\epsilon : R^{n+1} \to R^{n+1}$ be the map defined by $\eta_\epsilon(x) = \epsilon \cdot x, x \in R^{n+1}$. Then it is well-known (cf. [SL1] that, for any sequence of $\epsilon_n \to 0$, there is a subsequence $\{\epsilon'_n\}$ such that $\eta_{\epsilon'_n} \# M$ converges, locally in R^{n+1} as integral currents in the flat norm, to an area-minimizing cone $\mathbb{C}$ in R^{n+1}.

Then by the Sobolev's inequality on minimal submanifolds (see [SL4]) and regularity theory for minimal surfaces see [SL1], and also Poincare's inequality on area-minimal hypersurface, see [BG], one easily concludes that $\Delta_{\epsilon'_n} \overset{G}{\Longrightarrow} \Delta_{\mathbb{C}}$. There $\Delta_{\mathbb{C}}, \Delta_{\epsilon'_n}$ are Laplace-Beltrami operators on $\mathbb{C}$ and on $\eta_{\epsilon'_n} \# M$. In the case that the tangent cone of M at the infinity is unique (see [SL2]), then $\Delta_\epsilon \overset{G}{\Longrightarrow} \Delta_{\mathbb{C}}$ when $\epsilon \to 0^+$. In the latter situation, one has Δ_M is asymptotically conic. The proof of theorem 2.13 can carry through without any difficulties for Δ_M and we obtain in general the following:

Theorem 3.1. *Let M be a complete area-minimizing hypersurface in $\mathbb{R}^{n+1}$. Suppose M has an unique tangent cone at the infinity. Then polynomial growth harmonic functions, for a given growth rate, form a finite dimensional space. In fact, the dimension of this space can be calculated precisely in terms of the growth rate and eigenvalues of $\Delta_\Sigma, \Sigma = \mathbb{C} \cap \mathbb{S}^n$.*

Remark 3.2. The conclusion of the above theorem remains true if M is a complete minimal submanifold with an unique multiplicity one, smooth tangent cone at the infinity (cf. [SL2]).

In the case that both Sobolev and Poincare inequalities are valid on M (e.g. M is an area-minimizing hypersurface, see [BG]), then any harmonic function on M is Hölder continuous with some exponent $\alpha > 0$. Thus minimal growth of harmonic functions on M has to be at least like a power of the distance function. This also implies that the minimal nonzero eigenvalue of $-\Delta_\Sigma$ is strictly positive. It is an open problem proposed by S.T. Yau [Y, problem section] that if this minimal eigenvalue is $N - 1 = \dim \Sigma$.

Next let us consider a complete n-dimensional Riemannian manifold M with nonnegative Ricci curvature. It is an old theorem of S.T. Yau [Y2] that any nonconstant harmonic function u on M has to grow, at least, linearly. It is an open problem if polynomial growth harmonic functions (for a given growth rate) from a finite dimensional space.

When M has the Euclidian volume growth, i.e., $\mathrm{Vol}(B_R(a)) \geq C_0 R^n$, for some $C_0 > 0$, and for large R, when $B_R(a)$ is a geodesic ball of radius R in M. Then M has a metric cone structure at the infinity (see [CC]). When M has, in addition, quadratic curvature decay, then M has smooth cone-structures at the infinity. In our terminology, one has, in particular, that Δ_M is weakly asymptotically conic. In the recent work [CT], a large class of such manifolds were shown to have an unique cone structure at the infinity. In other words, Δ_M is asymptotically conic.

Theorem 3.3. *Suppose M is a complete Riemannian manifold with nonnegative Ricci curvature, quadratic curvature decay and Euclidian volume growth. If, in addition, M has an unique cone structure at the infinity. Then the spaces of polynomial growth harmonic functions are finite dimensional.*

When the cone structures at the infinity are not unique, we have the following result concerning those strictly polynomial growth harmonic functions.

Corollary 3.4. *Let P_β be the space of strictly polynomial growth harmonic functions with the growth rate $\beta > 0$ on a complete Riemannian manifold M. Then $\dim P_\beta < \infty$ if either (i) M is an area-minimizing hypersurface in $\mathbb{R}^{n+1}$ or (ii) M has nonnegative Ricci curvature, quadratic curvature decay and Euclidian Volume growth.*

We should also point out theorem 3.1, theorem 3.3 and the above corollary remain valid if we replace Δ_M by an operator $L_M = \mathrm{div}_M(A(x)\nabla_M)$. When $A(x)$ is non-matric-valued function on M with $\lambda I \leq A(x) \leq \Lambda I$ and with L_M asymptotically G-converging to $\Delta_{\mathbb{C}}$. The additional condition on the quadratic curvature decay in Theorem 3.3 and Corollary 3.4 can be easily removed.We add this condition so that readers do not have to go out the classical catagary and arguments. As we mentioned in the introduction that the proof in [AL] used the unique continuation theorem. Here we shall give an application of these arguments.

First we need the following

Lemma 3.5. *Let M be a complete Riemannian manifold with a pole and, has nonnegative sectional curvatures. Then for any harmonic function $u \neq 0$ on M, the quantity $N(r), 0 < r < \infty$, is monotone nondecreasing. Where $N(r) = rD(r)/H(r), H(r) = \int_{\partial B_r(a)} u^2, D(r) = \int_{B_r(a)} |\nabla_M u|^2$, and here a is a pole of M and $B_r(a)$ is the geodesic ball of radius r centered at a.*

Proof. Let $g = dr^2 + r^2 g_{ij}(r,\theta) d\theta^i d\theta^j$ be the metric on M in the polar-coordinates with the pole at $a \in M, 0 \leq r < \infty, \theta = (\theta_0^1, \cdots, \theta^{n-1}) \in \mathbb{S}^{n-1}$.

Since $H(r) = \int_{\mathbb{S}^{n-1}} u^2(r,\theta) r^{n-1} \sqrt{g(r,\theta)} d\theta$, by a calculation in [GL], one has

$$\frac{d}{dr} H(r) = \frac{n-1}{r} H(r) + 2D(r) - \epsilon_1(r).$$

Here $g(r,\theta) = \det(g_{ij}(r,\theta))$, and

$$\epsilon_1(r) = -\int_{\mathbb{S}^{n-1}} u^2(r,\theta) r^{n-1} \frac{d}{dr} \sqrt{g(r,\theta)} d\theta \geq 0$$

as Rie $(\partial_r, \partial_r) \geq 0$ implies that $\frac{d}{dr}\sqrt{g(r,\theta)} \leq 0$.

Also, since $D(r) = \int_0^r \int_{\mathbb{S}^{n-1}} [u_\rho^2(\rho,\theta) + \frac{1}{\rho^2} g^{ij}_{(\rho,\theta)} \frac{\partial u}{\partial \theta^i} \frac{\partial u}{\partial \theta^j}] \sqrt{g(\rho,\theta)} \rho^{n-1} d\theta \rho$, one has

$$\frac{d}{dr} D(r) = 2 \int_{\partial B_r(a)} u_\rho^2 + \frac{n-2}{r} D(r) + \epsilon_2(r).$$

where

$$\epsilon_2(r) = -\frac{1}{r} \int_{B_r(a)} \frac{\frac{d}{d\rho}\sqrt{g(\rho,\theta)}}{\sqrt{g(\rho,\theta)}} \cdot \rho |\nabla_M u|^2$$
$$+ \frac{1}{r} \int_{B_r} \frac{1}{\rho} \frac{\partial g^{ij}}{\partial \rho}(\rho,\theta) \frac{\partial u}{\partial \theta^i} \frac{\partial u}{\partial \theta^j}.$$

By a direct computation,

$$\frac{1}{\rho}\frac{\partial}{\partial\rho}g^{ij}(\rho,\theta)\frac{\partial u}{\partial\theta^i}\frac{\partial u}{\partial\theta^j} = \frac{2}{\rho^2}g^{ij}(\rho,\theta)\frac{\partial u}{\partial\theta^i}\frac{\partial u}{\partial\theta^j} - \frac{2}{\rho^3}g^{ij}\prod_{ij} g^{jt}\frac{\partial u}{\partial\theta^s}\frac{\partial u}{\partial\theta^t},$$

here $(\prod_{ij})$ is the second fundamental form of $\partial B_\rho(a)$ in M, and hence

$$\left(g^{iS}\prod_{ij} g^{it}\right) \leq (\rho g^{st}),$$

by the fact that the sectional curvature is nonnegative. Indeed $(\prod_{ij}) \leq \frac{1}{\rho}\cdot(\rho^2 g_{ij})$.

Therefore $\epsilon_2(r) \geq 0$.

Finally

$$\frac{N'(r)}{N(r)} = \frac{1}{r} + \frac{D'(r)}{D(r)} - \frac{H'(r)}{H(r)}$$
$$= \frac{2\int_{\partial B_r} u_\rho^2}{D(r)} - \frac{2D(r)}{H(r)} + \frac{\epsilon_1(r)}{H(r)} + \frac{\epsilon_2(r)}{D(r)} \geq 0.$$

Here we have used also the fact that $D(r) = \int_{\partial B_r(a)} uu_\rho$. We also note that

$$\frac{d}{dr}\log\frac{H(r)}{r^{n-1}} \leq \frac{2N(r)}{r}.$$

□

By using a bit knowledge from the theorey of Allexandrov spaces,the hypothesis in the the above Lemma 3.5 that M possess a pole can be also be droped.But we shall not elabrate this further here.

Theorem 3.6. *Let M be as in Lemma 3.5. Suppose M has, in addition, the quadratic decay in curvatures and Euclidian volume growth. Then polynomial growth harmonic functions (with any given growth rate) form a finite dimensional space.*

Proof. Let $\beta > 0$ be given and let H_β be defined as before. Then for any $U \in H_\beta$, there is a sequence of $R_j, j = 1, 2, \dots, R_j \to \infty$ as $j \to \infty$, such that

$$S(2R_j) \leq 2^{2N} S(R_j), \text{ for } j = 1, 2, \dots,$$

and $N = \beta + \epsilon, \epsilon > 0$. Here

$$S(\rho) = \rho^{-n} \int_{B_\rho(a)} U^2.$$

Since $\triangle_M$ is weakly asymptotically conic by our assumptions on M, we see that

$$N(R_j) = \frac{R_j D(R_j)}{H(R_j)} \leq C_0 N,$$

for j sufficiently large, and for some $C_0(M) > 0$.

Indeed, let $M_j = (M, R_j^{-2} g)$, and $V_j(x) = \frac{U(R_j x)}{\sqrt{S(R_j)}}$. Then $\triangle_{M_j} V_j = 0$. By G-convergence properties (iv), one has $M_j \to \mathbb{C}, \triangle_{M_j} \overset{G}{\Longrightarrow} \triangle_{\mathbb{C}}, V_j \to V$ in $L^2_{\text{loc}}(B_2)$ and weakly in $H^1_{\text{loc}}(B_2)$.

$$\triangle_{\mathbb{C}} V = 0 \text{ in } B_2 = \{(r, \theta) : 0 \leq r < 2\} \subseteq \mathbb{C}.$$

Moreover,

$$\int_{B_1 \cap \mathbb{C}} V^2 = 1, 2^{-n} \int_{B_2 \cap \mathbb{C}} V^2 \leq 2^{2N}.$$

For cone $\mathbb{C}$ and the function V, one has, by Lemma 2.12, that ([cf.GL]) $N(1) \leq C_1 N$. Since

$$\int_{\partial B_1} |V_j|^2 \to \int_{\partial B_1} V^2, \int_{B_1} |\nabla_{M_j} V_j|^2 \to \int_{B_1} |\nabla_{\mathbb{C}} V|^2,$$

by property (iii) of G-convergence, thus $N(R_j) \leq C_0 N$ for U.

Now by the monotonicity of $N(r)$, one has thus proved: for any $U \in H_\beta$, the corresponding function $N(r), 0 \leq r < \infty$, is uniformly bounded by $C_* \beta$.

Suppose now there are k-linearly independent harmonic functions $u_1, \dots, u_k$, in H_β. We choose a vector $(\alpha_1, \dots, \alpha_k) \in S^k$ such that $U = \alpha_1 u_1 + \dots + \alpha_k u_k \in H_\beta$ and that $U(a) = 0, |\nabla U(a)| = 0, \dots, |\nabla^Q U(a)| = 0$ where $Q > C_* \beta$. We note that whenever $k \geq k_0(n, C_* \beta)$, the above is always possible. Since $U \in H_\beta$, we have

$$N(0^+) = \lim_{r \downarrow 0} \frac{r D(r)}{H(r)} \leq C_* \beta \text{ for } U.$$

This implies the vanishing order of $U \leq C_* \beta$ by [GL]. We thus obtain a contradiction. □

Remark 1. The hypothesis on the quadratic curvature decay in the Theorem 3.6 can be omitted. However, one then has to deal with nonnegatively curved singular spaces (Alexandroff-spaces). To avoid such additional machinery, we stated the theorem in the classical version. Some further study on Alexandroff-spaces shall be discussed in our forth coming work. In particular, we can show both the Sobolev and Poincare inequalities for such spaces (see [Lin]).

To conclude the paper, we remark that we have discussed here only those elliptic equations of the form (1.1)-(1.2). Our main results can be also generalized to nondivergent form equations and certain elliptic systems. As an example, we consider the following problem:

$$A_{ij}(x) U_{x_i x_j}(x) = 0 \text{ in } \mathbb{R}^n, \tag{3.7}$$

with

$$\lambda I \leq (A_{ij}(x)) \leq \Lambda I. \tag{3.8}$$

Theorem 3.9. *Suppose $A^{ij}(x)$ converge to δ_{ij} uniformly as $(x) \to \infty$, for $i, j = 1, 2, \dots, n$. Then solutions u of the problem (3.7) with (3.8) and with polynomial growth, say, $u \in H_\beta$, for some $\beta > 0$, form a finite dimensional space. Moreover, $\dim H_\beta = Q(\beta) =$ dimension of the space of harmonic polynomials of degree $\leq [\beta]$.*

The proof of the above theorem is similar to that of theorem 2.13. Except here we may have to use the Krylov's estimates for equations of type (3.7)-(3.8), see [GT]. We also note that the assumption that $A^{ij}(x) \to \delta_{ij}$ uniformly as $|x| \to \infty$ can be weakened to, say, $R^{-n} \int_{B_R} |A^{ij}(x) - \delta_{ij}| dx \to 0$ as $R \to \infty$. We leave all these details to the reader.

Added in the proofs. We just saw an announcement [CM] of T.Colding and W.Minicozzi II,in which they stated the space of polynomial growth harmonic functions on a complete, Riemannian manifold with nonnegative Ricci curvature and Euclidian volume growth is finite dimensional.Their definition of polynomial growth is stronger than our's definition,but weak than our's definition of strict polynomial growth.We shouldh also refer to several other works done by P.Li, J.P.Wang,and the work of Cheeger-Colding and Minicozzi ,which contained in the reference of their article and not in the citations here. Finally we note that the Sobolev and Poincare inequalities are the main points that we can generalize the notion of G-convergence of Laplacian operators on manifolds.

References.

[AL] M. Avellaneda and F.H. Lin, Une thèorème de Liouville pour des èquations elliptique à coefficients pèriodiques; compt. Rendus Acad. Sci. Paris, 309, 1989, pp. 245-250.

[BG] E. Bombieri and E. Giusti, Harnack's inequality for Elliptic Differential Equations on Minimal Surfaces, Invent. Math. 15, 1972 pp. 24-46.

[CC] J. Cheeger and T. Colding, Lower bounds on Ricci curvature and the almost rigidity of warped products, Preprint.

[CM] T. Colding and W. Minicozzi II, *On Function Theory on spaces with lower Ricci curvature bound*, preprint, 1995.

[CT] J. Cheeger and G. Tian, On the cone structure at infinity of Ricci flat manifolds with Euclidian volume growth and quadratic curvature decay, Invent. Math. 118 1994, pp. 493-571.

[DGS] DeGiorgi and S. Spagnolo, Γ-Convergenza e G-Convergenza, Boll. Un. Mat. Italia A(5) 14, 1977, pp. 213-220.

[F] H. Federer, Geometric measure theory, Berlin-Heidelberg-New York, Springer-Verlag, 1969.

[GL] N. Garofalo and F.H. Lin, Monotonicity properties of variational integrals, Ap - weights, and unique continuation, Indiana Univ. Math. Journal 35, 1986, pp. 245-267.

[GT] D. Gilberg and N. Trudinger, Elliptic partial differential equations of second order, Berlin-Heidelberg-New York, Springer-Verlag, 1977.

[GW] R. Greene and H. Wu, Embedding of open Riemannian manifolds by harmonic functions, Ann. Inst. Fourier, Grenoble, 25, 1, 1975, pp. 215-235.

[L] P. Lax, A stability theorem for abstract differential equations and its application to the study of the local behavior of solutions of elliptic equations, Comm. Pure and Appl. Math., 9, 1956, pp. 747-766.

[Lin] F.H. Lin, *Analysis on singular spaces*, preprint, 1995.

[LN] F.H. Lin and W.M. Ni, On the least growth harmonic functions and the boundary behavior of Riemann Mapping, Comm. in P.D.E. 10, 1985, pp. 767-786.

[MS] J. Moser and M. Struwe, On a Liouville-Type Theorem for Linear and Nonlinear Elliptic Differential Equations on a Torus.

[SL1] L. Simon, Lectures on Geometric Measure Theory, Proc. of C.M.A., Australian National Univ., 3, 1983.

[SL2] L. Simon, Asymptotics for a class of non-linear evolution equations, with applications to geometric problems, Annals of Math. 118, 1983, pp. 525-572.

[Y] S.T. Yau, Survey on Partial Differential Equations in Differential Geometry, In "Seminars on Differential Geometry" Edited by S.T. Yau, Annals of Math. Studies 102, 1982, Princeton Univ. Press.

[Y2] S.T. Yau, Harmonic functions on complete Riemannian manifolds, Comm. Pure and Appl. Math., 28, 1975, pp. 201-228.

[ZKOK] V.V. Zhikov, S.M. Kozlov, O.A. Olzinik and Khatén Ngoan, Averaging and G-convergence of Differential operators, Russian Math. Surveys, §4, # 5, 1979, pp. 69-147.

RECEIVED MARCH 21, 1995.

Attainment results for the two-well problem by convex integration

STEFAN MÜLLER[1] AND VLADIMIR ŠVERÁK

Models for solid-solid phase transitions lead to the problem of finding Lipschitz continuous maps $u : \Omega \subset \mathbf{R}^n \to \mathbf{R}^m$ which satisfy $\nabla u \in K$ (for a given set K of $m \times n$ matrices) as well as suitable boundary conditions. We show how Gromov's method of convex integration can be used to construct such solutions and give an application to the two-well problem where $K = \mathrm{SO}(2)A \cup \mathrm{SO}(2)B$.

1. Introduction.

Let $\Omega \subset \mathbf{R}^n$ be an open and bounded set and let $K \subset M^{m\times n}$ be a set of $m \times n$ matrices. We study the question whether there exist (weakly differentiable) maps $u : \Omega \to \mathbf{R}^m$ that satisfy

$$\nabla u \in K \quad \text{a.e.} \tag{1}$$

as well as suitable boundary conditions, the simplest one being

$$u(x) = Fx \quad \text{on} \quad \partial\Omega, \tag{2}$$

where F is an $m \times n$ matrix.

Our interest in this question arises from applications to solid-solid phase transitions. In this case $n = m = 3$ and the matrices in the set K correspond to those affine deformations of a (reference) crystal lattice which represent the lowest energy states (see Ball and James [BJ 1], [BJ 2] for further details and references). The set K typically has a finite number of components each of which is invariant under the action of SO(3).

[1]Partially supported by SFB 256 at the University of Bonn, by NSF grant DMS-9002679 and DFG grant Mu-1067/1-1. This work was begun during a stay of V.S. at SFB 256 and was finished while both authors enjoyed the hospitality of the Institute for Advanced Study, which was made possible by grant 93-6-6 from the Sloan foundation.

Gromov has developed a very powerful method, called convex integration, for the construction of C^1 solutions to partial differential relations like (1). His work generalizes and unifies earlier approaches such as the striking constructions of Nash [Na] and Kuiper [Ku] of nontrivial isometric immersions. We refer to Gromov's treatise [Gr] for a detailed exposition and many further references.

For the applications we have in mind one is interested in Lipschitz continuous solutions and while this introduces certain additional difficulties, at the same time many technicalities that are necessary for the C^1 case are no longer required. This fact is already mentioned in [Gr], p. 218 and the purpose of this note is, on one hand, to spell out those details which Gromov leaves to the reader and, on the other hand, to provide a short self-contained introduction to convex integration in a special case of practical importance.

To illustrate the results which can be obtained by this method, we mention the following result for the so called two-well problem in two dimensions. Let

$$K = \mathrm{SO}(2) \cup \mathrm{SO}(2) \begin{pmatrix} \lambda & 0 \\ 0 & \mu \end{pmatrix}, \quad 0 < \lambda < 1 < \mu.$$

Suppose that $\lambda\mu > 1$ and let

$$\begin{aligned}\tilde{K} =& \{F \in M^{2\times 2} : F = \begin{pmatrix} y_1 & -y_2 \\ y_2 & y_1 \end{pmatrix} + \begin{pmatrix} z_1 & -z_2 \\ z_2 & z_1 \end{pmatrix}\begin{pmatrix} \lambda & 0 \\ 0 & \mu \end{pmatrix}, \quad \text{where} \\ & |y| \le 1 - (\det F - 1)/(\lambda\mu - 1), \quad \text{and} \quad |z| \le (\det F - 1)/(\lambda\mu - 1)\}.\end{aligned}$$

Theorem 1.1. *Let $F \in \operatorname{int} \tilde{K}$. Then* (1), (2) *has a solution.*

For the reader familiar with the jargon we remark that $\tilde{K}$ is exactly the lamination convex hull of K (see Definition 4 below) which in this particular case agrees with the rank-1 convex, quasiconvex and polyconvex hull of K (see [Sv]).

We remark that the case $K = O(2)$ (i.e. $\lambda = -\mu = 1$) or, more generally, $K = O(n)$ has been studied in connections with isometric maps (see [Gr]) and with a model for the folding of paper (see [JK] and also [Po]). In this case more explicit constructions are available, mainly due to the fact that the lamination convex hull of $O(n)$ is the set of all linear maps with Lipschitz constant less than 1 (see [JK] for $n = 2$, [CP] for $n = 3$ and also [Gr]).

The remainder of this paper is organized as follows. In Sections 2 to 4 we describe the procedure of convex integration (for the case at hand) in three steps. In Section 2 we consider the case where K is a neighbourhood of just two matrices. In Section 3 we iterate the construction derived in Section

2 and introduce the lamination convex hull (Gromov's P-convex hull) and are thus able to deal with open sets K. In our opinion the heart of the matter is Section 4. Following Gromov we approximate (bounded) sets K by open sets and show strong convergence of the corresponding approximate solutions. In Section 5 we verify that the method applies to the situation in Theorem 1.

Acknowledgement

This work was inspired by a seminar on convex integration at the University of Bonn. It is a pleasure to thank W. Ballmann, U. Hamenstädt, J. Lohkamp, and all the other participants for making this seminar possible, and for stimulating discussions. We would also like to thank G. Friesecke for very helpful suggestions.

2. The basic construction.

In the following Ω will denote a bounded domain (i.e. an open and connected set). To avoid technicalities, we will assume that Ω is sufficiently smooth. It suffices to assume that Ω is piecewise C^1, or the weaker condition that Ω is strongly Lipschitz in the sense of Morrey ([Mo], Definition 3.4.1). Under this assumption the class $W^{1,\infty}(\Omega;\mathbf{R}^m)$ of maps whose distributional derivative satisfies $|\nabla u| \le C$ a.e. agrees with the class of Lipschitz functions. In particular such maps have a continuous extension to the closure $\bar{\Omega}$.

We first construct maps whose gradient lies in a small neighbourhood of two values.

Definition 1. We say that a map $u : \Omega \to \mathbf{R}^m$ is piecewise linear if u is continuous and if there exist *(*finitely of countably many*)* mutually disjouint *(*strongly Lipschitz*)* domains Ω_i such that

$$u|_{\Omega_i} \quad \text{is affine and} \quad \operatorname{meas}(\Omega \setminus \cup_i \Omega_i) = 0.$$

Lemma 2.1. *Let A and B be $m \times n$ matrices and suppose that*

$$\operatorname{rank}(B - A) = 1. \tag{3}$$

Let

$$C = (1-\lambda)A + \lambda B, \quad \textit{where} \quad \lambda \in (0,1).$$

Then, for any $\delta > 0$, there exists a piecewise linear map $u_\delta : \Omega \to \mathbf{R}^m$ such that

$$\operatorname{dist}(\nabla u_\delta, \{A, B\}) \leq \delta \quad \textit{a.e. in} \quad \Omega, \tag{4}$$

$$\sup_\Omega |u_\delta(x) - Cx| \leq \delta, \quad \textit{and} \tag{5}$$

$$u_\delta(x) = Cx \quad \textit{on} \quad \partial\Omega. \tag{6}$$

Remarks. 1. If ∇u takes exactly the values A and B (a.e.), then it is not possible to satisfy the boundary condition (6). Also, if $\operatorname{rank} A - B \geq 2$ and if δ is sufficiently small, then it is not possible to satisfy (4) and (5) or (4) and (6). We refer to [BJ 1], Propositions 1 and 2 for the proofs.

2. Note that ∇u_δ has to oscillate very rapidly in order to satisfy simultaneously (4) and (5).

Proof. We will first construct a solution for a special domain U. The argument will then be finished by an application of the Vitali covering theorem.

By an affine change of variables we may assume without loss of generality that

$$A = -\lambda a \otimes e_n, \quad B = (1 - \lambda) a \otimes e_n, \quad C = 0, \quad \text{and} \quad |a| = 1.$$

Let $\varepsilon > 0$, let

$$V = (-1, 1)^{n-1} \times ((\lambda - 1)\varepsilon, \lambda\varepsilon)$$

and define $v : V \to \mathbf{R}^m$ by

$$v(x) = -\varepsilon\lambda(1 - \lambda)a + \begin{cases} -\lambda a x_n & \text{if } x_n < 0, \\ (1 - \lambda) a x_n & \text{if } x_n \geq 0. \end{cases}$$

Then $\nabla v \in \{A, B\}$ and $v = 0$ at $x_n = \varepsilon(\lambda - 1)$ and $x_n = \varepsilon\lambda$, but v does not vanish on the whole boundary ∂V. Next let

$$h(x) = \varepsilon\lambda(1 - \lambda)a \sum_{i=1}^{n-1} |x_i|.$$

Then h is piecewise linear and $|\nabla h| = \varepsilon\lambda(1 - \lambda)\sqrt{n - 1}$. Set

$$\tilde{u} = v + h.$$

Note that $\tilde{u} \geq 0$ on ∂V and let

$$U = \{x \in V : \tilde{u}(x) < 0\}.$$

Then

$$\begin{aligned} \tilde{u}|_U \quad \text{is piecewise linear,} \quad \tilde{u}|_{\partial U} &= 0, \\ \operatorname{dist}(\nabla \tilde{u}, \{A, B\}) &\le \varepsilon\lambda(1-\lambda)\sqrt{n-1}, \qquad \text{and} \\ |\tilde{u}| &\le \varepsilon\lambda(1-\lambda). \end{aligned}$$

By the Vitali covering theorem one can exhaust Ω by disjoint scaled copies of U. More precisely there exist $x_i \in \mathbf{R}^n$ and $r_i > 0$ such that the sets

$$U_i = x_i + r_i U$$

are mutually disjoint and $\operatorname{meas}(\Omega \setminus \cup_i U_i) = 0$. Define u by

$$u(x) = \begin{cases} r_i \tilde{u}(r_i^{-1}(x - x_i)) & \text{if} \quad x \in U_i, \\ 0 & \text{else.} \end{cases}$$

Note that

$$\nabla u(x) = \nabla \tilde{u}(r_i^{-1}(x - x_i)), \quad \text{if} \quad x \in \Omega_i.$$

It follows that u is piecewise linear, that $u|_{\partial\Omega} = 0$ and that u satisfies (4) for a suitable choice of ε. Moreover by choosing $r_i \le 1$ one can also satisfy (5). □

3. Iteration of the basic construction for open relations.

Let K be an open subset of the $m \times n$ matrices. We seek maps $u : \Omega \to \mathbf{R}^m$ such that $\nabla u \in K$ a.e. in Ω and $u(x) = Fx$ on $\partial\Omega$, where F is an $m \times n$ matrix. From the basic construction we see that the above problem can be solved for $F = C$ provided that $A, B \in K$, $\operatorname{rank} A - B = 1$ and $C = (1-\lambda)A + \lambda B$, where $\lambda \in (0,1)$.

We will see that this procedure can be iterated. Specifically, if $A', B' \in K$ with $\operatorname{rank} A' - B' = 1$, if $C' = (1-\lambda')A' + \lambda' B'$, and if $\operatorname{rank} C - C' = 1$, then the above problem can also be solved for $F = (1-\mu)C + \mu C'$. This motivates the following definition which no longer assumes that K is open.

Definition 2. Let $K \subset M^{m\times n}$. We say that K is stable under lamination (or *lamination convex)* if for all $A, B \in K$ which satisfy $\operatorname{rank} A - B = 1$ and all $\lambda \in (0,1)$ one has $(1-\lambda)A + \lambda B \in K$.

The *lamination convex hull* K^L is defined as the smallest lamination convex set that contains K.

Remarks. 1. Gromov uses the term P-convex hull. Our terminology is motivated by the theory of composite materials.

2. Some authors use the term rank-1 convex hull for K^L. We follow the convention that rank-1 convex sets are defined as cosets of rank-1 convex functions (see, e.g. [Sv]). With this terminology K^L is contained in the rank-1 convex hull, but the inclusion may be strict (see [BFJK]).

Proposition 3.1. *The lamination convex hull K^L is given by successively adding rank-1 segments. More precisely*

$$K^L = \cup_i K^{(i)}, \tag{7}$$

where $K^{(0)} = K$ and

$$\begin{aligned} K^{(i+1)} = \{C \in M^{m\times n} : \exists A, B \in K^{(i)},\ \lambda \in (0,1), \quad \textit{such that} \\ C = (1-\lambda)A + \lambda B,\ \operatorname{rank} B - A = 1\} \cup K^{(i)}. \end{aligned} \tag{8}$$

If, moreover, K is open then all the sets $K^{(i)}$ are open.

Proof. Clearly K^L must contain each $K^{(i)}$ as K^L is lamination closed. On the other hand, if $A, B \in \cup_i K^{(i)}$, then $A, B \in K^{(i)}$ for some i. If now $\operatorname{rank} A - B = 1$ then $(1-\lambda)A + \lambda B \in K^{(i+1)}$. Hence $\cup_i K^{(i)}$ is lamination convex and therefore contained in K^L.

Suppose now that K is open. By induction it suffices to show that $K^{(1)}$ is open. Let $C = (1-\lambda)A + \lambda B$, where $A, B \in K$ and $\operatorname{rank} B - A = 1$. Then $C + D \in K^{(1)}$ for sufficiently small $|D|$ since $\operatorname{rank}((A+D) - (B+D)) = 1$ and $A + D, B + D \in K$. □

Lemma 3.2. *Suppose that $K \subset M^{m\times n}$ is open and bounded. Let $v : \Omega \to \mathbf{R}^m$ be a piecewise linear map that satisfies*

$$\nabla v \in K^L \quad a.e.$$

Then, for each $\delta > 0$, there exists a piecewise linear map $u_\delta : \Omega \to \mathbf{R}^m$ which satisfies

$$\nabla u_\delta \in K \quad a.e.,$$

$$\sup_\Omega |u_\delta - v| \le \delta, \quad \textit{and}$$

$$u_\delta = v \quad \textit{on} \quad \partial\Omega.$$

Proof. We may assume that v is affine as otherwise we can argue on each subset Ω_i where v is affine. Normalizing additive constants we assume

$$v(x) = Fx, \quad F \in K^L.$$

By Proposition 5 we have $F \in K^{(i)}$ for some i. We now argue by induction. If $i = 1$ it suffices to apply Lemma 2.1. Suppose the result was proved for $i \leq j$ and let $F \in K^{(j+1)} \setminus K^{(j)}$. There exist $A, B \in K^{(j)}$, $\lambda \in (0,1)$ such that $\operatorname{rank} B - A = 1$ and $F = (1-\lambda)A + \lambda B$. Since $K^{(j)}$ is open, Lemma 2.1 implies that there exists a piecewise linear map w such that

$$\sup_\Omega |w - v| < \delta/2, \quad w = v \quad \text{on} \quad \partial\Omega,$$

and

$$\nabla w \in K^{(j)} \quad \text{a.e.}$$

Let Ω_k denote the subdomains on which w is piecewise affine. By the induction hypothesis there exist piecewise linear maps u_k such that

$$\sup_{\Omega_k} |u_k - w| < \delta/2, \quad u_k = w \quad \text{on} \quad \partial\Omega_k,$$

and

$$\nabla u_k \in K \quad \text{a.e.}$$

Finally, we let $u_\delta = u_k$ on $\overline{\Omega}_k$. □

4. The heart of the matter — closed relations.

We now consider more general bounded sets $K \subset M^{m\times n}$ (in applications K is often compact). To construct solutions of $\nabla u \in K$, we approximate K by open sets U_i. This leads to approximate solutions u_i which satisfy $\nabla u_i \in U_i$. Somewhat surprisingly, it turns out that by a careful choice of u_i one can achieve strong (or a.e) convergence of ∇u_i as $i \to \infty$, despite the fact that the u_i develop increasingly faster spatial oscillations. The idea, which appears already in the work of Nash [Na] and Kuiper [Ku], is to superimpose at each step oscillations which are much faster than all the oscillations that occurred in the previous step. The actual analytic formulation is quite simple (see equations (12) to (15) below).

Before we state the main result, we have to clarify in which sense the sets U_i have to approximate the original set K. We would like to use Lemma 3.2 to obtain the approximation u_{i+1} with $\nabla u_{i+1} \in U_{i+1}$ from u_i. This motivates the following definition.

Definition 3. ([Gr], p. 218) Let $K \subset M^{m\times n}$. A sequence of open sets $U_i \subset M^{m\times n}$ is called an *in-approximation* of K if the following three conditions are satisfied.

1. $U_i \subset U_{i+1}^L$;

2. the sets U_i are uniformly bounded;

3. if a sequence $F_i \in U_i$ converges to $F \in M^{m\times n}$ as $i \to \infty$ then $F \in K$.

In the scalar case $m = 1$ lamination convexity is the same as ordinary convexity and a typical example of an in-approximation is

$$K = S^{n-1}, \quad U_i = \{x \in \mathbf{R}^n : 2^{-(i+1)} < 1 - |x| < 2^{-i}\}.$$

Theorem 4.1. ([Gr], p. 218) *Suppose that $K \subset M^{m\times n}$ admits an in-approximation by open sets U_i in the sense of* Definition 3. *Let $v \in C^1(\Omega)$ and suppose that*

$$\nabla v \in U_1.$$

Then, for every $\delta > 0$, there exists a Lipschitz map u_δ such that

$$\nabla u_\delta \in K \quad a.e., \tag{9}$$

$$u_\delta = v \quad on \quad \partial\Omega, \qquad and \tag{10}$$

$$\sup_\Omega |u_\delta - v| < \delta. \tag{11}$$

Remarks. 1. It follows from the proof that instead of $v \in C^1$ one can assume that v is piecewise linear. We do not know whether for a general Lipschitz v one can find a Lipschitz u_δ such that (9) to (11) hold. The question arises already in Lemma 3.2.

2. The theorem continues to hold if one replaces in Definition 2 the lamination convex hull by the rank-1 convex hull. One can also handle constraints on the Jacobian det ∇u where an in-approximation by open sets is not possible. These refinements will be discussed in a forthcoming paper.

Proof. We first construct a sequence of piecewise linear maps u_i that satisfy

$$\nabla u_i \in U_i \quad \text{a.e},$$
$$\sup |u_{i+1} - u_i| < \delta_{i+1}, \quad u_{i+1} = u_i \quad \text{on} \quad \partial\Omega,$$
$$\sup |u_1 - v| < \delta/4, \quad u_1 = v \quad \text{on} \quad \partial\Omega.$$

To construct u_1 note that if Ω' is open and $\Omega' \subset\subset \Omega$ (i.e $\bar\Omega' \subset \Omega$) then $\text{dist}(\nabla v(x), \partial U_1) \geq c(\Omega') > 0$ for all $x \in \Omega'$. Hence it is easy to obtain

$u_1|_{\Omega'}$ by introducing a sufficiently fine triangulation. Now exhaust Ω by an increasing sequence of sets $\Omega_i \subset\subset \Omega$.

To construct u_{i+1} and δ_{i+1} from u_i and δ_i we proceed as follows. Let

$$\Omega_i = \{x \in \Omega : \mathrm{dist}(x, \partial\Omega) > 2^{-i}\}.$$

Let ϱ be a usual mollifying kernel, i.e. let ϱ be smooth with support in the unit ball and $\int \varrho = 1$. Let

$$\varrho_\varepsilon(x) = \varepsilon^{-n}\varrho(x/\varepsilon).$$

Since the convolution $\varrho_\varepsilon * \nabla u_i$ converges to u_i in $L^1(\Omega_i)$ as $\varepsilon \to 0$ we can choose $\varepsilon_i \in (0, 2^{-i})$ such that

$$||\varrho_{\varepsilon_i} * \nabla u_i - \nabla u_i||_{L^1(\Omega_i)} < 2^{-i}. \tag{12}$$

Let

$$\delta_{i+1} = \delta_i \varepsilon_i. \tag{13}$$

Use Lemma 3.2 to obtain u_{i+1} such that $\nabla u_{i+1} \in U_{i+1}$, $u_{i+1} = u_i$ on $\partial\Omega$ and

$$\sup_\Omega |u_{i+1} - u_i| < \delta_{i+1}. \tag{14}$$

Since $\delta_{i+1} \leq \delta_i/2$, we have

$$\sum_{i=1}^{\infty} \delta_i \leq \delta/2.$$

Thus

$$u_i \to u_\infty \quad \text{uniformly},$$

and u_∞ is Lipschitz since the u_i are uniformly Lipschitz (by (ii) in Definition 7). Taking $u_\delta = u_\infty$ we obtain (10) and (11).

It only remains to show that $\nabla u_\infty \in K$. The key point is to ensure strong convergence of ∇u_i. Since $||\nabla\varrho_\varepsilon||_{L^1} \leq C/\varepsilon$, we deduce from (14) and (13)

$$\begin{aligned} ||\varrho_{\varepsilon_k} * (\nabla u_k - \nabla u_\infty)||_{L^1(\Omega_k)} &= ||\nabla\varrho_{\varepsilon_k} * (u_k - u_\infty)||_{L^1(\Omega_k)} \\ &\leq \frac{C}{\varepsilon_k} \sup |u_k - u_\infty| \leq \frac{C}{\varepsilon_k} \sum_{j=k+1}^{\infty} \delta_j \\ &\leq 2\frac{C}{\varepsilon_k}\delta_{k+1} \leq C'\delta_k. \end{aligned} \tag{15}$$

Taking into account (12), it follows that

$$||\nabla u_k - \nabla u_\infty||_{L^1(\Omega)} \le C'\delta_k + 2^{-k} + ||\varrho_{\varepsilon_k} * \nabla u_\infty - \nabla u_\infty||_{L^1(\Omega_k)} + ||\nabla u_k - \nabla u_\infty||_{L^1(\Omega\setminus\Omega_k)}.$$

Since ∇u_k and ∇u_∞ are bounded, we obtain $\nabla u_k \to \nabla u_\infty$ in $L^1(\Omega)$. Therefore there exists a subsequence u_{k_j} such that

$$\nabla u_{k_j} \to \nabla u_\infty \quad \text{a.e.}$$

It follows from the definition of an in-approximation that

$$\nabla u_\infty \in K \quad \text{a.e.}$$

□

5. Application to the two-well problem.

To give an idea of the results that can be obtained by convex integration we consider sets of the form

$$K = \mathrm{SO}(2)A \cup \mathrm{SO}(2)B, \tag{16}$$

where

$$0 < \det A < \det B. \tag{17}$$

The problem of finding maps $u : \Omega \subset \mathbf{R}^2 \to \mathbf{R}^2$ that satisfy

$$\nabla u \in K \quad \text{a.e.} \tag{18}$$

if often referred to as the two-well problem. A short calculation shows that, depending on the choice of A and B, each matrix in K may be rank-1 connected to two, one or no other matrices in K (we say that F and G are rank-1 connected if $\operatorname{rank} G - F = 1$). Denote by $K(A, B)$ the lamination convex hull of K,

$$K(A,B) = \Big(\mathrm{SO}(2)A \cup \mathrm{SO}(2)B\Big)^L.$$

By $\lambda_1(BA^{-1}) \le \lambda_2(BA^{-1})$ we denote the singular values of BA^{-1}, i.e. the eigenvalues of $\Big((BA^{-1})^T(BA^{-1})\Big)^{1/2}$.

Lemma 5.1. *Suppose that* $\lambda_1(BA^{-1}) < 1 < \lambda_2(BA^{-1})$.

(i) *Then the lamination convex hull is given by*

$$K(A,B) = \Bigg\{ F : F = \begin{pmatrix} y_1 & -y_2 \\ y_2 & y_1 \end{pmatrix} A + \begin{pmatrix} z_1 & -z_2 \\ z_2 & z_1 \end{pmatrix} B,$$
$$|y| \leq \frac{\det B - \det F}{\det B - \det A},$$
$$|z| \leq \frac{\det F - \det A}{\det B - \det A} \Bigg\}.$$

(ii) *Moreover, the interior of* $K(A,B)$ *is obtained by replacing the inequalities for* $|y|$ *and* $|z|$ *by strict inequalities.*

Proof. After an affine change of variables we may suppose $A = \mathrm{Id}$, $B = \begin{pmatrix} \lambda_1 & \\ & \lambda_2 \end{pmatrix}$. Then the formula for $K(A,B)$ was obtained in Šverak [Sv]. One easily checks that $\det A \leq \det F \leq \det B$ for all $F \in K(A,B)$. If strict inequalities hold at a matrix F_0 then F_0 is in the interior of K since the coordinate map $F \mapsto (y,z)$ is nonsingular as $\lambda_1 \neq \lambda_2$.

Suppose conversely that for $F_0 = (y_0, z_0)$

$$|z_0| = \frac{\det F_0 - \det A}{\det B - \det A}.$$

If F_0 was an interior point of K then one would have

$$\det(F_0 + tY) \geq \det F_0 \quad \text{for all} \quad Y \in SO(2)$$

and all sufficiently small t (since $z(F_0 + tY) = z_0$). It follows that

$$0 = F_0 : \operatorname{cof} Y = F_0 : Y \quad \forall Y \in SO(2).$$

Thus

$$F_0 = \begin{pmatrix} a & b \\ b & -a \end{pmatrix}$$

which implies $\det F_0 \leq 0$. This contradiction shows that F_0 must be a boundary point of K. The same argument applies if equality holds in the inequality for $|y|$.

Corollary 5.2. *Let* $V \subset M^{2\times 2} \times M^{2\times 2}$ *be the set*

$$V = \Big\{ (A,B) : \lambda_1(BA^{-1}) < 1 < \lambda_2(BA^{-1}), \quad 0 < \det A < \det B \Big\}.$$

Then the map $(A,B) \to K(A,B)$ is continuous on V in the following sense: if $(A_j,B_j) \to (A,B)$ then each compact set $G \subset \operatorname{int} K(A,B)$ is contained in $K(A_j,B_j)$ for sufficiently large j.

Proof. This follows from (ii) of Lemma 5.1 and the continuity of the coordinate map $(A,B,F) \mapsto (y,z)$ on $V \times M^{2\times 2}$.

Theorem 5.3. *Suppose* $A = \mathrm{Id}$, $B = \begin{pmatrix} \lambda_1 & \\ & \lambda_2 \end{pmatrix}$, $\lambda_1 < 1 < \lambda_2$, $\lambda_1\lambda_2 > 1$. *Let* $F_0 \in \operatorname{int} K(A,B)$. *Then there exists a solution of the two well problem* (18) *that satisfies*

$$u = F_0 x \quad on \quad \partial\Omega.$$

Proof. By Theorem 4.1 it suffices to construct an in-approximation. The existence of such an approximation is a direct consequence of Corollary 5.2. One possible construction is as follows.
Let $U_1 \subset\subset \operatorname{int} K(A,B)$ be open neighbourhood of F_0 and let $\delta > 0$. By Corollary 5.2 there exist $A_2, B_2 \in \operatorname{int} K(A,B)$ such that

$$\overline{U}_1 \subset K(A_2,B_2) = \left(\mathrm{SO}(2)A_2 \cup \mathrm{SO}(2)B_2\right)^L.$$

We may suppose that $|A_2 - A| < \delta/2$, $|B_2 - B| < \delta/2$. Let U_2 be a $\delta/2$ neighbourhood of $\mathrm{SO}(2)A_2 \cup \mathrm{SO}(2)B_2$ then

$$U_1 \subset U_2^L,$$

and

$$\operatorname{dist}(F, \mathrm{SO}(2)A \cup \mathrm{SO}(2)B) < \delta \quad \forall F \in U_2.$$

Proceeding inductively we obtain the desired in-approximation. □

References.

[BJ 1] J.M. Ball and R.D. James, *Fine phase mixtures as minimizers of energy,* Arch. Rat. Mech. Anal. **100** (1987), 13–52.

[BJ 2] J.M. Ball and R.D. James, *Proposed tests of a theory of fine microstructure and the two-well problem,* Phil. Trans. R. Soc. London **A 388** (1992), 389–450.

[BFJK] K. Bhattacharya, N. Firoozye, R.D. James and R.V. Kohn, *Restrictions on microstructure,* Proc. Roy. Soc. Edinburgh **124 A** (1994), 843–878.

[CP] A. Celina and S. Perrotta, *On a problem of potential wells,* preprint SISSA.

[Gr] M. Gromov, *Partial differential relations,* Springer, 1986.

[JK] R.D. James and R.V. Kohn, personal communication.

[Ku] N.H. Kuiper, *On C^1 isometric embeddings,* I. Proc. Könikl. Nederl. Ak. Wet., **A 58** (1955), 545–556.

[Mo] C.B. Morrey, *Multiple integrals in the calculus of variations,* Springer, 1966.

[Na] J. Nash, *C^1 isometric embeddings,* Ann. Math. **60**, 383–396.

[Po] Y. Pomeau, Papier froissé, C.R.A.S. Paris **320** (1995), 975–979.

[Sv] V. Šverák, *On the problem of two wells*, in: *Microstructure and phase transitions,* IMA Vol. Appl. Math. **54** (eds. J. Eriksen, R.D. James, D. Kinderlehrer, M. Luskin), Springer, 1993, pp. 183–189.

RECEIVED AUGUST 10, 1995.

MATHEMATISCHES INSTITUT
ALBERSTR. 23B
UNIVERSITÄT FREIBURG
D–79104 FREIBURG, GERMANY

AND

SCHOOL OF MATHEMATICS
UNIVERSITY OF MINNESOTA
MINNEAPOLIS, MN 55455, USA

On the existence and uniqueness of constant mean curvature hypersurfaces in hyperbolic space[1]

BARBARA NELLI AND JOEL SPRUCK

1. Introduction.

Let Γ be an embedded codimension one submanifold of $\partial_\infty \mathbb{H}^{n+1}$ (the boundary at infinity of hyperbolic space). We study the problem of finding a constant mean curvature hypersurface of $\mathbb{H}^{n+1}$ with prescribed asymptotic boundary Γ. To state precisely our results, we must first give some definitions.

Consider the halfspace model for hyperbolic space, i.e.

$$\mathbb{H}^{n+1} = \{ (x_1, \ldots, x_{n+1}) \in \mathbb{R}^{n+1} \mid x_{n+1} > 0 \}$$

with the *hyperbolic metric* $ds^2 = \sum_{i=1}^{n} \frac{dx_i^2}{x_{n+1}^2}$. In the halfspace model, we view $\Gamma = \partial\Omega$ as a codimension one submanifold of euclidean space $\mathbb{R}^n$, with Ω a bounded $C^{2,\alpha}$ domain. We denote by H_Γ the mean curvature of Γ with respect to the interior normal vector, computed in the Euclidean metric. We say that Γ is *mean convex* if $H_\Gamma > 0$ at each point of Γ. We remark that mean convexity is not an intrinsic hyperbolic notion.

We shall prove that any such mean convex Γ is the asymptotic boundary of a complete embedded hypersurface of $\mathbb{H}^{n+1}$ of constant mean curvature H, for each $H \in (0,1)$ (here and in the following the mean curvature is computed with respect to the upward normal vector). We construct the desired M as a limit of constant mean curvature graphs over a fixed compact domain in a horosphere, for constant boundary data. By graphs, we mean graphs in the system of coordinates defined as follows: at a point p on the horosphere $\{x_{n+1} = c\} = L(c)$ we associate a point on the geodesic passing by p and orthogonal to the horosphere (i.e. the vertical geodesic passing by p). Thus an important part of our study concerns the existence and uniqueness of constant mean curvature hypersurfaces which are graphs over

[1]Dedicated to Stefan Hildebrandt on his sixtieth birthday

a bounded domain in a horosphere, whose boundary is mean convex. For such graphs, we are able to prove existence and uniqueness for $H \in (0,1)$. This leads to the following

Theorem 1.1. *Let Ω be a bounded domain in $L(c)$, respectively $\partial_\infty \mathbb{H}^{n+1}$ such that $\Gamma = \partial\Omega$ is of class $C^{2,\alpha}$ and mean convex. Then for each $H \in (0,1)$ there exists a complete embedded hypersurface M of $\mathbb{H}^{n+1}$ of constant mean curvature H with $\partial M = \Gamma$, respectively $\partial_\infty M = \Gamma$. Moreover, M can be represented as a graph $x_{n+1} = u(x)$ over Ω with $u \in C^{2,\alpha}(\bar{\Omega})$ and there is a unique such graph.*

If Γ is mean convex and bounds a star-shaped domain Ω, we have a stronger uniqueness result. We say that Γ is the *asymptotic homological boundary* of a hypersurface M in $\mathbb{H}^{n+1}$ if, for each c sufficiently small, $M \cap L(c) = \Gamma(c)$, where $\Gamma(c) \to \Gamma$ as $c \to 0$ and $\Gamma(c)$ is homologous to 0 in M. We denote the asymptotic homological boundary of M by $\partial_\infty M$.

Theorem 1.2. *Let Ω be a bounded domain in $L(c)$, respectively $\partial_\infty \mathbb{H}^{n+1}$ such that $\Gamma = \partial\Omega$ is of class $C^{2,\alpha}$ and mean convex. Let M be an embedded hypersurface of constant mean curvature $H \in (0,1)$ such that $\partial M = \Gamma$, respectively $\partial_\infty M = \Gamma$ and such that the mean curvature vector at the highest point of M points upward. Then M is the unique graph constructed in Theorem 1.1.*

We remark that an embedded hypersurface of constant mean curvature bigger than one, with asymptotic homological boundary a codimension one embedded submanifold of $\partial_\infty \mathbb{H}^{n+1}$ does not exist; this follows easily by comparing such a hypersurface with horospheres.

The study of minimal hypersurfaces, $H = 0$, with prescribed asymptotic boundary was initiated by Anderson [A] using methods of geometric measure theory. The boundary regularity of these solutions was studied by Hardt and Lin [HL] and Lin [L]. The extension of these results to constant $H \in (0,1)$ is due to Tonegawa [T] who makes a detailed study of boundary regularity, using the methods of [L]. Our Theorem 1.2 says that the geometric measure theory solution studied by Tonegawa is actually a topological disk when Γ is mean convex and star-shaped.

In the case of intrinsic Gauss curvature between -1 and 0, Rosenberg and Spruck [RS] completely answered the question of existence and uniqueness of graph type solutions. They proved that any embedded codimension one submanifold of $\partial_\infty \mathbb{H}^{n+1}$ is the asymptotic homological boundary of a complete embedded hypersurface of constant intrinsic Gauss curvature K, for each $K \in (-1,0)$; furthermore they proved that in $\mathbb{H}^3$, there are exactly two such surfaces, each of one is a graph over one of the two components of $\partial_\infty \mathbb{H}^{n+1} \setminus \Gamma$. Our approach follows the spirit of [RS]. It is possible that the graphical solutions of Theorem 1.1 always exist for any smooth Ω without convexity condition (this is false for $H=0$) but this is far from clear.

The first author wishes to thank her thesis advisor, Harold Rosenberg, for his continuous support and interest in the present work.

2. Constant mean curvature graphs.

Theorem 2.1. *Let Ω be a $C^{2,\alpha}$ subdomain of a horosphere $L(c)$ such that $\Gamma = \partial\Omega$ is mean convex. Then, for each $H \in (0,1)$ there exists a $C^{2,\alpha}$ graph M over $\bar{\Omega}$ of constant mean curvature H such that $\partial M = \Gamma$. Furthermore the graph M is unique.*

We claim that if $u\colon \Omega \to \mathbb{R}$ is a $C^{2,\alpha}$ solution of the following Dirichlet problem

$$\begin{cases} F[u] = \operatorname{div}\left(\dfrac{\nabla u}{W_u}\right) - \dfrac{n}{u}\left(h - \dfrac{1}{W_u}\right) & \text{in } \Omega, \\ u = c \quad \text{on } \Gamma, \end{cases} \tag{A}$$

where $W_u = \sqrt{1+|\nabla u|^2}$, then the graph of u is a hypersurface of constant mean curvature H with boundary equal to Γ. In fact, the hyperbolic metric in the halfspace model is conformally equivalent to the Euclidean metric with coefficient of conformality x_{n+1}^{-2}, so the principal curvatures of M in $\mathbb{H}^{n+1}$ are given by

$$k = x_{n+1}k_e + n_{n+1},$$

where k_e is the Euclidean principal curvature and n_{n+1} is the last component of the unit (in the Euclidean metric) normal vector to M. So, if we denote by H_e the Euclidean mean curvature, we have

$$H = x_{n+1}H_e + n_{n+1}$$

The claim follows, if we substitute in the previous equality the well known formula for the Euclidean mean curvature of a graph

$$H_e = \frac{1}{n}\operatorname{div}\left(\frac{\nabla u}{W_u}\right).$$

Proof of Theorem 2.1. Set

$$S = \{\, t \in [0,1] \mid \exists\, u^t \text{ admissible solution of } (\mathrm{A}^t) \,\}.$$

Ry Remark 2.3, $0 \in S \neq \emptyset$ so if we prove that S is open and closed, we have $S = [0,1]$ and the admissible solution u_1 is a solution of the Dirichlet problem (A).

(i) First, we prove that we can solve (A^t) in a neighborhood of $t = 0$ in $[0,1]$ with the aid of the Implicit Function Theorem.

Consider the linear operator $\mathfrak{L}^t_u\colon C^{2,\alpha}(\bar\Omega) \to C^{0,\alpha}(\bar\Omega)$ defined by

$$\mathfrak{L}^t_u h = F^t_{ij}(x) h_{ij} + b^t_i(x) h_i + c^t(x) h$$

where

$$F^t_{ij}(x) = \frac{\partial F^t}{\partial u_{ij}}(x,u,\nabla u)$$
$$b^t_i(x) = \frac{\partial F^t}{\partial u_i}(x,u,\nabla u)$$
$$c^t(x) = \frac{\partial F^t}{\partial u}(x,u,\nabla u) = \frac{n}{u^2 W_t}(tHW_t - 1)$$

$\mathfrak{L}^0_{u^0}$ is invertible since $c^0(x) = -\frac{n}{(u^0)^2 W_{u^0}} < 0$ (cf. Theorem 6.14 [GT]).

Then, by the Implicit Function Theorem (cf. Theorem 17.6 [GT]), there exists $t_0 > 0$ such that for each $t \in [0, t_0)$ there exists a solution u^t of (A^t) and such that each u^t varies continuously in t in the norm $C^{2,\alpha}(\bar\Omega)$. In particular, there exists a positive constant M such that

$$|W_{u^t} - W_{u^0}| \le M t_0$$

hence

$$tHW_{u^t} - 1 < t_0(W_{u^0} + Mt_0)H - 1$$

and the second term is negative if t_0 is small enough.

So, for each $t \in [0, t_0)$, we have found an admissible solution of (A^t).

To prove that S is open and closed, we can assume $t \ge t_0 > 0$.

(ii) S is open in $[0,1]$. Let $t_1 \in S$, $t_1 \ge t_0$, and let u^{t_1} be an admissible solution of (A^{t_1}). The linear operator $\mathfrak{L}^t_u$ defined in (i) is invertible in B^t as $c^t(x) \le 0$ for each $u \in B^t$, for each $t \ge t_0$.

Then, by the Implicit Function Theorem, there exists ϵ, $0 < \epsilon < t_0$, such that for each $t \in (t_1 - \epsilon, t_1 + \epsilon)$ there exists a solution u^t of (A^t) and such that u^t varies continuously in t in the $C^{2,\alpha}(\bar{\Omega})$ norm. By the same argument used in (i), we obtain that $(t_1 - \epsilon, t_1 + \epsilon) \subset S$, hence S is open.

(iii) S is closed in $[0,1]$. Let $t \in \bar{S}$, $t \geq t_0$, and let $\{t_m\} \subset S$ be a sequence such that $t_m \geq t_0$ for each m, and $t_m \to t$. Let $\{u^m\}$ be the corresponding sequence of admissible solutions of (A^{t_m}).

For each m, $t_m W_{u^m} H - 1 < 0$, hence $W_{u^m} < (t_0 H)^{-1}$; so the set $\{u^m\}$ is $C^{2,\alpha}(\bar{\Omega})$ bounded by a constant not depending on m. Up to a subsequence there exists $u^t \in C^{2,\alpha}(\bar{\Omega})$ such that $u^m \to u^t$ in $C^{2,\alpha}(\bar{\Omega})$. By continuity u^t is a solution of (A^t) and $tW_{u^t}H - 1 \leq 0$.

To prove that u^t is an admissible solution, we have to show that $tW_{u^t}H - 1 < 0$.

First we prove that the maximum of $|\nabla u^t|$ is on the boundary Γ for each solution of (A^t) such that $tW_{u^t}H - 1 \leq 0$.

By differentiating equation $F^t[u] = 0$ with respect to x_k, $k \leq n$, we obtain that $v = u_k^t$ satisfies a linear differential equation of the form

$$a_{ij}(x)v_{ij} + b_i(x)v_i + c(x)v = 0, \tag{1}$$

where $c(x) = tHW_{u^t} - 1 \leq 0$. By the maximum principle, v attains its maximum at the boundary and hence

$$\sup_{\Omega} |\nabla u^t| = \sup_{\Gamma} |\nabla u^t|.$$

Now, we evaluate the maximum of W_{u^t} (i.e. $|\nabla u^t|$) on the boundary Γ.

Let $0 \in \Gamma$ be a point of maximum of $|\nabla u^t|$ and choose coordinates on $L(c)$ so that the positive x_n-axis is the interior normal to Γ at 0 (i.e. $u_n^t(0) = |\nabla u^t|(0)$ and $u_{nn}(0) \leq 0$). Near 0, we can represent Γ as a graph $x_n = \rho(x')$, where $x' = (x_1, \ldots, x_{n-1})$, $\rho(0) = \rho_\alpha(0) = 0$, $\alpha < n$.

Consider the constant function $\underline{u} = c$; as $F^t[\underline{u}] = -\frac{n}{c}(tH - 1) > 0$, $\underline{u}$ is a subsolution of (A^t), hence $u^t \geq c$ in Ω and $u_n^t > 0$ on Γ by the Hopf boundary point lemma.

Since $u^t(x', \rho(x')) = c$, differentiating with respect to $\alpha, \gamma < n$, we have

$$u_{\alpha\gamma}^t(0) = -u_n^t(0)\rho_{\alpha\gamma}(0).$$

Substituting in $F^t[u^t] = 0$ gives

$$u_{nn}^t(0) - W_{u^t}^2 u_n^t(0) \sum_{\alpha<n} \rho_{\alpha\alpha}(0) - \frac{n}{u} W_u^2 (tHW_{u^t} - 1) = 0.$$

Since $\sum_{\alpha<n} \rho_{\alpha\alpha} = (n-1)H_\Gamma$ and $u_{nn}^t \leq 0$, we obtain

$$u_n^t(0)(n-1)H_\Gamma + \frac{n}{c}(tHW_{u^t} - 1) \leq 0.$$

As $H_\Gamma > 0$ and $u_n^t > 0$ at each point of Γ, the last inequality implies at 0

$$tHW_{u^t} - 1 < 0$$

hence

$$\max_\Omega W_{u^t} \le \max_\Gamma W_{u^t} < \frac{1}{tH}.$$

Thus u^t is an admissible solution and S is closed.

We have proved the existence part of Theorem 2.1.

(iv) Uniqueness. Let u be an admissible solution of the Dirichlet problem (A) and v be an arbitrary solution; by Remark 2.2 $u \ge c$, $v \ge c$. Let $x^0 \in \Omega$ be such that the function $w = v - u$ takes on its maximum at x^0 (x^0 is interior); hence $\nabla w(x^0) = \nabla v(x^0) - \nabla u(x^0) = 0$. As u is admissible, in a neighborhood of x^0 we have $HW_v - 1 = HW_u - 1 < 0$. Thus w also satisfies a linear equation of the form (1) and so, by the maximum principle $w \le 0$. By reversing the role of u and v we find $w \equiv 0$.

Theorem 2.1 is proved. □

Theorem 2.5. *Let Ω be a $C^{2,\alpha}$ subdomain of $\partial_\infty \mathbb{H}^{n+1}$ such that $\Gamma = \partial\Omega$ is mean convex. Then for each constant $H \in (0,1)$ there exists a graph*

$$M = \{x_{n+1} = u(x),\ u \in C^\infty(\Omega) \cap C^{0,1}(\bar{\Omega})\}$$

over Ω of constant mean curvature H such that $\partial_\infty M = \Gamma$. Furthermore the graph M is unique.

Proof. Let $\Gamma(c)$, $\Omega(c)$ be the vertical translations of Γ and Ω to $L(c)$. By Theorem 2.1, $\Gamma(c)$ is the boundary of a graph of a function u_c over $\Omega(c)$ of constant mean curvature H.

To prove the theorem we will pass to the limit as $c \to 0$ for the Dirichlet problems

$$\begin{cases} F[u] = 0 & \text{in } \Omega(c), \\ u = c & \text{on } \Gamma(c). \end{cases}$$

The sequence $\{u_c\}$ is decreasing with c. In fact, if $c' < c$

$$u_c|_{\Gamma(c)} - u_{c'}|_{\Gamma(c')} = c - c' > 0$$

hence, by the maximum principle ($HW - 1 < 0$)

$$u_c - u_{c'} \geq 0 \quad \text{in } \Omega.$$

Furthermore, u_c has a positive lower bound independent of c in Ω. In fact equidistant spheres (i.e. the set of equidistant points from a hyperbolic hyperplane) with asymptotic boundary in Ω, of constant mean curvature H, whose mean curvature vector points upward are lower barriers for u_c for all c. Therefore, by Schauder estimates, we have uniform bounds for u_c in $C^{0,1}(\bar{\Omega}) \cap C^{2,\alpha}(\Omega')$ for any compact subdomain Ω' strictly contained in Ω, independent of c. So, we can pass to the limit (up to subsequence) for $c \to 0$ and obtain a solution $u \in C^{0,1}(\bar{\Omega}) \cap C^{2,\alpha}(\Omega)$ of the Dirichlet problem

$$\begin{cases} F[u] = 0 & \text{in } \Omega, \\ u = 0 & \text{on } \Gamma. \end{cases} \tag{A$^\infty$}$$

By standard elliptic regularity $u \in C^\infty(\Omega)$.

Uniqueness is obtained as in (iv) of Theorem 2.1. □

3. Higher regularity.

The graph M of Theorem 2.5 was obtained by constructing a solution $u \in C^\infty(\Omega) \cap C^{0,1}(\bar{\Omega})$ of the degenerate Dirichlet problem

$$\begin{cases} F[u] = \operatorname{div}\left(\dfrac{\nabla u}{W_u}\right) - \dfrac{n}{u}\left(h - \dfrac{1}{W_u}\right) & \text{in } \Omega, \\ u = c \quad \text{on } \Gamma, \end{cases} \tag{A$^\infty$}$$

where $W_u = \sqrt{1 + |\nabla u|^2} \leq H^{-1}$.

We will show in fact that $u \in C^{2,\alpha}(\bar{\Omega})$. Define $v(x) = \lambda d(x)$, where $\lambda = \frac{\sqrt{1-H^2}}{H}$ and $d(x)$ is the distance from x to $\partial\Omega$.

Lemma 3.1. *$u(x) < v(x)$ in Ω.*

Proof. Suppose $T = \sup_\Omega (u - v) > 0$ and $u(x_0) - v(x_0) = T$, $x_0 \in \Omega$. Let $y_0 \in \partial\Omega$ be a closest point to x_0 i.e. $|x_0 - y_0| = d(x_0)$, and choose coordinates $(y_1, \ldots, y_n)$ at y_0 such that e_n is the interior unit normal and $\partial\Omega$ is locally represented as a graph $y_n = \rho(y_1, \dots, y_{n-1})$ with $\rho(0) = \rho_\alpha(0) = 0$ and $\rho_{\alpha\beta}(0) = \kappa_\alpha \delta_{\alpha\beta}$, $\alpha, \beta < n$, where κ_α are the principal curvatures of $\partial\Omega$ at y_0 with respect to the normal e_n. Then $u(x) - u(x_0) \leq v(x) - v(x_0) \leq \lambda|x - x_0|$, hence $|\nabla u(x_0)| \leq \lambda$. On the other hand,

$$u(y_0 + se_n) \leq v(y_0 + se_n) + T \leq \lambda s + T, \quad 0 \leq s \leq d(x_0)$$

and

$$u(y_0 + d(x_0)e_n) = u(x_0) = v(x_0) + T = \lambda d(x_0) + T.$$

Thus

$$|\nabla u(x_0)| = u_n(x_0) = \lambda, \quad u_{nn}(x_0) \leq 0.$$

Consider now the level set

$$\Gamma_0 = \{ x \in \Omega \mid u(x) = \lambda d(x_0) + T \}$$

passing through x_0; since $|\nabla u(x_0)| = \lambda$, Γ_0 is smooth near x_0 and also $d(x) \geq d(x_0)$ on Γ_0. Hence we can find a small ball $B_\epsilon(z_0)$ (hence also $d(x) > d(x_0)$ on $B_\epsilon(z_0)$). According to the geometric meaning of $d(x)$, the ball of radius $d(x_0) + \epsilon$ centered at z_0 is contained in $\bar{\Omega}$ (for otherwise there exists $z \in B_\epsilon(z_0)$ with $d(z) < d(x_0)$; a contradiction). This implies that $z_0 = y_0 + (d(x_0) + \epsilon)e_n$ and moreover

$$1 - \kappa_i(y_0)d(x_0) \geq 1 - \frac{d(x_0)}{d(x_0) + \epsilon} = \frac{\epsilon}{d(x_0) + \epsilon} > 0, \quad i = 1, \dots, n-1.$$

It follows that $d(x)$ is actually C^2 near x_0 and satisfies

$$\begin{aligned} |\nabla d| &\equiv 1, \\ d_i d_j d_{ij} &= 0 \end{aligned}$$

near x_0.

$$\Delta d(x_0) = -\sum_{i=1}^{n-1} \frac{\kappa_i(y_0)}{1 - \kappa_i(y_0)d(x_0)} \leq -(n-1)H_\Gamma(y_0) < 0.$$

Therefore $v - \lambda d(x)$ satisfies

$$\operatorname{div}\left(\frac{\nabla v}{W_v}\right) < 0, \quad HW_v - 1 = 0$$

at x_0.

Since $u - v$ has its maximum at x_0

$$\operatorname{div}\left(\frac{\nabla u}{W_u}\right) \leq \operatorname{div}\left(\frac{\nabla_v}{W_v}\right) < 0, \quad HW_u - 1 = HW_v - 1 = 0$$

at x_0.

This gives $F[u] < 0$ at x_0; a contradiction. □

Remark 3.2. The above argument really shows that $v = \lambda d(x)$ is a strict viscosity supersolution of (A^∞).

Now fix δ_0 so small that each point $P \in \Gamma$ can be touched by an interior tangent ball B_{δ_0} of radius δ_0. Choosing P as origin and introducing coordinates $(x_1, \ldots, x_n)$ with e_n the interior normal to Γ at 0, there is an equidistant sphere solution $w(x) \leq u(x)$ of (A^∞) which is a graph over $B_{\delta_0}(\delta_0 e_n)$ given by

$$w(x) = -RH + \sqrt{R^2 - \sum_{\alpha<n} x_\alpha^2 - (x_n - \delta_0)^2}, \quad R = \frac{\delta_0}{\sqrt{1-H^2}}.$$

Expanding w and $v = \lambda d(x)$ in a Taylor series about the origin, we find (with $\lambda = \frac{\sqrt{1-H^2}}{H}$)

$$w(x) = \lambda x_n + O(|x|^2),$$
$$v(x) = \lambda x_n + O(|x|^2).$$

Since $w \leq u \leq v$ in $B_{\delta_0}(\delta_0 e_n)$ this gives

Lemma 3.3. *Let $x_0 = \delta e_n$, $\delta \leq \delta_0$ and let $x \in B_{\frac{\delta}{2}}(x_0)$. Then*

$$|u(x) - \lambda x_n| \leq C\delta^2$$

with C independent of δ.

Now observe that $l(x) = \lambda x_n$ is also a solution of $F = 0$. Since homothety from $0 \in \Gamma$ is a hyperbolic isometry, the rescaled functions

$$u^\delta(x) = \frac{1}{\delta} u(\delta x), \quad l^\delta(x) = \frac{1}{\delta} l(\delta x) = l(x)$$

are solutions of $F = 0$ in $B_{\frac{1}{2}}(e_n)$. By standard interior estimates (since $|\nabla u^\delta| \leq \frac{1}{H}$, $\frac{\lambda}{10} \leq u^\delta \leq 10\lambda$ in $B_{\frac{1}{2}}(e_n)$) all derivatives of u^δ are uniformly bounded in $B_{\frac{1}{4}}(e_n)$. Therefore the difference $u^\delta - l^\delta$ satisfies a uniformly elliptic equation with nice coefficients. This implies that

$$\sup_{B_{\frac{1}{8}}(e_n)} |\nabla(u^\delta - l^\delta)| \le C(H) \sup_{B_{\frac{1}{2}}(e_n)} |u^\delta - l^\delta| \le C\delta,$$

$$\sup_{B_{\frac{1}{8}}(e_n)} |\nabla^2(u^\delta - l^\delta)| \le C(H) \sup_{B_{\frac{1}{2}}(e_n)} |u^\delta - l^\delta| \le C\delta$$

by Lemma 3.3. Returning to the original variables, this gives

$$|\nabla u(x_0) - \lambda e_n| \le C\delta, \quad |\nabla^2 u(x_0)| \le C$$

with C independent of δ. Thus we have proved

Theorem 3.4. *Let $u \in C^\infty(\Omega) \cap C^{0,1}(\bar\Omega)$ be a solution of (A^∞). Then $u \in C^{1,1}(\bar\Omega)$ and $W_u = H^{-1}$ on $\partial\Omega$.*

We will now utilize the work of Tonegawa (cf. [T]) to show that $u \in C^{2,\alpha}(\bar\Omega)$. Let $0 \in \partial\Omega$ and represent $\partial\Omega$ near 0 as a graph $x_n = \rho(x)$, $\rho(0) = \rho_\alpha(0) = 0$, $\alpha < n$ with e_n the interior normal to $\partial\Omega$ at 0.

We flatten $\partial\Omega$ near 0 by the transformation

$$y = (x', u(x)).$$

Since $u_n(0) = \lambda > 0$, there is a local inverse map $x_n = \psi(y)$ (the zeroth order Legendre transform in terminology of [KNS]) defined on a small upper half-ball $B_\delta^+(0)$. Using the transformation rules

$$u_\alpha(x) = -\frac{\psi_\alpha(y)}{\psi_n(y)}, \quad \alpha < n,$$

$$u_n(x) = \frac{1}{\psi_n(y)},$$

$$\frac{\partial}{\partial x_\alpha} = \frac{\partial}{\partial y_\alpha} - \frac{\psi_\alpha}{\psi_n}\frac{\partial}{\partial y_n}, \quad \alpha < n,$$

$$\frac{\partial}{\partial x_n} = \frac{1}{\psi_n}\frac{\partial}{\partial y_n}$$

it follows that $\psi \in C^{1,1}(\bar B_\delta^+(0))$ satisfies

$$\begin{cases} \left(\delta_{ij} - \dfrac{\psi_i\psi_j}{W_\psi^2}\right)\psi_{ij} = \dfrac{n}{y_n}(\psi_n - HW_\psi) & \text{in } B_\delta^+(0), \\ \psi(y',0) = \rho(y') & \text{on } \{y_n = 0\}. \end{cases} \tag{$*$}$$

Geometrically, this is just the representations of the graph $x_{n+1} = u(x)$ over the vertical plane passing through the tangent plane to $\partial\Omega$ at the origin and $(*)$ is just the equation of constant H in these coordinates.

This is precisely the situation studied in [T]. Applying his Theorem 2.14, we obtain that $\psi \in C^{2,\alpha}(\bar{B}_\delta^+(0))$. This gives

Theorem 3.5. *Let $u \in C^\infty(\Omega) \cap C^{0,1}(\bar{\Omega})$ be a solution of* (A^∞). *Then* $u \in C^{2,\alpha}(\bar{\Omega})$.

For further regularity results see [T].

4. A uniqueness result.

Let M be an embedded hypersurface of $\mathbb{H}^{n+1}$ such that $\partial M = \Gamma \subset L(c)$, $c \geq 0$; let Ω be the compact domain in $L(c)$ such that $\partial\Omega = \Gamma$. By Remark 2.2, if M has constant mean curvature $H \in (0,1)$, then it lies above $L(c)$; hence M divides $\mathbb{H}^{n+1} \cap \{x_{n+1} \geq c\}$ in two connected components. We denote by $\mathfrak{B}$ the component of $(\mathbb{H}^{n+1} \setminus M) \cap \{x_{n+1} \geq c\}$ that does not contain Ω. As H is constant, the mean curvature vector of M points towards the same component at each point of M.

Theorem 4.1. *Let Ω be a star-shaped (with respect to some interior point) subdomain of a horosphere $L(c)$ (respectively $\partial_\infty\mathbb{H}^{n+1}$) such that $\Gamma = \partial\Omega$ is mean convex. Let M be an embedded hypersurface of constant mean curvature $H \in (0,1)$ such that $\partial M = \Gamma$ (respectively $\partial_\infty M = \Gamma$), with mean curvature vector that points towards $\mathfrak{B}$. Then M is a graph over Ω, so M is the unique disk given by Theorem 2.1 (respectively 2.5).*

Proof. We start by proving uniqueness for a hypersurface with boundary at infinity. By Theorem 2.5, there exists a graph S over Ω of constant mean curvature H, with asymptotic boundary Γ. Denote by 0 the point with respect to which Ω is star-shaped and consider the family of hyperbolic isometries $\{H_t\}_{t\in\mathbb{R}}$ generated by translations along the vertical geodesic passing by 0; each H_t is an Euclidean homothety and we can choose the parameter t such that $H_1 = Id$. The fact that Ω is star-shaped with respect to 0 guarantees that, if $t \neq 1$, then $H_t(\Gamma) \cap M = \emptyset$. For t big enough, $H_t(S) \cap M = \emptyset$ and $H_t(S)$ is above M. Then, decrease t until we have a first point of contact between $H_t(S)$ and M; by the maximum principle, the first point of contact must be on the boundary or $M = S$. Hence, S is above M or equal to it.

Now, let t be small enough to have $H_t(S) \cap M = \emptyset$ and $H_t(S)$ below M. By increasing t, we find a first point of contact, that cannot be interior by the maximum principle. Hence S is below M. So, $M = S$.

Now let M have compact boundary Γ on $L(c)$. Consider the family $\{h_t\}_{t\in\mathbb{R}}$ of horizontal homotheties about the point $0 \in \Omega$, such that $h_1 = Id$. As Ω is star-shaped with respect to 0, the family $\{h_t(\Gamma)\}_{t\in\mathbb{R}} = \{C_t\}_{t\in\mathbb{R}}$ is a foliation of $L(c)$ by mean convex codimension one submanifolds that are the boundaries of star-shaped domains Ω_t in $L(c)$.

By Theorem 2.1, for each t, there exists a unique graph N_t over Ω_t of constant mean curvature H, such that $\partial N_t = C_t$. We can choose $\tau, \sigma \in \mathbb{R}$, $\tau < 1$, $\sigma > 1$, such that N_τ is below M and N_σ is above M (Figure 1).

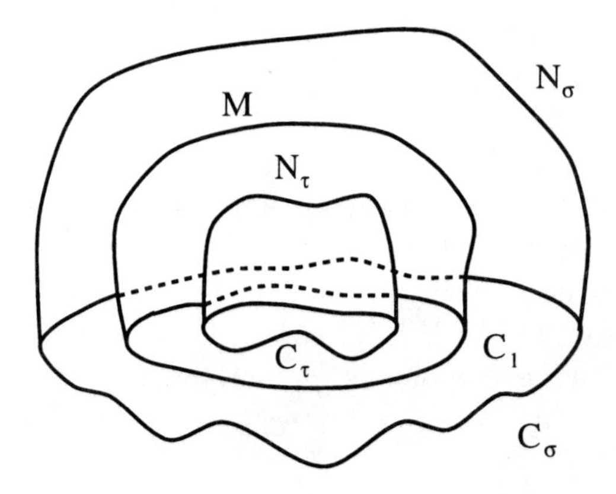

Figure 1.

In fact we can find an equidistant spheres S_1 of constant mean curvature H with constant mean curvature vector that points upward, lying below M and one such equidistant sphere S_2 lying above M. Let Γ_1 and Γ_2 be the codimension one spheres in $L(c)$ such that $\Gamma_1 = \partial S_1$ and $\Gamma_2 = \partial S_2$.

Choose τ and σ such that C_τ is interior to Γ_1 and C_σ is exterior to Γ_2. Then N_τ must lie below S_1, hence below M and N_σ must lie above S_2, hence above M (by the argument used at the beginning of the proof of this theorem).

We claim that the family $\mathfrak{F} = \{N_t\}$, $t \in [\tau, \sigma]$, is a foliation of the region (that contains M) bounded by $N_\tau \cup N_\sigma \cup [(\Omega_\sigma \setminus \Omega_\tau) \cap L(c)]$. In fact, by the maximum principle, two distinct N_t's cannot intersect, so we have only to prove that N_t varies continuously with t.

Fix $t_0 \in \mathbb{R}$; for ϵ small enough, for each $t \in [t_0 - \epsilon, t_0 + \epsilon]$, there exists a diffeomorphism $f_t \colon \Omega_{t_0} \to \Omega_t$ with the property $\|f_t\|_{C^2(\Omega_{t_0})} \leq \epsilon$.

Let F be defined as in Section 2 and for each $t \in [t_0 - \epsilon, t_o + \epsilon]$ consider the two families of equivalent Dirichlet problems

$$\begin{cases} F[u \circ f_t] = 0 & \text{in } \Omega_{t_0}, \\ u = c & \text{on } C_{t_0}, \end{cases} \tag{D^t}$$

$$\begin{cases} F[u] = 0 & \text{in } \Omega_t, \\ u = c & \text{on } C_t, \end{cases} \tag{E^t}$$

By Theorem 2.1, there exists a unique admissible solution u^{t_0} of (E^{t_0}); as u^{t_0} is admissible, $HW_{u^{t_0}} - 1 < 0$. Then, as $\|f_t\|_{C^2(\Omega_{t_0})} \leq \epsilon$, we have

$$HW_{u^{t_0} \circ f^{t_0}} - 1 < 0,$$

so the linearized operator associated to (D^{t_0}) as in Theorem 2.1, is invertible. Hence, by Implicit Function Theorem, there exists δ, $0 < \delta < \epsilon$, such that for each $t \in [t_0 - \delta, t_0 + \delta]$ there exists a solution v^t of (D^t) that depends continuously on t. Thus $u^t = v^t \circ (f^t)^{-1}$ is the unique solution of (E^t) and depends continuously on t and so $\mathfrak{F} = \{N_t\}$ is a foliation.

Now, using the foliation $\mathfrak{F}$, we prove that $M = N_1$, hence it is a graph and it is unique by Theorem 2.1.

For $t < 1$ no N_t intersect M. Otherwise, for the smallest such t, N_t is on one side of M at an intersection point (necessarily interior) and their mean curvature vectors both point towards $\mathfrak{B}$, so N_t would be equal to M by the maximum principle. This is impossible as $\partial N_t \neq \partial M$ for $t < 1$. Thus M is above N_1 (or equal to it). We repeat the same argument starting with N_σ and decreasing to N_1 and, as before, we conclude that M is below N_1. So $M = N_1$ and the Theorem 3.1 is proved. □

References.

[A] M. T. Anderson, *Complete minimal varieties in hyperbolic space*, Invent. Math. **69** (1982), 477–494.

[GT] D. Gilbarg, N. S. Trudinger, *Elliptic partial differential equations of second order*, Springer–Verlag, 1983.

[HL] R. Hardt, F. H. Lin, *Regularity at infinity for area-minimizing hypersurfaces in hyperbolic space*, Invent. Math. **88** (1987), 217–224.

[KNS] D. Kinderlehrer, L. Nirenberg, J. Spruck, *Regularity in elliptic free boundary problems I*, Journal D'Analyse Math. **34** (1978), 86–119.

[L] F. H. Lin, *On the Dirichlet problem for minimal graphs in hyperbolic space*, Invent. Math. **96** (1989), 593–612.

[NR] B. Nelli, H. Rosenberg, *Some remarks on embedded hypersurfaces in hyperbolic space of constant mean curvature and spherical boundary*, Ann. Global Anal. Geom. **13** (1995), 23–30.

[RS] H. Rosenberg, J. Spruck, *On the existence of convex hypersurfaces of constant Gauss curvature in hyperbolic space*, J. of Diff. Geometry **40** (1994), 379–409.

[T] Y. Tonegawa, *Existence and regularity of constant mean curvature hypersurfaces in hyperbolic space*, to appear in Math. Zeitschrift.

Received March 14, 1995.

Départament de Mathématiques
Université de Paris VII
2, Place Jussieu — 75251 Paris, France
E-mail address: nelli@mathp7.jussieu.fr

and

Departament of Mathematics
John Hopkins University
Baltimore, MD 21218, USA
E-mail address: js@chow.mat.jhu.edu

Homoclinics for a singular Hamiltonian system

Paul H. Rabinowitz [1]

Dedicated to Stefan Hildebrandt for his 60th birthday

1. Introduction.

This paper studies the existence of homoclinic solutions for a family of singular Hamiltonian systems which are periodically forced. Consider

$$\text{(HS)} \qquad \ddot{q} + V_q(t,q) = 0$$

where $q \in \mathbb{R}^2$ and the potential V satisfies:

(V_1) There is a $\xi \in \mathbb{R}^2 \backslash \{0\}$ such that $V \in C^1(\mathbb{R} \times (\mathbb{R}^2 \backslash \{\xi\}), \mathbb{R})$ and is T periodic in t,

(V_2) $\lim_{x \to \xi} V(t,x) = -\infty$ uniformly in t,

(V_3) There is a neighborhood, $\mathcal{N}$, of ξ and $U \in C^1(\mathcal{N} \backslash \{\xi\}, \mathbb{R})$ such that $|U(x)| \to \infty$ as $x \to \xi$ and

$$|U'(x)|^2 \leq -V(t,x) \quad \text{for } x \in \mathcal{N} \backslash \{\xi\} \text{ and } t \in \mathbb{R}$$

(V_4) $V(t,x) \leq 0$ and $V(t,x) = 0$ if and only if $x = 0$,

(V_5) There is a constant $V_0 < 0$ such that

$$\overline{\lim_{x \to \infty}}\, V(t,x) \leq V_0 \quad \text{for all } t.$$

[1] This research was sponsored by the National Science Foundation under Grant#MCS-8110556 and by the U.S. Army under contract #DAAL03-87-12-0043.

For simplicity, T will be set equal to 1 for what follows. Hypothesis (V_3) is the so-called "strong force" condition introduced by Gordon [1] and implies that $W^{1,2}_{\text{loc}}$ collisions are not possible for (HS), i.e. no $W^{1,2}_{\text{loc}}$ solution of (HS) can enter the singularity ξ in finite time. Hypotheses (V_4)–(V_5) show that the equilibrium solution $q \equiv 0$ is the unique global maximum of the potential V, although if e.g. $V \in C^2$, 0 may be a highly degenerate maximum.

The goal of this paper is to find solutions of (HS) which are homoclinic to 0. In §2, elementary minimization arguments, exploiting the topology of the plane, show (HS) has a pair of solutions which are homoclinic to 0. They wind respectively about ξ in a positive and a negative sense. When V is autonomous, these two solutions may possibly be represented by the same orbit modulo time reversal. Moreover we show there is such a solution given by a simple curve. The effect of dropping (V_3) will also be studied in §2. It is then proved that (HS) has a "generalized homoclinic solution" which may contain collision points, i.e. points at which $q(t) = \xi$.

Some criteria for there to exist more homoclinic solutions will be given in §3. This will be the case if e.g. it is easier, in a sense to be made precise later, for an orbit to wind around ξ than to go from 0 to ξ. Then in §4 a more general situation in which V possesses multiple singularities will be treated. Suppose there are strong force singularities of V at $\xi_1, \dots, \xi_k$ and (V_1)–(V_3) are modified to reflect this fact. Then it will be shown that (HS) has at least k geometrically distinct solutions homoclinic to 0. Moreover if V is autonomous, there are k homoclinics whose orbits are simple curves.

There does not seem to be much work on homoclinics for singular Hamiltonian systems in the literature. In [1], Gordon studied the existence of periodic solutions of singular Hamiltonian systems when $n = 2$ and $V = V(q)$ somewhat in the spirit of the approach used here. When $n > 2$ and $V = V(q)$, the existence of homoclinics under slightly stronger hypotheses than (V_1)–(V_5) was shown by Tanaka [2]. He used a variant of a minimax argument from Bahri-Rabinowitz [3] to get subharmonic solutions of (HS) and then passed to a limit with the aid of appropriate estimates to obtain the homoclinics. In a different setting, Bessi [4] found multiplicity results for homoclinics when (HS) is singular and V satisfies a pinching condition. Homoclinics have also been obtained for autonomous Hamiltonian systems which are not singular under hypotheses like (V_1)–(V_5) by Bolotin and Kozlov [5] and Rabinowitz and Tanaka [6].

In a sequel to this paper, we will show how multibump homoclinics can be obtained for (HS) when $n \geq 2$ under some further assumptions.

2. A pair of homoclinics.

A simple geometrical variational argument will be used to get a pair of homoclinic solutions of (HS) when V satisfies (V_1)–(V_5). Let

$$E = \{q \in W^{1,2}_{\mathrm{loc}}(\mathbb{R}, \mathbb{R}^2) \mid \int_{\mathbb{R}} |\dot{q}|^2 dt < \infty\}.$$

Then E is a Hilbert space under the norm

$$\|q\|^2 = \int_{\mathbb{R}} |\dot{q}|^2 dt + |q(0)|^2.$$

Let

$$\Lambda = \{q \in E \mid q(t) \neq \xi \quad \text{for all } t \in \mathbb{R}\}, \tag{2.1}$$

the set of curves in E which avoid ξ. For $q \in \Lambda$, set

$$I(q) = \int_{\mathbb{R}} \left(\tfrac{1}{2}|\dot{q}|^2 - V(t,q)\right) dt. \tag{2.2}$$

The following simple result will be used several times.

Proposition 2.3. *If V satisfies (V_1)–(V_5) and $q \in \Lambda$ with $I(q) < \infty$, then $|q(t)| \to 0$ as $|t| \to \infty$.*

Proof. The proof is the same as that of Proposition 3.11 of [7].

The continuous embedding of $C(\mathbb{R}, \mathbb{R}^2)$ in E and asymptotic behavior of $q(t)$ given by Proposition 2.3 imply that whenever $I(q) < \infty$, q can be considered to be a closed curve in $\mathbb{R}^2$. Moreover after a reparametrization, q can be considered to be the continuous image of S^1. Since $q(t) \neq \xi$ for all $t \in \mathbb{R}$, associated with the curve is its Brouwer degree with respect to ξ, $d(q)$. The degree in turn equals the winding number of the curve about ξ.

Let

$$\Gamma = \{q \in E \mid d(q) \neq 0\}. \tag{2.4}$$

Writing $\Gamma = \Gamma^+ \cup \Gamma^-$ where

$$\Gamma^{\pm} = \{q \in \Gamma \mid \pm d(q) > 0\}, \tag{2.5}$$

define

$$c^{\pm} = \inf_{q \in \Gamma^{\pm}} I(q). \tag{2.6}$$

The main result of this section is

Theorem 2.7. *If V satisfies (V_1)–(V_5), then $c^{\pm} > 0$ and there exists $q^{\pm} \in \Gamma^{\pm}$ such that $I(q^{\pm}) = c^{\pm}$. Moreover $q^{\pm}$ is a classical homoclinic solution of (HS).*

Proof. The proof is the same for the $\pm$ cases so it will be carried out for the $+$ case. Let (q_m) be a minimizing sequence for (2.6). For $q \in E$ and $j \in \mathbb{Z}$, set

$$\tau_j q(t) = q(t-j). \tag{2.8}$$

By (V_1), if $q \in \Gamma^+$, so is $\tau_j q$ for all $j \in \mathbb{Z}$ and

$$I(\tau_j q) = I(q). \tag{2.9}$$

Thus I possesses a natural $\mathbb{Z}$ symmetry. Let $\mathcal{S}$ denote the ray joining ξ to infinity given by

$$\mathcal{S} = \{\theta\xi \mid \theta \geq 1\}. \tag{2.10}$$

If $q \in \Gamma$, since $d(q) \neq 0$,

$$q(\mathbb{R}) \cap \mathcal{S} \neq \phi. \tag{2.11}$$

By (2.9) and (2.11), the minimizing sequence can be normalized so that q_m intersects $\mathcal{S}$ for the first time for some $t \in [0,1)$. Thus

$$\begin{cases} q_m(-\infty, 0) \cap \mathcal{S} = \phi \\ q_m([0,1)) \cap \mathcal{S} \neq \phi. \end{cases} \tag{2.12}$$

This normalization shows any L^∞_{loc} limit of (q_m) is nontrivial.

Next some bounds will be obtained for (q_m). By (V_4) and (2.6), it can be assumed that

$$0 \leq I(q_m) \leq c+1 \tag{2.13}$$

where $c = c^+$. Therefore

$$\|\dot{q}_m\|^2_{L^2} \leq 2(c+1). \tag{2.14}$$

Let $B_r(a)$ denote an open ball of radius r about $a \in \mathbb{R}^2$. By Lemma 3.6 of [7], for any $\delta > 0$, there is a $\beta(\delta) > 0$ such that if $q \in E$, $I(q) < \infty$, and $q(t) \notin B_\delta(0)$ for $t \in [s, \sigma]$, then

$$I(q) \geq \beta(\delta)|q(s) - q(\sigma)|. \tag{2.15}$$

Choosing e.g. $\delta = |\xi|/2$, s so that $|q_m(s)| = \delta$ and σ so that $q_m(\sigma) = \|q_m\|_{L^\infty}$, (2.13) and (2.15) show there is a $K > 0$ satisfying

$$\|q_m\|_{L^\infty} \leq K \tag{2.16}$$

for all $m \in \mathbb{N}$. Thus by (2.14) and (2.16), (q_m) is bounded in the Hilbert space E. Therefore, along a subsequence, q_m converges weakly in E and strongly in L^∞_{loc} to $Q \in E$. By (V_4), for all $\ell > 0$,

$$\int_{-\ell}^{\ell} \left(\tfrac{1}{2}|\dot{Q}|^2 - V(t,Q)\right) dt \leq \varliminf_{m\to\infty} I(q_m) = c. \tag{2.17}$$

Hence

$$I(Q) \leq c. \tag{2.18}$$

Consequently by Proposition 2.3, $|Q(t)| \to 0$ as $|t| \to \infty$. Moreover by (2.15),

$$I(q_m) \geq \beta\left(\frac{|\xi|}{2}\right)\frac{|\xi|}{2} \tag{2.19}$$

so

$$c \geq \beta\left(\frac{|\xi|}{2}\right)\frac{|\xi|}{2}. \tag{2.20}$$

To complete the proof, it remains to show that (i) $Q \in \Gamma^+$ and (ii) Q is a solution of (HS). Proving (i) requires showing that (iii) $Q \in \Lambda$ and (iv) $d(Q) > 0$. To verify (iii), one argues as in [1] using (V_2)–(V_3). If $q \in \Lambda$ and $q(t)$ is near ξ for $t \in [\sigma, \mu]$, by (V_3)

$$\begin{aligned} |U(q(\mu))| &\leq |U(q(\sigma))| + \left| \textstyle\int_\sigma^\mu \frac{d}{dt} U(q(t))dt \right| \\ &\leq |U(q(\sigma))| + \left(\textstyle\int_\sigma^\mu |U'(q)|^2 dt\right)^{1/2} \left(\textstyle\int_\sigma^\mu |\dot{q}|^2 dt\right)^{1/2} \\ &\leq |U(q(\sigma))| + \left(\textstyle\int_\sigma^\mu -V(t,q)dt\right)^{1/2} \left(\textstyle\int_\sigma^\mu |\dot{q}|^2 dt\right)^{1/2}. \end{aligned} \tag{2.21}$$

If further,

$$I(q) \leq c + 1, \tag{2.22}$$

the form of I shows

$$|U(q(\mu))| \leq |U(q(\sigma))| + 2^{1/2}(c+1). \tag{2.23}$$

Choose σ so that

$$|q(\sigma) - \xi| \leq |\xi|/2. \tag{2.24}$$

Then by (2.23), $|U(q(\mu))|$ is bounded from above. Therefore by (V_2), $q(\mu)$ can not get too close to ξ (independently of q satisfying (2.22) and (2.24)). In particular there is an $\epsilon > 0$ so that

$$q_m(t) \cap B_\epsilon(\xi) = \phi \tag{2.25}$$

for all $m \in \mathbb{N}$ and $t \in \mathbb{R}$. The L^∞_{loc} convergence of q_m to Q then implies

$$Q(t) \cap B_\epsilon(\xi) = \phi \tag{2.26}$$

for all $t \in \mathbb{R}$, i.e. (iii) holds.

An indirect argument will be employed to obtain (iv). Suppose

$$d(Q) \leq 0. \tag{2.27}$$

Let $\delta \in \left(0, \frac{|\xi|}{2}\right)$. Since $I(Q) < \infty$, there is a $T = T(\delta)$ such that $Q(t) \in B_{\delta/2}(0)$ for $|t| \geq T$. Now $q_m \to Q$ in L^∞_{loc} along a subsequence, which for notational convenience we take to be all of $\mathbb{N}$. Therefore for large m, $q_m(t) \in B_\delta(0)$ for $t \in [-T-1, -T] \cup [T, T+1]$. By (2.25)–(2.26), q_m and Q can be expressed in polar coordinates relative to ξ:

$$q_m(t) = \rho_m(t)e^{i\varphi_m(t)}; Q(t) = \rho(t)e^{i\varphi(t)}$$

where ρ_m, ρ, φ_m, φ are continuous and $\rho_m(t), \rho(t) \geq \epsilon$. Moreover $\rho_m(t) \to \rho_m(\pm\infty) = \rho(\pm\infty)$ and $\varphi_m(t) \to \varphi_m(\pm\infty) = \varphi(\pm\infty)$ as $|t| \to \infty$.

Now

$$d(Q) = \tfrac{1}{2\pi}(\varphi(\infty) - \varphi(-\infty)), \tag{2.28}$$

the winding number of Q about ξ. By the normalization made for q_m,

$$\varphi_m(-T) - \varphi_m(-\infty) \simeq 0 \tag{2.29}$$

for small δ, i.e. the net change of φ_m over $(-\infty, -T)$ is near 0 for small δ. Similarly

$$\varphi(\infty) - \varphi(T), \;\; \varphi(-T) - \varphi(-\infty) \simeq 0 \tag{2.30}$$

for small δ. Since q_m is L^∞ close to Q for $t \in [-T, T]$, by (2.28)–(2.30),

$$\tfrac{1}{2\pi}(\varphi_m(T) - \varphi_m(-T)) - d(Q) \simeq 0. \tag{2.31}$$

Therefore

$$\tfrac{1}{2\pi}(\varphi_m(\infty) - \varphi_m(T)) \simeq d(q_m) - d(Q) > 0 \tag{2.32}$$

via (2.27) and $q_m \in \Gamma^+$.

Define a new sequence $(\hat{q}_m) \subset \Gamma^+$ as follows:

$$\begin{aligned} \hat{q}_m(t) &= 0, \qquad t \le T \\ &= (t-T)q_m(T+1) \qquad T \le t \le T+1 \\ &= q_m(t) \qquad t \ge T+1. \end{aligned} \tag{2.33}$$

Since $\hat{q}_m$ is near 0 for $t \le T+1$ and then coincides with q_m, by (2.32),

$$d(\hat{q}_m) = d(q_m) - d(Q) > 0. \tag{2.34}$$

Consequently $\hat{q}_m \in \Gamma^+$. Consider

$$\begin{aligned} I(q_m) - I(\hat{q}_m) &= \textstyle\int_{-\infty}^{T+1} \left(\frac{1}{2}|\dot{q}_m|^2 - V(t,q_m)\right)dt \\ &\quad - \textstyle\int_{T}^{T+1} \left(\frac{1}{2}|q_m(T+1)|^2 - V(t,\hat{q}_m)\right)dt. \end{aligned} \tag{2.35}$$

By (2.15),

$$\int_{-\infty}^{T+1} \left(\tfrac{1}{2}|\dot{q}_m|^2 - V(t,q_m)\right)dt \ge \beta\left(\frac{|\xi|}{2}\right)|\xi|. \tag{2.36}$$

Since $\hat{q}_m(t) \in B_\delta(0)$ for $t \in [T, T+1]$ and m large,

$$\int_{T}^{T+1} \left(\tfrac{1}{2}|q_m(T+1)|^2 - V(t,\hat{q}_m)\right)dt \le \tfrac{1}{2}\delta^2 + \chi(\delta) \tag{2.37}$$

where $\chi(\delta) \to 0$ as $\delta \to 0$. Thus for δ sufficiently small,

$$I(\hat{q}_m) \le I(q_m) - \beta\left(\frac{|\xi|}{2}\right)|\xi| \tag{2.38}$$

for all large m and

$$\varliminf_{m\to\infty} I(\hat{q}_m) \le c - \tfrac{1}{2}\beta\left(\frac{|\xi|}{2}\right)|\xi| < c \tag{2.39}$$

contrary to (2.6). It follows that $Q \in \Gamma^+$. Therefore by (2.18),

$$I(Q) = c. \tag{2.40}$$

To complete the proof of Theorem 2.7, we must show that Q is a classical solution of (HS) and $\dot{Q}(\pm\infty) = 0$. The first fact follows in a standard way. Let $\psi \in C(\mathbb{R}, \mathbb{R}^2)$ have compact support. Then for α small, $Q + \alpha\psi \in \Gamma^+$. Consider $I(Q + \alpha t)$. This is a C^1 function of α for α near 0 with a minimum at $\alpha = 0$. Therefore

$$\frac{d}{d\alpha} I(Q + \alpha\psi)\big|_{\alpha=0} = 0 = \int_{\mathbb{R}} \left(\tfrac{1}{2}\dot{Q} \cdot \dot{\psi} - V_q(t, Q) \cdot \psi\right) dt \tag{2.41}$$

for all such ψ. Equation (2.41) implies Q is a weak solution of (HS). Standard arguments then show that it is a classical solution.

Finally to prove that $\dot{Q}(\pm\infty) = 0$, by an interpolation inequality from [8], for any $\epsilon > 0$,

$$\|\dot{Q}\|_{L^\infty[a,b]} \le \epsilon \|\ddot{Q}\|_{L^\infty[a,b]} + k(\epsilon)\|Q\|_{L^\infty[a,b]} \tag{2.42}$$

where k depends on $|b - a|$. Note that by (HS) and (V_4), as $|t| \to \infty$,

$$|\ddot{Q}(t)| = |V_q(t, Q(t))| \to 0.$$

Hence taking $[a, b] = [j, j+1]$ in (2.42) and letting $|j| \to \infty$ shows $\dot{Q}(\pm\infty) = 0$. The proof of Theorem 2.7 is complete.

Remark 2.43. The full strength of (V_2)–(V_3) is not needed for Theorem 2.7. It suffices that V is sufficiently negative at ξ. Suppose e.g. that for some $\rho > 0$ and all $x \in B_\rho(\xi) \setminus B_{\rho/2}(\xi)$ and $t \in \mathbb{R}$,

$$|V(t, x)| \ge \tfrac{1}{2} \min_{t \in [0,1]} |V(t, \xi)|. \tag{V_2'}$$

If $q(\sigma) \in \partial B_{\rho/2}(\xi)$, $q(s) \in \partial B_\rho(\xi)$ and $q(t) \in B_\rho(\xi) \setminus B_{\rho/2}(\xi)$ for $t \in [\sigma, s)$, then

$$\begin{aligned} \rho/2 &\le |q(s) - q(\sigma)| = \left| \textstyle\int_\sigma^s \dot{q}(t)dt \right| \\ &\le |s - \sigma|^{1/2} \left(\textstyle\int_\sigma^s |\dot{q}|^2 dt \right)^{1/2} \le |s - \sigma|^{1/2} (2I(q))^{1/2} \end{aligned} \tag{2.44}$$

Therefore

$$\begin{aligned} I(q) &\ge \left| \textstyle\int_\sigma^s V(t, q)dt \right| \ge \tfrac{1}{2} \min_{t \in [0,1]} |V(t, \xi)| |s - \sigma| \\ &\ge \frac{\rho^2}{16 I(q)} \min_{t \in [0,1]} |V(t, \xi)| \end{aligned}$$

or

$$I(q) \geq \frac{\varrho}{4} \min_{t\in[0,1]} |V(t,\xi)|^{1/2}. \tag{2.45}$$

Consequently if there is a $\hat{q} \in \Gamma^+$ such that

$$I(\hat{q}) < \frac{\varrho}{4} \min_{t\in[0,1]} |V(t,\xi)|^{1/2}, \tag{2.46}$$

then

$$c < \frac{\varrho}{4} \min_{t\in[0,1]} |V(t,\xi)|^{1/2} \tag{2.47}$$

and there exists a homoclinic solution of (HS) which winds around $B_{\varrho/2}(\xi)$.

Remark 2.48. If V is time reversible, i.e. $V(t,x) = V(-t,x)$, whenever $q \in \Gamma^+$, $p(t) = q(-t) \in \Gamma^-$ and $I(q) = I(p)$. It follows that $c^+ = c^- \equiv c$ and for each minimizer $Q \in \Gamma^+$ of $(2.6)^+$, $P(t) = Q(-t) \in \Gamma^-$ is a minimizer for $(2.6)^-$.

The next result improves Theorem 2.6 when V is autonomous provided that it is a bit smoother.

Theorem 2.49. *If V satisfies (V_1)–(V_5), is autonomous and V' is locally Lipschitz continuous for $x \neq \xi$, then any minimizer Q of Theorem 2.7 is a simple curve.*

Proof. If not, there are numbers $\sigma < s$ such that $Q(\sigma) = Q(s)$. Consider $\hat{Q} = Q\big|_{[\sigma,s]}$. Let P denote the function obtained by removing the loop $\hat{Q}$ from Q. Observe that $d(\hat{Q}) > 0$ for otherwise, $P \in \Gamma^+$ with $I(P) < I(Q)$, contrary to $(2.6)^+$. Note also that $d(P) = 0$, again since otherwise $P \in \Gamma^+$ and $I(P) < I(Q)$ contrary to Remark 2.48. Define a new function $R \in \Lambda$ as follows:

$$\begin{aligned} R(t) &= Q(t) & t &\leq \sigma \\ &= Q(s-t+\sigma) & t &\in [\sigma, s] \\ &= Q(t) & t &\geq s \end{aligned} \tag{2.50}$$

Then $R \in \Gamma^-$ and it is easy to see that $I(R) = I(Q) = c$. But minimizers of I in $\Gamma^\pm$ are solutions of (HS) and $R(t) = Q(t)$ for $t \notin [\sigma, s]$. Therefore, since V' is locally Lipschitz continuous, $R(t) \equiv Q(t)$. Consequently $R \in \Gamma^+ \cap \Gamma^-$ which is impossible. Hence Q must be a simple curve.

Remark 2.51. Since any minimizer, Q, of I in Γ is a simple curve, $d(Q) = 1$ or -1.

To conclude this section, the effect of dropping hypothesis (V_3) but retaining (V_2) will be studied. This is sometimes called the "weak force" case. Following arguments from [3], it will be shown that (HS) possesses a "generalized" homoclinic solution in the following sense: There is a function Q satisfying:

$$\begin{aligned}
&\text{(i)} \quad Q \in E \text{ and } I(Q) < \infty \\
&\text{(ii)} \quad Q(\pm\infty) = \dot{Q}(\pm\infty) = 0 \\
&\text{(iii)} \quad \text{Aside from a closed set } \mathcal{D} \subset \mathbb{R} \text{ of measure 0, } Q \text{ is a classical solution of (HS).} \qquad (2.52)\\
&\text{(iv)} \quad \text{If } V \text{ is independent of } t, \quad \tfrac{1}{2}|\dot{Q}(t)|^2 + V(Q(t)) = 0, \qquad t \in \mathbb{R}\backslash\mathcal{D}.
\end{aligned}$$

This notion parallels that of generalized T-periodic solution introduced in [3]. Dropping the hypothesis (V_3) allows the solution Q to enter the singularity, ξ, i.e. collisions are possible.

Theorem 2.53. *If V satisfies (V_1)–(V_2), (V_4)–(V_5), (HS) possesses a generalized homoclinic solution.*

Proof. An approximation argument based on [3] will be used. Let $\epsilon > 0$ and set

$$V_\epsilon(t,x) = V(t,x) - \frac{\epsilon(\varphi(|x-\xi|)}{|x-\xi|^4}, \tag{2.54}$$

where $\varphi(s) = 1$ if $s \le |\xi|/4$; $= 0$ if $s \ge |\xi|/2$ and is positive and continuous in between. Then V_ϵ satisfies (V_1)–(V_5). Hence by Theorem 2.7, there is a homoclinic solution, Q_ϵ, of

$$\text{(HS)}_\epsilon \qquad \ddot{q} + V_{\epsilon q}(t,q) = 0$$

where Q_ϵ minimizes

$$I_\epsilon(q) = I(q) + \epsilon \int_{\mathbb{R}} \frac{\varphi(|q-\xi|)}{|q-\xi|^4} dt \tag{2.55}$$

for $q \in \Gamma^+$. Bounds will be obtained for Q_ϵ. Then letting $\epsilon \to 0$, a subsequence of Q_ϵ converges to a generalized homoclinic solution, Q, of (HS). By (2.55),

$$c_\epsilon = \inf_{q \in \Gamma^+} I_\epsilon(q) \tag{2.56}$$

is a nondecreasing function of ϵ. Hence for $\epsilon \in (0,1)$, by (2.20),

$$c_1 \geq c_\epsilon \geq \beta\left(\tfrac{|\xi|}{2}\right)\tfrac{|\xi|}{2}. \tag{2.57}$$

Arguing as in (2.13)–(2.15) shows the functions Q_ϵ are bounded E and in L^∞. Consequently, choosing $\epsilon_m \to 0$, along a subsequence, $Q_m \equiv Q_{\epsilon_m}$ converges in L^∞_{loc} to $Q \in E$.

For any $\delta > 0$, let

$$S_\delta = \{t \in \mathbb{R} \mid |Q(t) - \xi| \geq \delta\}.$$

Then by (2.55) and (2.57),

$$c_1 \geq -\int_{S_\delta} V_{\epsilon_m}(t, Q_m)dt \geq -\int_{S_\delta} V(t, Q_m)dt. \tag{2.58}$$

Since Q_m converges to Q in L^∞_{loc}, (V_4) and (2.58) imply

$$c_1 \geq -\int_{S_\delta} V(t, Q)dt. \tag{2.59}$$

Letting $\delta \to 0$, the Monotone Convergence Theorem shows

$$c_1 \geq -\int_{\mathbb{R}} V(t, Q)dt. \tag{2.60}$$

Hence (i) of (2.52) is satisfied.

Let

$$\mathcal{D} = \{t \in \mathbb{R} \mid Q(t) = \xi\}.$$

Since $I(Q) < \infty$, $\mathcal{D}$ has measure 0. Moreover $\mathcal{D}$ is closed since Q is continuous. The L^∞_{loc} convergence of Q_m to Q and fact that Q_m satisfies $(\text{HS})_{\epsilon_m}$ imply that Q_m converges to Q in C^2_{loc} on $\mathbb{R}\backslash\mathcal{D}$. Therefore Q satisfies (HS) on $\mathbb{R}\backslash\mathcal{D}$ and (iii) of (2.52) holds. By Proposition 2.3, $I(Q) < \infty$ implies $Q(\pm\infty) = 0$. Arguing as in (2.42) shows $\dot{Q}(\pm\infty) = 0$ and (ii) of (2.52) is verified. Finally if V is independent of t,

$$\tfrac{1}{2}|\dot{Q}_\epsilon(t)|^2 + V_\epsilon(Q_\epsilon(t)) = 0 \tag{2.61}$$

since $Q_\epsilon(\pm\infty) = 0 = \dot{Q}_\epsilon(\pm\infty)$. Thus for $t \in \mathbb{R}\backslash\mathcal{D}$, letting $\epsilon = \epsilon_m \to 0$ in (2.61) yields (iv) of (2.52).

Remark 2.62. In recent work on periodic solutions of singular Hamiltonian systems, several authors, e.g. [9–12], have used Morse theoretic and blow-up arguments to show that certain generalized solutions obtained variationally have only a finite number or even no collisions. It seems likely that analogous results obtain here. On the other hand, suppose (V_1)–(V_2) and (V_3)–(V_4) are satisfied. Let $E_0 = \{q \in E \mid q(0) = \xi\}$. If there is a $v \in E_0$ such that

$$I^+(v) \equiv \int_0^\infty \left(\tfrac{1}{2}|\dot{v}|^2 - V(t,v)\right)dt < \infty,$$

then

$$\inf_{q \in E_0} I^+(q) < \infty$$

and a simpler version of our earlier minimization argument shows that there is a generalized solution q^+ of (HS) for $t \geq 0$ such that $q^+(0) = \xi$, $q^+(\infty) = 0 = \dot{q}^+(\infty)$. Similarly if there is a $w \in E_0$ such that

$$I^-(w) \equiv \int_{-\infty}^0 \left(\tfrac{1}{2}|\dot{w}|^2 - V(t,w)\right)dt < \infty,$$

then

$$\inf_{q \in E_0} I^-(q) < \infty$$

and there is a generalized solution q^- of (HS) for $t \leq 0$ satisfying $q^-(0) = \xi$, $q^-(-\infty) = 0 = \dot{q}^-(-\infty)$. Concatenating q^+ and q^- then produces a generalized homoclinic solution of (HS) which is in fact a collision orbit.

3. Some extensions.

The main result in this section is a criterion for there to exist more solutions of (HS) in $\Gamma^\pm$. By Theorem 2.7, the numbers

$$c^\pm = \inf_{q \in \Gamma^\pm} I(q)$$

correspond to homoclinic solutions of (HS). For $k \in \mathbb{N}$, set

$$\Gamma_k^\pm = \{q \in \Gamma^\pm \mid d(q) = \pm k\} \tag{3.1}$$

and define

$$c_k^\pm = \inf_{q \in \Gamma_k^\pm} I(q) \tag{3.2}$$

It is natural to ask whether there are solutions $q_k^\pm$ of (HS) in $\Gamma_k^\pm$ with $I(q_k^\pm) = c_k^\pm$. This need not be the case in general. E.g. if V is autonomous, $c_1^+ = c_1^- \equiv c_1$ and if $Q \in \Gamma^+$ satisfies $I(Q) = c_1^+$, it is easy to produce functions q_k near a k-fold concatenation of Q with itself having $d(q_k) = k$ and with $I(q_k)$ as close to kc_1 as desired. Thus if $c_k^\pm = kc_1^\pm$, it cannot be expected that (HS) has a solution $q_k^\pm$ in $\Gamma_k^\pm$.

To give a criterion guaranteeing more solutions of (HS) in $\Gamma^\pm$, we proceed a bit differently and define for $k \in \mathbb{N}$,

$$\hat{\Gamma}_k^\pm = \{q \in \Lambda \mid \pm d(q) \geq k\} \tag{3.3}$$

and

$$\hat{c}_k^\pm = \inf_{q \in \hat{\Gamma}_k^\pm} I(q)\,. \tag{3.4}$$

Then we have:

Theorem 3.5. *If V satisfies (V_1)–(V_5) and there is a smallest integer $k > 1$ such that*

$$\hat{c}_k^+ < k\hat{c}_1^+ \qquad (\textit{resp. } \hat{c}_k^- < k\hat{c}_1^-), \tag{3.6}$$

then there is a solution Q_k of (HS) with $Q_k \in \hat{\Gamma}_k^+$ (resp. $\hat{\Gamma}_k^-$) and $I(q_k) = \hat{c}_k^+$ (resp. $\hat{c}_k^-$).

The proof of Theorem 3.5 is based on analyzing minimizing sequences for (3.4). The key technical result, Proposition 3.7 will be stated next and used to prove the theorem. Then a sharper version of Theorem 3.5 for $V = V(x)$ will be given. The section will then conclude by studying when (3.6) is satisfied.

The key to Theorem 3.5 is the following Proposition. It is a variant in our setting of various results on the behavior of Palais-Smale sequences for a C^1 functional J, i.e. sequences such that (i) $J(u_m) \to b$ and (ii) $J'(u_m) \to 0$. (See e.g. P. L. Lions [13], Coti Zelati, Ekeland, Séré [14]). Here (ii) is replaced by the fact that we have a minimizing sequence.

Proposition 3.7. *If V satisfies (V_1)–(V_5), for any minimizing sequence, (q_m), of $(3.4)^+$, there is an $\ell \leq k$, functions $Q_1, \dots, Q_\ell \in E$ and sequences $(j_m^i) \subset \mathbb{Z}$, $1 \leq i \leq \ell$ such that*

$$Q_i[0,1] \cap \mathcal{S} \neq \phi, \tag{3.8}$$

$$1 \le d(Q_i), \tag{3.9}$$

$$\sum_1^\ell d(Q_i) \ge k, \tag{3.10}$$

$$I(Q_i) = \hat{c}^+_{d(Q_i)}, \tag{3.11}$$

$$\sum_1^\ell I(Q_i) = \hat{c}^+_k, \tag{3.12}$$

Along a subsequence, $(\tau_{j^i_m} q_m)$ *converges weakly in* E *and strongly in* L^∞_{loc} *to* Q_i, $1 \le i \le \ell$, (3.13)

$$|j^i_m - j^p_m| \to \infty \quad \textit{as } m \to \infty \textit{ if } i \ne p, \tag{3.14}$$

$$Q_i \in \Gamma^+_{d(Q_i)} \quad \textit{and is a solution of (HS).} \tag{3.15}$$

Assuming Proposition 3.7 for the moment, we can give the:

Proof of Theorem 3.5. The + case will be proved. Let (q_m) be a minimizing sequence for (3.4). If $\ell = 1$, by (3.9)–(3.10), $d(Q_1) \ge k$ so $Q_1 \in \hat{\Gamma}^+_k$. Moreover by (3.12), $I(Q_1) = \hat{c}^+_k$. Hence Q_1 is a solution of (HS) as in the proof of Theorem 2.7. Thus Theorem 3.5 holds for this case.

To complete the proof, it suffices to show that ℓ must be 1. If $\ell > 1$, by (3.11)–(3.12),

$$\hat{c}^+_k = \sum_1^\ell I(Q_i) = \sum_1^\ell \hat{c}^+_{d(Q_i)}.$$

Clearly $\hat{c}^+_{d(Q_i)} \le d(Q_i)\hat{c}^+_1$ and since k is the smallest integer greater than 1 so that (3.6) holds, $\hat{c}^+_{d(Q_i)} = d(Q_i)\hat{c}^+_1$. Consequently by (3.10),

$$\hat{c}^+_k = \Big(\sum_1^\ell d(Q_i)\Big)\hat{c}^+_1 \ge k\hat{c}^+_1$$

contrary to (3.6). Therefore $\ell = 1$ and the proof is complete.

Results related to Theorems 2.7 and 3.5 were obtained in [15] where Brezis and Coron found a harmonic map via a minimization argument. When an inequality in the spirit of (3.6) is satisfied, they were able to find a second solution by minimizing over a class of functions involving a degree constraint. The results of [15] answer a question raised by Giaquinta and Hildebrandt [16] who observed the two solutions for a special case. See also Jost [17].

Proof of Proposition 3.7. By the proof of Theorem 2.7, any normalized minimizing sequence (q_m) for (3.4) is bounded in E and therefore possesses a subsequence converging weakly in E and strongly in L^∞_{loc} to $Q_1 \in \Lambda$ such that

$$0 < \tfrac{|\xi|}{2}\beta(\tfrac{|\xi|}{2}) \le I(Q_1) \le \underline{\lim}_{m\to\infty} I(q_m) = \hat{c}_k^+ \tag{3.16}$$

and

$$Q_1([0,1]) \cap \mathcal{S} \ne \phi . \tag{3.17}$$

Remark 3.18. For future reference, note that if V is independent of t, the stronger normalization $Q_1(0) \in \mathcal{S}$ can be made.

If $d(Q_1) \ge k$, $Q_1 \in \hat{\Gamma}_k^+$ so (3.16) implies $I(Q_1) = \hat{c}_k^+$ and the argument of Theorem 2.7 shows Q_1 is a critical point of I and a solution of (HS). Thus Proposition 3.7 is proved with $\ell = 1$, $j_m^1 = 0$ for all m, and q_m replaced by the above subsequence. Thus suppose that $d(Q_1) < k$. Let $\delta > 0$. As in the proof of Theorem 2.7, there is a $T_1 = T_1(\delta) > 0$ such that $Q_1(t) \in B_\delta(0)$ for $|t| \ge T_1$ and for large m,

$$\tfrac{1}{2\pi}(\Phi_1(T_1) - \Phi_1(-\infty)) \simeq d(Q_i) \simeq \tfrac{1}{2\pi}(\varphi_m(T_1) - \varphi_m(-\infty)) \tag{3.19}$$

where $Q_1(t) = R_1(t)e^{i\Phi_1(t)}$. If $d(Q_1) \le 0$, by choosing δ sufficiently small, (q_m) can be replaced by a new sequence $(\hat{q}_m)$ where $\hat{q}_m(t) = q_m(t)$ for $t \ge T_1 + 1$, $\hat{q}_m(t) = 0$ for $t \le T_1$, $d(\hat{q}_m) \ge d(q_m) \ge k$, and

$$I(\hat{q}_m) \le I(q_m) - \tfrac{|\xi|}{4}\beta(\tfrac{|\xi|}{2}) . \tag{3.20}$$

By (3.20),

$$\overline{\lim_{m\to\infty}}\, I(\hat{q}_m) \le \hat{c}_k^+ - \tfrac{|\xi|}{4}\beta(\tfrac{|\xi|}{2}) . \tag{3.21}$$

But (3.21) contradicts that (q_m) is a minimizing sequence for (3.4). Therefore $d(Q_1) \ge 1$, i.e. (3.9) holds for $i = 1$.

Since $d(Q_1) < k$ and $q_m \in \hat{\Gamma}_k$, by (3.19), for large m, q_m must wind around ξ a nonzero number of times in the interval (T_1, ∞). For t near T_1, $q_m(t) \in B_{2\delta}(0)$. Therefore there is a smallest integer $t_m > T_1$ such that $q_m((t_m, t_m + 1)) \cap \mathcal{S} \ne \phi$. We claim $t_m \to \infty$ as $m \to \infty$. Otherwise there is a subsequence along which (t_m) is bounded and converges to $\hat{T} < \infty$. Then $q_m \to Q_1$ in $L^\infty([-T_1, \hat{T}]))$ and

$$\varphi(\hat{T}) - \varphi(T_1) \simeq \varphi_m(\hat{T}) - \varphi_m(T_1) \tag{3.22}$$

for large m. But

$$\varphi(\hat{T}) - \varphi(T_1) \simeq 0 \tag{3.23}$$

since $Q_1(t) \in B_\delta(0)$ for $t \geq T_1$ while

$$\varphi_m(\hat{T}) - \varphi_m(T_1) \simeq \pi \pmod{2\pi}. \tag{3.24}$$

Thus (3.23)–(3.24) are contrary to (3.22). Consequently $t_m \to \infty$ as $m \to \infty$.

Consider a new sequence of functions $(p_m(t)) = (\tau_{-t_m} q_m(t)) = (q_m(t + t_m))$. These functions are bounded in E by earlier arguments and therefore converge as in (3.13) to $Q_2 \in E$ with Q_2 satisfying (3.8) for $i = 2$ and

$$I(Q_2) \leq \hat{c}_k^+. \tag{3.25}$$

Note also that (3.14) holds with $j_m^1 = 0$ and $j_m^2 = -t_m$. By (3.25), $Q_2(t) \to 0$ as $|t| \to \infty$ and as earlier there is a $T_2 > 0$ such that $Q_2(t) \in B_\delta(0)$ for $|t| \geq T_2$ and if $Q_2(t) = R_2(t) e^{i\Phi_2(t)}$,

$$\Phi_2(T_2) - \Phi_2(-T_2) \simeq \varphi_m(t_m + T_2) - \varphi_m(t_m - T_2). \tag{3.26}$$

An argument as in (3.20)–(3.21) shows

$$d(Q_2) \geq 1. \tag{3.27}$$

Note that $q_m(t_m - T_2) \in B_\delta(0)$. Hence the definition of t_m implies that for large m, q_m cannot wind around 0 in $[T_1, t_m - T_2]$. In fact

$$\varphi_m(t_m - T_2) - \varphi_m(T_1) \simeq 0. \tag{3.28}$$

Observe also that for any $s > 0$,

$$\begin{aligned} I(q_m) \geq & \int_{-\infty}^{T_1+s} \left(\tfrac{1}{2}|\dot{q}_m|^2 - V(t, q_m)\right) dt \\ & + \int_{t_m - T_2 - s}^{t_m + T_2 + s} \left(\tfrac{1}{2}|\dot{q}_m|^2 - V(t, q_m)\right) dt. \end{aligned}$$

Hence

$$\begin{aligned} \hat{c}_k^+ \geq & \int_{-\infty}^{T_1+s} \left(\tfrac{1}{2}|\dot{Q}_1|^2 - V(t, Q_1)\right) dt \\ & + \int_{-T_2 - s}^{T_2 + s} \left(\tfrac{1}{2}|\dot{Q}_2|^2 - V(t, Q_2)\right) dt. \end{aligned} \tag{3.29}$$

and letting $s \to \infty$,

$$\hat{c}_k^+ \geq I(Q_1) + I(Q_2). \tag{3.30}$$

Now if

(3.31)
$$k < d(Q_1) + d(Q_2) \simeq \tfrac{1}{2\pi}(\Phi_1(T_1) - \Phi_1(-T_1)) + \tfrac{1}{2\pi}(\Phi_2(+T_2) - \Phi_2(-T_2))$$
$$\simeq \tfrac{1}{2\pi}(\varphi_m(t_m + T_2) - \varphi_m(-\infty)),$$

the above construction can be repeated to get $Q_3(t)$, etc. At the pth step, as in (3.30),

$$\hat{c}_k^+ \geq \sum_1^p I(Q_i). \tag{3.32}$$

Since

$$I(Q_i) \geq \tfrac{|\xi|}{2}\beta(\tfrac{|\xi|}{2}) \tag{3.33}$$

for each i, (3.32)–(3.33) show this process must terminate in a finite number of steps. Thus after e.g. ℓ steps, we have $Q_1, \dots, Q_\ell$ satisfying (3.8)–(3.10), (3.13)–(3.14) as well as (3.32) with $p = k$.

To verify (3.12), suppose that there is strict inequality in (3.32). Then there exists an $\epsilon > 0$ such that

$$\sum_1^\ell I(Q_i) < \hat{c}_k^+ - \epsilon. \tag{3.34}$$

Let $s > 0$ and set $P(s) = \sum_{i=1}^\ell \tau_{si} Q_i$. Then for s sufficiently large,

$$d(P(s)) = \Sigma d(Q_i) \geq k \tag{3.35}$$

via (3.10). Consequently $P(s) \in \hat{\Gamma}_k^+$ for large s. Moreover

$$I(P(s)) \to \sum_1^\ell I(Q_i) \tag{3.36}$$

as $s \to \infty$. Hence by (3.36) and (3.34),

$$I(P(s)) \leq \hat{c}_k^+ - \tfrac{\epsilon}{2} \tag{3.37}$$

for large s, contrary to (3.4). Thus (3.12) holds.

It remains to prove (3.11) and (3.15). The former implies the latter as in the proof of Theorem 2.7. To get (3.11), note that since $Q_i \in \hat{\Gamma}^+_{d(Q_i)}$,

$$I(Q_i) \geq \hat{c}^+_{d(Q_i)}. \tag{3.38}$$

If there is strict inequality in (3.38), there exists a $p \in \hat{\Gamma}^+_{d(Q_i)}$ and an $\epsilon > 0$ such that

$$I(p) < I(Q_i) - \epsilon\,. \tag{3.39}$$

Arguing again as in (3.20)–(3.21), (q_m) can be replaced by a new sequence of functions, (p_m), which equal p in an interval of length $2T_i$ centered at j^i_m; equal q_m outside of an interval of length $2(T_i+1)$ centered at j^i_m; have $d(p_m) \geq d(q_m)$, and for which $I(p_m) \leq I(q_m) - \frac{\epsilon}{2}$. But then (p_m) violates (3.4). Consequently (3.11) is satisfied and the proof of Proposition 2.7 is complete.

As in §2, if V is autonomous, a better result obtains:

Theorem 3.40. *If V satisfies (V_1)–(V_5), V is independent of t, and there is a smallest integer $k > 1$ so that (3.6) holds, then there is a solution Q_k of (HS) with $Q_k \in \Gamma^+_k$ (resp. Γ^-_k) and $I(Q_k) = c^+_k$ (resp. c^-_k).*

The proof of Theorem 3.40 relies on the following:

Proposition 3.41. *Under the hypotheses of Theorem 3.40, if $Q \in \hat{\Gamma}^+_k$ such that $I(Q) = \hat{c}^+_k$, there exist $\underline{s}, \overline{s} \in \mathbb{R}$ with $\underline{s} < \overline{s}$, $Q(\underline{s}) = Q(\overline{s})$, and $d(Q\big|_{[\underline{s},\overline{s}]}) =$ 1.*

Assuming Proposition 3.41 for the moment, we can give the:

Proof of Theorem 3.40. Let $Q \in \hat{\Gamma}^+_k$ such that $I(Q) = \hat{c}^+_k$. Suppose that $d(Q) > k$. Then by Proposition 3.41, a new function P can be defined as follows:

$$\begin{aligned} P(t) &= Q(t), & t &\leq \underline{s} \\ &= Q(t + \overline{s} - \underline{s}) & t &\geq \underline{s}. \end{aligned} \tag{3.42}$$

Note that $d(P) \geq k$ and $P \in \hat{\Gamma}^+_k$. But $I(P) < I(Q) = \hat{c}^+_k$, a contradiction. Therefore $d(Q) = k$.

Proof of Proposition 3.41. Since $d(Q) \geq 2$, there are points $\underline{t}, \overline{t} \in \mathbb{R}$ with $\underline{t} < \overline{t}$ and $Q(\underline{t}) = Q(\overline{t})$. If $d(Q\big|_{[\underline{t},\overline{t}]}) = 1$, the result obtains with $\underline{s} = \underline{t}$, $\overline{s} = \overline{t}$. If $d(Q\big|_{[\underline{t},\overline{t}]}) \neq 1$, either

$$d(Q\big|_{[\underline{t},\overline{t}]}) \leq 0 \tag{3.43}$$

or

$$d(Q\big|_{[\underline{t},\overline{t}]}) \geq 2. \tag{3.44}$$

If (3.43) occurs, a new function P can be defined as in (3.42) by deleting the closed loop corresponding to $[\underline{t}, \overline{t}]$ from Q obtaining a new function $P \in \hat{\Gamma}_k^+$ with $I(P) < I(Q)$, contrary to the definition of $\hat{c}_k^+$. Thus (3.43) is impossible. If (3.44) is satisfied, there are points $\underline{t}_1 < \overline{t}_1$ in $[\underline{t}, \overline{t}]$, with not both $\underline{t}_1 = \underline{t}$ and $\overline{t}_1 = \overline{t}$, and such that $Q(\underline{t}_1) = Q(\overline{t}_1)$. If $d(Q\big|_{[\underline{t}_1, \overline{t}_1]}) = 1$, we are through; if not, by the above reasoning, $d(Q\big|_{[\underline{t}_1, \overline{t}_1]}) \geq 2$. Repeating this argument, after a finite number of steps, we find $\underline{t}_j < \overline{t}_j$ in $[\underline{t}_{j-1}, \overline{t}_{j-1}]$ with $Q(\underline{t}_j) = Q(\overline{t}_j)$ and $d(Q\big|_{[\underline{t}_j, \overline{t}_j]}) = 1$ or there are infinitely many distinct pairs $\underline{r} < \overline{r}$ in $[\underline{t}, \overline{t}]$ with $Q(\underline{r}) = Q(\overline{r})$ and $d(Q\big|_{[\underline{r}, \overline{r}]}) \geq 2$. In the latter event, let

$$\mathcal{R} = \{\underline{r} \in [\underline{t}, \overline{t}) \mid \text{there is an } \overline{r} \in (\underline{t}, \overline{t}] \text{ with } Q(\underline{r}) = Q(\overline{r}) \text{ and } d(Q\big|_{[\underline{r}, \overline{r}]}) \geq 2\} \tag{3.45}$$

and

$$\underline{T} = \sup\{\underline{r} \mid \underline{r} \in \mathcal{R}\}. \tag{3.46}$$

By its definition, there is a sequence $\underline{t}_m \to \underline{T}$ as $m \to \infty$, where $\underline{t}_m \leq \underline{T}$ and $\underline{t}_m \in \mathcal{R}$. For each m, there is a $\overline{t}_m \in (\underline{t}_m, \overline{t}]$ such that $Q(\underline{t}_m) = Q(\overline{t}_m)$ and $d(Q\big|_{[\underline{t}_m, \overline{t}_m]}) \geq 2$. Writing $Q(t) = r(t)e^{i\varphi(t)}$ where r and φ are continuous,

$$d(Q\big|_{[\underline{t}_m, \overline{t}_m]}) = \tfrac{1}{2\pi}(\varphi(\overline{t}_m) - \varphi(\underline{t}_m)) \geq 2. \tag{3.47}$$

Letting $m \to \infty$, there is a $\overline{T} \in [\underline{t}, \overline{t}]$ such that along a subsequence, $\overline{t}_m \to \overline{T}$, $Q(\underline{T}) = Q(\overline{T})$, and

$$\tfrac{1}{2\pi}(\varphi(\overline{T}) - \varphi(\underline{T})) = d(Q\big|_{[\underline{T}, \overline{T}]}) \geq 2 \tag{3.48}$$

via (3.47). Consequently $\underline{T} \in \mathcal{R}$.

Let

$$\hat{T} = \inf\{\sigma \in (\underline{T}, \overline{t}] \mid Q(\sigma) = Q(\underline{T}) \ \text{ and } \ d(Q\big|_{[\underline{T}, \sigma]}) \geq 2\}. \tag{3.49}$$

If $\hat{T} = \underline{T}$, there is a sequence $\sigma_m > \underline{T}$ such that $\sigma_m \to \underline{T}$ as $m \to \infty$, $Q(\sigma_m) = Q(\underline{T})$, and $d(Q\big|_{[\underline{T}, \sigma_m]}) \geq 2$. Thus Q winds around $B_\epsilon(\xi)$ in time $\sigma_m - \underline{T} \to 0$ as $m \to \infty$ which implies $|\dot{Q}(\underline{T})| = \infty$. Since this is not possible, $\hat{T} > \underline{T}$ and

$$d(Q\big|_{[\underline{T}, \hat{T}]}) \geq 2. \tag{3.50}$$

Hence there are points $\underline{s} < \overline{s}$ in $[\underline{T}, \hat{T}]$ with not both $\underline{s} = \underline{T}$ and $\overline{s} = \hat{T}$ such that $Q(\underline{s}) = Q(\overline{s})$. An earlier argument shows that $d(Q|_{[\underline{s},\overline{s}]}) \le 0$ is impossible and $d(Q|_{[\underline{s},\overline{s}]}) \ge 2$ contradicts the maximality of $\underline{T}$ or minimality of $\hat{T}$. Thus $d(Q|_{[\underline{s},\overline{s}]}) = 1$ and the Proposition is proved.

Remark 3.51. See also Whitney [18] for a topological result finding a degree 1 subloop of a loop of degree > 1 in a generic setting. If V is locally Lipschitz continuous as in Theorem 2.49, a stronger variant of Proposition 3.41 can be obtained.

Corollary 3.52. *If under the hypotheses of Proposition 3.41, V' is also locally Lipschitz continuous for $x \neq \xi$ then $Q|_{[\underline{s},\overline{s}]}$ is a simple closed curve.*

Proof. If not, there are points $\underline{\sigma} < \overline{\sigma}$ in $[\underline{s}, \overline{s}]$ (other than $\underline{s}, \overline{s}$) such that $Q(\underline{\sigma}) = Q(\overline{\sigma})$. By an earlier argument, $d(Q|_{[\underline{\sigma},\overline{\sigma}]}) > 0$. It can be assumed that

$$d(Q|_{[\underline{\sigma},\overline{\sigma}]}) = 1 \tag{3.53}$$

for otherwise the argument of Proposition 3.41 yields a subloop of $Q|_{[\underline{\sigma},\overline{\sigma}]}$ with this property and then relabeling its endpoints $\underline{\sigma}, \overline{\sigma}$ gives (3.53). Now,

$$\int_{\underline{s}}^{\underline{\sigma}} \left(\tfrac{1}{2}|\dot{Q}|^2 - V(Q)\right)dt = \int_{\overline{\sigma}}^{\overline{s}} \left(\tfrac{1}{2}|\dot{Q}|^2 - V(Q)\right)dt \tag{3.54}$$

for otherwise suppose e.g. the integral on the left is smaller. Set

$$\begin{aligned} \hat{Q}(t) &= Q(t), & t &\le \overline{\sigma} \\ &= Q(\underline{\sigma} - (t - \overline{\sigma})), & \overline{\sigma} &\le t \le \overline{\sigma} + (\underline{\sigma} - \underline{s}) \\ &= Q(\overline{s} + t - \overline{\sigma} - (\underline{\sigma} - \underline{s})), & t &\ge \overline{\sigma} + (\underline{\sigma} - \underline{s}). \end{aligned} \tag{3.55}$$

Then $\hat{Q} \in \hat{\Gamma}_k^+$ and $I(\hat{Q}) < I(Q)$, a contradiction. Therefore (3.54) is satisfied. But then with $\hat{Q}$ as in (3.55), $\hat{Q} \in \hat{\Gamma}_k^+$ and $I(\hat{Q}) = I(Q)$. Therefore $\hat{Q}$ is a solution of (HS) which coincides with Q in $(-\infty, \overline{\sigma})$. Therefore $\hat{Q}(t) = Q(t)$ for all $t \in \mathbb{R}$. In particular

$$Q(t) = Q(t + \overline{s} - \overline{\sigma} - (\underline{\sigma} - \underline{s})) \tag{3.56}$$

for all large t. Since Q is not a periodic function, (3.56) implies $\overline{s}-\overline{\sigma}=\underline{\sigma}-\underline{s}$. It also shows $Q(t)$ moves back and forth along the segment $Q([\underline{s},\underline{\sigma}]) = Q([\overline{\sigma},\overline{s}])$. Set $\overline{Q}(t)=Q\big(t-(\frac{\underline{\sigma}+\overline{\sigma}}{2})\big)$ so this segment is traversed during time intervals $S,-S$ symmetric about $t=0$. Let $P(t)=\overline{Q}(-t)$. Then P is a solution of (HS) which equals $\overline{Q}(t)$ on $S\cup(-S)$. Therefore $P(t)\equiv\overline{Q}(t)$. But P and $\overline{Q}$ traverse the closed loop $Q[\underline{\sigma},\overline{\sigma}]$ in opposite directions. Therefore $Q([\underline{s},\overline{s}])$ must be a simple curve.

Remark 3.57. We suspect that in the setting of Corollary 3.52, Q has exactly $k-1$ self intersections.

It is natural to ask when the hypotheses (3.6) is satisfied. The next result gives a sufficient condition for this to be the case. For simplicity we will assume V is autonomous. For $L>0$, let $W_L^{1,2}$ denote the subset of functions $q\in W^{1,2}([0,L],\mathbb{R}^2)$ such that q is L periodic.

Proposition 3.58. *Suppose V satisfies (V_1)-(V_5) and is autonomous. If there is an $L>0$ and $p\in W_L^{1,2}$ such that*

$$d(p\big|_{[0,L]})=1 \tag{3.59}$$

and

$$\int_0^L \left(\tfrac{1}{2}|\dot{p}|^2-V(p)\right)dt\equiv A<I(Q) \tag{3.60}$$

where Q is as in Proposition 3.41, then there exists $k\in\mathbb{N}\backslash\{1\}$ such that $\hat{c}_k^+<k\hat{c}_1^+$, i.e. (3.6) holds.

Proof. By (3.59), there is an $s\in[0,L)$ such that $p(s)\in\mathcal{S}$ where $\mathcal{S}$ is as in (2.10). It can be assumed that $s=0$. Recall that if $Q\in\Gamma^+$ satisfies $I(Q)=\hat{c}^+$, $Q(0)\in\mathcal{S}$. For each $j\in\mathbb{N}\backslash\{1\}$, define $p_j\in E$ as follows:

$$\begin{aligned} p_j(t) &= Q(t) \qquad t\le 0\\ &=\tfrac{t}{L}p(0)+\tfrac{(1-t)}{L}Q(0),\quad t\in[0,L]\\ &=p(t),\qquad t\in[L,jL]\\ &=\tfrac{(t-j-L)}{L}Q(0)+\tfrac{((j+1)L-t)}{L}p(0)\quad t\in[jL,(j+1)L]\\ &=Q(t-(j+1)L)\quad t\ge(j+1)L. \end{aligned}\tag{3.61}$$

Then $p_j \in \hat{\Gamma}_j^+$ and

$$I(p_j) = I(Q) + (j-1)A + B \tag{3.62}$$

where

$$B = \int_0^1 \mathcal{L}(p_j)dt + \int_{1+(j-1)L}^{2+(j-1)L} \mathcal{L}(p_j)dt \tag{3.63}$$

and $\mathcal{L}(q) = \frac{1}{2}|\dot{q}|^2 - V(q)$. Choose j so large that

$$A + \tfrac{B}{j-1} < I(Q) = \hat{c}_1^+. \tag{3.64}$$

This is possible via (3.60). Therefore by (3.62) and (3.64),

$$\hat{c}_j^+ \le I(p_j) < j\,\hat{c}_1^+ \tag{3.65}$$

and there is a minimal k so that (3.6) is verified.

Remark 3.66. With a bit more care, a similar construction can be made if $V = V(t,x)$. Under the hypotheses of Proposition 3.57, for each $L > 0$, simpler variants of the arguments of Theorems 2.7 and 2.49 show there is an L-periodic solution, P_L, of (HS) which winds once around ξ. (See also [1]). This is a natural candidate for p in Proposition 3.58. Thus heuristically if (3.60) does not hold, the 'cost' as measured by I of winding around ξ in the class of all periodic functions is least for a function of infinite period, namely the homoclinic solution. It is not difficult to show that as $L \to \infty$, a subsequence of the function P_L converge in C^2_{loc} to the homoclinic solution, Q. Since the family of functions P_L will not intersect each other via earlier comparison arguments, if (3.60) fails, this suggests that the functions P_L lie either inside or outside the curve Q and there may be another such family on the other side of Q.

Finally, to conclude this section, it is easy to give concrete examples of when (3.59)–(3.60) are satisfied. Indeed considering ξ as a parameter, by (2.20), the 'cost' of Q winding around ξ is proportional to $|\xi|$. On the other hand, one can construct V so that A in (3.60) is independent of $|\xi|$. Thus (3.60) will hold for $|\xi|$ large.

4. Multiple singularities.

Suppose V has singularities of strong force type at $\xi_1, \dots, \xi_k$, i.e. (V_3) holds with ξ replaced by $\{\xi_1, \dots, \xi_k\}$. We will show that (HS) then has at least k geometrically distinct homoclinics. To formulate matters more precisely, assume V satisfies

$(\overline{V}_1)$ There exist k distinct points $\xi_1, \dots, \xi_k \in \mathbb{R}^2 \backslash \{0\}$ such that

$V \in C^1(\mathbb{R} \times (\mathbb{R}^2 \backslash \{\xi_1, \dots, \xi_k\}), \mathbb{R})$ and is T periodic in t,

$(\overline{V}_2)$ $\lim_{x \to \xi_i} V(t,x) = -\infty$ uniformly in t, $1 \le i \le k$,

$(\overline{V}_3)$ there is a neighborhood $\mathcal{N}_i$ of ξ_i and $U_i \in C(\mathcal{N}_i \backslash \{\xi_i\}, \mathbb{R})$ such that $|U_i(x)| \to \infty$ as $x \to \xi_i$ and

$$|U_i'(x)|^2 \le -V(t,x) \quad \text{for } x \in \mathcal{N}_i \backslash \{\xi_i\} \text{ and } t \in \mathbb{R},\ 1 \le i \le k$$

as well as (V_4)–(V_5). Again set $T = 1$. Let

$$\Lambda_k = \{q \in E \mid q(t) \notin \{\xi_1, \dots, \xi_k\} \text{ for all } t \in \mathbb{R}\}. \tag{4.1}$$

For $q \in \Lambda_k$, the Brouwer degree of q with respect to ξ_i is defined and will be denoted by $d(q, \xi_i)$, $1 \le i \le k$. Set $D(q) = (d(q,\xi_1), \dots, d(q,\xi_k))$. Therefore $D(q) \in \mathbb{Z}^k$. For $\gamma \in \mathbb{Z}^k \backslash \{0\}$, set

$$\Gamma(\gamma) = \{q \in \Lambda_k \mid D(q) = \gamma\} \tag{4.2}$$

and define

$$c(\gamma) = \inf_{q \in \Gamma(\gamma)} I(q). \tag{4.3}$$

Now we can state:

Theorem 4.4. *If V satisfies $(\overline{V}_1)$–$(\overline{V}_3)$, (V_4)–(V_5), there are at least k geometrically distinct solutions of (HS) in Λ_k.*

Remark 4.5. As in §2, if q is a solution of (HS), so is $\tau_j q$ for any $j \in \mathbb{Z}$. Thus by distinct solutions in Theorem 4.4, we mean as distinct elements of $E/\mathbb{Z}$.

The proof of Theorem 4.4 is based on a variant of Proposition 3.7 for the current setting which will be stated next.

Proposition 4.6. *If V satisfies the hypotheses of Theorem 4.4 and (q_m) is a minimizing sequence for (4.3), there is an $\ell = \ell(\gamma) \in \mathbb{N}$, functions $v_1, \dots, v_\ell \in \Lambda_k$, and sequences $(j_m^i) \subset \mathbb{Z}$, $1 \le i \le \ell$ such that*

$$D(v_i) \neq 0, \quad 1 \le i \le \ell, \quad \textit{and} \quad \sum_1^\ell D(v_i) = \gamma, \tag{4.7}$$

$$I(v_i) = \inf_{\Gamma(D(v_i))} I, \tag{4.8}$$

$$\sum_1^\ell I(v_i) = c(\gamma). \tag{4.9}$$

$$\begin{aligned} &\textit{Along a subsequence } \tau_{-j_m^i} q_m \to v_i, \quad 1 \le i \le \ell, \\ &\textit{weakly in } E \textit{ and strongly in } L^\infty_{\text{loc}}, \end{aligned} \tag{4.10}$$

$$|j_m^i - j_m^p| \to \infty \quad \textit{as } m \to \infty \textit{ if } i \neq p, \tag{4.11}$$

$$v_i \quad \textit{is a solution of (HS).} \tag{4.12}$$

The proof of Proposition 4.6 is close to that of Proposition 3.7. However more care must be taken with normalizations. The proof will be given at the end of this section.

Next two proofs of Theorem 4.4 will be given. Both are based on Proposition 4.6. The first is brief and indirect.

Proof 1 *of Theorem* 4.4. Choosing successively $\gamma = e_1, \dots, e_k$, the usual orthonormal basis in $\mathbb{R}^k$, Proposition 4.6 yields a total of e.g. s geometrically distinct homoclinic solutions $Q_1, \dots, Q_s$ of (HS). By (4.7),

$$e_i = \sum_1^s \alpha_{ij} D(Q_j), \quad 1 \le i \le k \tag{4.13}$$

where $\alpha_{ij} \in \mathbb{Z}$. Thus the span of $D(Q_1), \dots, D(Q_s)$ over $\mathbb{Z}$ is all of $\mathbb{Z}^k$. Hence $s \ge k$.

This proof provides no variational characterization of the homoclinic solutions. The next one has such characterizations. It is related to arguments of Bolotin-Kozlov [5] and of [7].

Proof 2 *of Theorem* 4.4. An inductive argument will be used to define k minimization values of I which correspond to k distinct solutions of (HS).

Set $G = \bigcup_{\gamma \neq 0} \Gamma(\gamma)$ and define

$$c_1 = \inf_{q \in G} I(q). \tag{4.14}$$

Let (q_m) be a minimizing sequence for (4.14). Each q_m has a corresponding $\gamma_m \in \mathbb{Z}^k \backslash \{0\}$ such that $q_m \in \Gamma(\gamma_m)$. The cost, as measured by I, of winding around a singularity goes to infinity as $|\gamma| \to \infty$, i.e. if $q \in \Gamma(\gamma)$,

$$I(q) \geq \psi(\gamma) \tag{4.15}$$

where $\psi(\gamma) \to \infty$ as $|\gamma| \to \infty$. Thus (4.15) shows that only finitely many values of γ are significant for the minimization (4.14) and therefore it can be assumed that $(q_m) \subset \Gamma(\gamma_1)$ for some fixed $\gamma_1 \in \mathbb{Z}^k \backslash \{0\}$. Thus

$$c_1 = \inf_{q \in \Gamma(\gamma_1)} I(q) = c(\gamma_1). \tag{4.16}$$

By Proposition 4.6, there is an $\ell(\gamma_1) \in \mathbb{N}$ and solutions $v_1, \dots, v_\ell$ of (HS) such that along a subsequence, (4.10)–(4.11) holds. Note that by (4.7), $D(v_1) \neq 0$ and if $\ell > 1$, by (4.9),

$$\inf_{q \in G} I(q) \leq I(v_1) < c_1, \tag{4.17}$$

a contradiction to (4.14). Thus $\ell = 1$, $D(v_1) = \gamma_1$ by (4.7), and $c_1 = I(v_1)$. Now setting $Q_1 = v_1$, yields one homoclinic solution of (HS) in Λ_k.

Next define

$$c_2 = \inf_{D(q) \in G \backslash \operatorname{span}_{\mathbb{Z}} \gamma_1} I(q). \tag{4.18}$$

As with c_1, there is a $\gamma_2 \in G \backslash \operatorname{span}_{\mathbb{Z}} \gamma_1$ such that

$$c_2 = \inf_{q \in \Gamma(\gamma_2)} I(q) = c(\gamma_2). \tag{4.19}$$

Let (q_m) now be a minimizing sequence for (4.19). As above, there is an $\ell(\gamma_2)$ and functions $v_1, \dots, v_\ell$ satisfying (4.7)–(4.12). Note that $D(v_i) \notin \operatorname{span}_{\mathbb{Z}} \gamma_1$ for at least one $i \in [1, \ell] \cap \mathbb{N}$; otherwise by (4.7), $\sum_1^\ell D(v_i) = \gamma_2 \in \operatorname{span}_{\mathbb{Z}} \gamma_1$ contrary to the construction of γ_2. If $\ell > 1$, as above,

$$\inf_{D(q) \in G \backslash \operatorname{span}_{\mathbb{Z}} \gamma_1} I(q) \leq I(v_i) < \sum_1^\ell I(v_j) = c(\gamma_2) \tag{4.20}$$

contrary to (4.19). Thus again, $\ell = 1$, $D(v_i) = \gamma_2$, and $c_2 = I(v_i)$. Setting $Q_2 = v_2$, it is a second distinct homoclinic solution of (HS). Continuing inductively, once γ_p, c_p, and Q_p have been obtained, we get their $(p+1)$ analogues from

$$c_{p+1} = \inf_{D(q)\in G\setminus \text{span}_{\mathbb{Z}}\{\gamma_1,\dots,\gamma_p\}} I(q) \tag{4.21}$$

and Proposition 4.6 if $p+1 \le k$. Proof 2 is complete.

As usual, if V is autonomous, a better result than Theorem 4.4 obtains.

Theorem 4.22. *If $V = V(x)$ satisfies $(\overline{V}_1)$–$(\overline{V}_3)$, (V_4)–(V_5) and V' is locally Lipschitz continuous for $x \ne \xi_1, \dots, \xi_k$, then (HS) has at least k solutions which are simple curves.*

Proof. It suffices to show the functions $Q_1, \dots, Q_k$ obtained in the second proof of Theorem 4.4 represent simple curves. Consider first Q_1, obtained from (4.14). Suppose Q_1 has a self intersection: $Q_1(\sigma) = Q_1(s)$ with $-\infty < \sigma < s < \infty$. If

$$D(Q_1|_{[\sigma,s]}) = D(Q_1), \tag{4.23}$$

a new function, P, can be defined as in (2.50) by reversing Q_1 along the loop $Q_1[\sigma, s]$:

$$P(t) = \begin{cases} Q_1(t) & t \le \sigma \\ Q_1(s - t + \sigma) & \sigma \le t \le s \\ Q_1(t) & t \ge s \end{cases} \tag{4.24}$$

and $P \in G$ with $I(P) = I(Q_1) = c$. Consequently P is a solution of (HS) which coincides with Q_1 for $t \notin (\sigma, s)$. Therefore $P(t) \equiv Q_1(t)$ while $D(P) = -D(Q_1)$ via (4.23). Thus (4.23) is false:

$$D(Q_1|_{[\sigma,s]}) \ne D(Q_1). \tag{4.25}$$

Deleting the loop $Q_1[\sigma, s]$ from Q_1 produces a new function $R \in G$ (via (4.25)) and $I(R) < I(Q_1) = c_1$ contrary to the definition of Q_1. Thus Q_1 has no self intersections.

The proof of the same result for Q_j, $2 \le j \le k$ follows similar lines. If Q_j has a self intersection at σ and s and (4.23) holds (with Q_1 replaced by Q_j), then the above argument using P again leads to a contradiction. Therefore

(4.25) must be satisfied (with Q_1 replaced by Q_j). Delete $Q_j[\sigma, s]$ from Q_j producing a new function R with

$$I(R) < I(Q_j) = c_j \,. \tag{4.26}$$

If $D(Q_j\big|_{[\sigma,s]}) = 0$, $D(R) = D(Q_j)$ and (4.26) contradicts the definition of c_j. Hence

$$D(Q_j\big|_{[\sigma,s]}) \neq 0. \tag{4.27}$$

Note that $D(R) \neq 0$ via (4.25). Thus by (4.26) and the definition of c_j

$$D(R) \in \text{span}\{D(Q_1), \dots, D(Q_{j-1})\}. \tag{4.28}$$

Reversing the loop $Q_j[\sigma, s]$ produces P as in (4.24) with $D(P) = D(R) - D(Q_j\big|_{[\sigma,s]})$. Note that

$$D(Q_j\big|_{[\sigma,s]}) \notin \text{span}\{D(Q_i), \dots, D(Q_{j-1})\} \tag{4.29}$$

for otherwise

$$D(Q_j) = D(R) + D(Q_j\big|_{[\sigma,s]}) \in \text{span}\{D(Q_i), \dots, D(Q_{j-1})\} \tag{4.30}$$

via (4.28), contrary to $D(Q_j) \in \mathbb{Z}^k \backslash \text{span}\{D(Q_1), \dots, D(Q_{j-1})\}$. By (4.28)–(4.29),

$$D(P) \notin \text{span}\{D(Q_1), \dots, D(Q_{j-1})\}. \tag{4.31}$$

Since $I(P) = I(Q_j) = c_j$, it follows that P is also a minimizer of I over

$$\{q | D(q) \in \mathbb{Z}^k | \{D(q_1), \dots, D(Q_{j-1})\}\}$$

and therefore in a solution of (HS). But as earlier since P coincides with Q_j for $t \notin [\sigma, s]$, $P \equiv Q_j$ while on the other hand,

$$D(P) - D(Q_j) = -2D(Q_j\big|_{[\sigma,s]}) \neq 0 \tag{4.32}$$

via (4.27). It follows that Q_j is a simple curve.

Lastly we give the:

Proof of Proposition 4.6. Let (q_m) be a minimizing sequence for (4.3). Since $D(q_m) = \gamma \neq 0$, (q_m) can be normalized as follows: Let $\rho > 0$ be small. Then there is a smallest $z_m \in \mathbb{Z}$ and $s_m \in [z_m, z_m + 1)$ such that

$$q_m(s_m) \in \partial B_\rho(0) \tag{4.33}$$

and the winding number of $q_m\big|_{[-\infty,s_m]}$ with respect to at least one ξ_j, $1 \leq j \leq k$, is (nearly) in $\mathbb{Z}\backslash\{0\}$. (The error depends on the smallness of ρ.) Due to the analogue of (2.9) for the current setting, it can be assumed that $z_m = 0$ for all m. Now as in the proof of Proposition 3.7, (q_m) is bounded in E and therefore possesses a subsequence converging weakly in E and strongly in L^∞_{loc} to v_1 satisfying

$$0 < \min_{1\leq i\leq k} \tfrac{|\xi_i|}{2}\beta\left(\tfrac{|\xi_i|}{2}\right) \leq I(v_1) \leq c(\gamma) \tag{4.34}$$

and

$$v_1([0,1]) \cap \partial B_\rho(0) \neq \phi. \tag{4.35}$$

If $D(v_1) = \gamma$, $v_1 \in \Gamma(\gamma)$ and (4.7)–(4.12) readily follow with $\ell = 1$. Thus suppose that $D(v_1) \neq \gamma$.

Observe that $D(v_1) \neq 0$ for otherwise by arguments from §2 or 3, (q_m) would not be a minimizing sequence for (4.3). The function v_1 can be expressed in polar coordinates with respect to ξ_j, $1 \leq j \leq k$ via $v_1(t) = R_{1j}(t)e^{\Phi_{1j}(t)}$. Hence as earlier, for $0 < \delta < \rho$, there is a $T_1 = T_1(\delta)$ such that $Q_1(t) \in B_\delta(0)$ for $|t| \geq T_1$ and if now $\Phi_1 = (\Phi_{1i}, \dots, \Phi_{1k})$, as in (3.19),

$$\tfrac{1}{2\pi}(\Phi_1(T_1) - \Phi_1(-\infty)) \simeq D(Q_1). \tag{4.36}$$

Moreover if φ_m is the analogue for q_m of Φ_1 for Q_1,

$$D(Q_1) \simeq \tfrac{1}{2\pi}(\varphi_m(T_1) - \varphi_m(-\infty)). \tag{4.37}$$

Indeed $q_m(t) \to Q_1(t)$ uniformly for $|t| \leq T_1$ along a subsequence and

$$\tfrac{1}{2\pi}(\varphi_m(-T_1) - \varphi_m(-\infty)) \simeq 0$$

due to the normalization made on q_m.

Since $D(Q_1) \neq \gamma$ and q_m satisfies (4.37), $D(q_m\big|_{[T_1,\infty]}) \neq 0$. Therefore $q_m(t)$ near 0 for t near T_1 implies there is a smallest integer $t_m > T_1$ and $y_m \in [t_m, t_m+1)$ such that $q_m(y_m) \in \partial B_\rho(0)$ and the winding number of $q_m\big|_{[T_1,y_m]}$ with respect to some ξ_i, $1 \leq i \leq j$ is (nearly) a nonzero integer. This normalization leads to the existence of a subsequence of $\tau_{-t_m}q_m$ converging to Q_2. Continuing in this fashion essentially as in the proof of Proposition 3.7 but with the above modifications leads in a finite number of steps to (4.7)–(4.12).

References.

[1] Gordon, W. B., *Conservative dynamical systems involving strong forces*, Trans. A.M.S., **204**, (1975), 113–135.

[2] Tanaka, K., *Homoclinic orbits for a singular second order Hamiltonian system*, Ann. Inst. H. Poincaré, Analyse non lineaire, **7**, (1990), 427–438.

[3] Bahri, A. and P. H. Rabinowitz, *A minimax method for a class of Hamiltonian systems with singular potentials*, J. Funct. Analysis, **82**, (1989), 412–428.

[4] Bessi, U., *Multiple homoclinic orbits for autonomous singular potential*, Proc. Roy. Soc. Edinburg Sec. A, **124** (1994), 785–802.

[5] Kozlov, V.V., *Calculus of variations in the large and classical mechanics*, Russ. Math. Surv., **40**, (1985), 37–71.

[6] Rabinowitz, P. H. and K. Tanaka, *Some results on connecting orbits for a class of Hamiltonian systems*, Math. Z., **206**, (1991), 473–499.

[7] Rabinowitz, P. H., *Periodic and heteroclinic orbits for a periodic Hamiltonian system*, Ann. Inst. H. Poincaré, Analyse non lineaire, **6**, (1989), 331–346.

[8] Friedman, A., *Partial Differential Equations*, Holt, Rinehard, and Winston, 1969.

[9] Ambrosetti, A. and V. Coti Zelati, *Periodic Solutions of Singular Lagrangian Systems*, Birkhaüser, 1993.

[10] Coti Zelati, V. and E. Serra, *Some properties of collision and noncollision orbits for a class of singular dynamical system*, Rend. Lincei Ser. 9, **3**, (1992) 217-222.

[11] Serra, E. and S. Terracini, *Noncollision solutions to some singular minimization problems with Keplerian-like potentials*, Nonlinear Analysis T.M.A., **22**, (1994) 45–62.

[12] Tanaka, K., *Non-collision solutions for a second order singular Hamiltonian system with weak force*, Ann. Inst. H. Poincaré, Analyse non lineaire, **10** (1993), 215–238.

[13] Lions, P. L., *The concentration compactness principle in the calculus of variations: The locally compact case, Part 2*, Ann. Inst. H. Poincaré, Analyse non lineaire, **1**, (1994) 223–283.

[14] Coti Zelati, V., I. Ekeland, and E. Séré, *A variational approach to homoclinic orbits in Hamiltonian systems*, Math. Ann. **288**, (1990) 133-160.

[15] Brezis, H. and J.-M. Coron, *Large solutions for harmonic maps in two dimensions*, Comm. Math. Physics, **92**, (1983) 203–215.

[16] Giaquinta, M. and S. Hildebrandt, *A priori estimates for harmonic mappings*, J. Reine Angew. Math., **336**, (1982) 124–164.

[17] Jost, J., *The Dirichlet problem for harmonic maps from a surface with boundary into a two sphere with nonconstant boundary values*, J. Diff. Geom., **19**, (1984) 393–401.

[18] Whitney, H., *On regular closed curves in the plane*, Compos. Math., **4**, (1937), 276–284.

RECEIVED JANUARY 25, 1995.

DEPARTMENT OF MATHEMATICS
UNIVERSITY OF WISCONSIN
MADISON, WI 53706

Uniqueness of Plateau's problem for certain contours with a one–to–one, nonconvex projection onto a plane

FRIEDRICH SAUVIGNY

Dedicated to Professor Stefan Hildebrandt on the occasion of his sixtieth birthday

§ 1. The Result.

Let $\Omega \subset \mathbb{R}^2$ be a simply connected, bounded domain with the regular C^2–Jordan–curve $\partial\Omega$ as its boundary. Furthermore, let $g = g(x, y) : \partial\Omega \to \mathbb{R} \in C^0(\partial\Omega)$ be a continuous function such that the contour

$$\Gamma := \big\{(x, y, z) \in \mathbb{R}^3 \mid (x, y) \in \partial\Omega,\ z = g(x, y)\big\} \tag{1.1}$$

is a Jordan–curve with a one–to–one projection onto the x, y–plane. Due to classical investigations of T. Rado, J. Douglas and R. Courant (see [1], [2], [8]) one can construct a parametric minimal surface $\mathbf{x}$ bounded by the contour Γ, minimizing Dirichlet's integral. We take the unit disc

$$B := \{w = u + iv \in \mathbb{C} \mid |w| < 1\}$$

in the complex plane and three different points $\pi_k = (x_k, y_k, g(x_k, y_k)) \in \Gamma$ such that the points $(x_k, y_k) \in \partial\Omega$; $k = 0, 1, 2$; induce the positive orientation on $\partial\Omega$. Then a *solution* $\mathbf{x} \in \mathbf{P}(\Gamma)$ *of Plateau's problem* has the following properties:

$$\mathbf{x} = \mathbf{x}(u, v) = (x(u, v), y(u, v), z(u, v)) : \overline{B} \to \mathbb{R}^3 \in C^2(B) \cap C^0(\overline{B}) \tag{1.2}$$

$$\Delta\mathbf{x}(u, v) = 0, \quad |\mathbf{x}_u| = |\mathbf{x}_v|, \quad \mathbf{x}_u \cdot \mathbf{x}_v = 0 \quad \text{in } B , \tag{1.3}$$

$$\mathbf{x} : \partial B \to \Gamma \text{ is a topological mapping} , \tag{1.4}$$

$$\mathbf{x}(\exp(2\pi ik/3)) = \pi_k, \quad k = 0, 1, 2. \tag{1.5}$$

Now the question arises, which additional conditions on Γ guarantee that a parametric solution $\mathbf{x} \in \mathbf{P}(\Gamma)$ has a representation as a minimal graph solving the following *Dirichlet– Problem for the minimal surface equation* $\mathbf{P}(\Omega, g)$:

$$
\begin{aligned}
z &= \varsigma(x,y) : \overline{\Omega} \to \mathbb{R} \in C^2(\Omega) \cap C^0(\overline{\Omega}), & (1.6)\\
L(\varsigma) &:= (1+\varsigma_y^2)\varsigma_{xx} - 2\varsigma_x\varsigma_y\varsigma_{xy} + (1+\varsigma_x^2)\varsigma_{yy} = 0 \quad \text{in } \Omega, & (1.7)\\
\varsigma(x,y) &= g(x,y) \quad \text{on } \partial\Omega. & (1.8)
\end{aligned}
$$

Here L is called the *minimal surface operator.*

The question above can be answered in the affirmative for convex domains Ω by a celebrated result of H. Kneser and T. Rado (see [6] and [9]). For nonconvex domains Ω this question cannot always be answered positively, since then the Dirichlet–Problem $\mathbf{P}(\Omega, g)$ does not have a solution for arbitrary continuous boundary data $g : \partial\Omega \to \mathbb{R}$.

For an arbitrary bounded C^2–domain $\Omega \subset \mathbb{R}^2$ as above we call $(x_0, y_0) \in \partial\Omega$ a *convex boundary point* , if a support function $\psi(x,y) := ax + by + c$ with the real parameters a, b, c satisfying $a^2 + b^2 > 0$ exists, such that

$$\psi(x,y) \le 0 \text{ for all } (x,y) \in \partial\Omega \text{ and } \psi(x_0,y_0) = 0 \tag{1.9}$$

holds true. Now we distinguish between the *convex boundary*

$$\partial^+\Omega := \{(x_0,y_0) \in \partial\Omega \,|\, (x_0,y_0) \text{ is a convex boundary point } \}$$

and the *concave boundary* $\partial^-\Omega := \overline{(\partial\Omega \backslash \partial^+\Omega)}$. By $\nu = \nu(x,y) : \partial\Omega \to \mathbb{R}^2$ we denote the exterior unit normal to the domain Ω .

Definition 1. The *domain Ω fulfills an exterior ray condition* if the property

$$(x_0,y_0) + t\nu(x_0,y_0) \notin \overline{\Omega} \text{ for all } (x_0,y_0) \in \partial^-\Omega \text{ and all } t \in (0,+\infty) \tag{1.10}$$

is satisfied.

For these domains Ω with an exterior ray condition we shall give quantitative conditions on the boundary distributions $g : \partial\Omega \to \mathbb{R}$ which guarantee that Plateau's problem $\mathbf{P}(\Gamma)$ has exactly one solution for the associated contour

$$\Gamma := \left\{(x,y,g(x,y)) \in \mathbb{R}^3 \,|(x,y) \in \partial\Omega\right\} .$$

In the following we need the family of *Scherk's functions*

$$\Phi^{\alpha,\beta}(t) := -\frac{1-\alpha}{\beta}\log\cos(\beta t) , \qquad |t| < \frac{\pi}{2\beta} \tag{1.11}$$

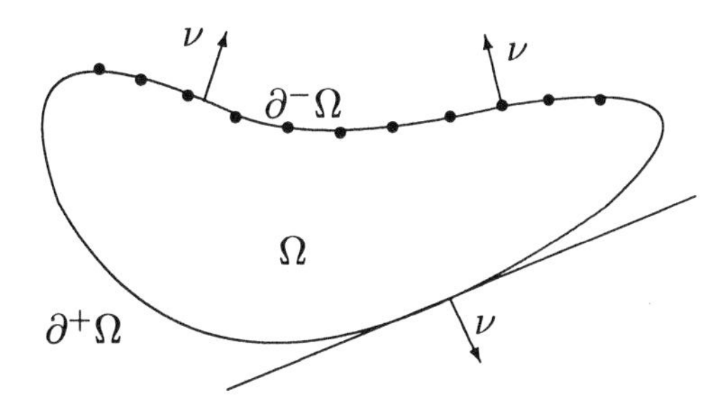

Figure 1: Domain with an exterior ray condition

for all $\alpha \in [0,1)$ and $\beta \in (0,+\infty)$. We easily verify

$$\begin{cases} \phi^{\alpha,\beta}(-t) = \phi^{\alpha,\beta}(t) \ \text{for} \ |t| < \frac{\pi}{2\beta}, \\ \phi^{\alpha,\beta}(0) = 0, \qquad \lim_{t \to \pm \frac{\pi}{2\beta}} \phi^{\alpha,\beta}(t) = +\infty. \end{cases} \tag{1.12}$$

Furthermore, we derive

$$\begin{cases} \frac{d}{dt}\phi^{\alpha,\beta}(t) = (1-\alpha)\tan(\beta t) \\ \frac{d^2}{dt^2}\phi^{\alpha,\beta}(t) = \frac{(1-\alpha)\beta}{\cos^2(\beta t)} \end{cases}, \qquad \text{for} \ |t| < \frac{\pi}{2\beta} \, . \tag{1.13}$$

For the arc $t \longmapsto (t, \Phi^{\alpha,\beta}(t))$ we determine the oriented curvature

$$\kappa(t) = \frac{(1-\alpha)\beta\cos(\beta t)}{\sqrt{1+\alpha(\alpha-2)\sin^2(\beta t)}^{\,3}}\,, \qquad |t| < \frac{\pi}{2\beta} \, . \tag{1.14}$$

In the case $\alpha = 0$ these curves appear as principal lines of curvature in the Scherk minimal surface, which will serve as a comparison surface in our investigations. We denote by

$$D(\alpha,\beta) := \left\{ (x,y) \,\middle|\, x \in (-\frac{\pi}{2\beta}, +\frac{\pi}{2\beta}), \ y > \Phi^{\alpha,\beta}(x) \right\}$$

the domain above the graph of $\Phi^{\alpha,\beta}$. If we rotate the domain $D(\alpha,\beta)$ about the origin such that the vector $(0,1)$ is transformed into an arbitrary unit vector $\nu_0 = (\nu_x, \nu_y) \in \mathbb{R}^2$ and if we subsequently apply a translation of the

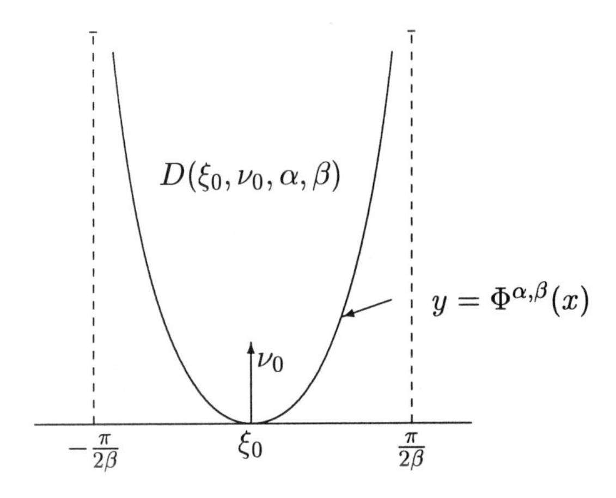

Figure 2.

plane, mapping the origin $(0, 0)$ into an arbitrary point $\xi_0 = (x_0, y_0) \in \mathbb{R}^2$, we obtain the domain $D(\xi_0, \nu_0, \alpha, \beta)$.

Now we shall quantify the degree of concavity of our domains admitted.

Definition 2. Let $\Omega \subset \mathbb{R}^2$ be a bounded C^2–domain as above with the nonvoid concave boundary $\partial^-\Omega$. Then Ω satisfies a *uniform concavity condition* with the parameters $\alpha \in [0, 1)$, $\beta \in (0, +\infty)$, $\gamma \in (0, \operatorname{diam} \Omega]$, if the following conditions are satisfied:

i) For each point $\xi_0 = (x_0, y_0) \in \partial^-\Omega$ with the unit normal $\nu_0 = \nu(x_0, y_0)$ we have

$$\Omega \cap D(\xi_0, \nu_0, \alpha, \beta) = \emptyset \; ; \tag{1.15}$$

ii) For each point $\xi_0 \in \partial^-\Omega$ the set

$$\omega_0 = \omega(\xi_0) := \partial\Omega \cap D(\xi_0 - \gamma\nu_0, \nu_0, 0, \beta)$$

can be represented as a graph above an open interval contained in the straight line through the point ξ_0 orthogonal to ν_0 .

Remarks.

1.) Considering the normalized situation $\xi_0 = (0,0)$ and $\nu_0 = (0,1)$, we obtain

$$\omega_0 = \{(t, y(t)) \mid t \in (t_-, t_+)\} \tag{1.16}$$

with the function $y = y(t) \in C^2(t_-, t_+)$ satisfying $y(0) = 0 = y'(0)$. Here we have

$$t_- = t_-(\xi_0) \in (-\frac{\pi}{2\beta}, 0) \qquad \text{and} \qquad t_+ = t_+(\xi_0) \in (0, +\frac{\pi}{2\beta}).$$

2.) For C^2–domains with an exterior ray condition and for given $\alpha \in [0,1)$ one can determine the quantities $\beta = \beta(\alpha) \in (0, +\infty)$ and $\gamma = \gamma(\alpha, \beta) \in (0, \operatorname{diam} \Omega]$ such that Ω satisfies a uniform concavity condition for these parameters.

We can now formulate our main result, namely the following

Theorem. Assumptions.

I. *Let $\Omega \subset \mathbb{R}^2$ be a bounded C^2–domain as above satisfying a uniform concavity condition with the parameters $\alpha \in [0,1)$, $\beta \in (0, +\infty)$, $\gamma \in (0, diam\Omega]$. We introduce the* associated height

$$h = h(\Omega) := \frac{1}{2\beta} \arccos \exp(-\beta\gamma)$$

and the associated Lipschitz constant $l = l(\Omega) := \alpha$.

II. *Let $g : \partial\Omega \to \mathbb{R}$ be a continuous function satisfying the maximum estimate*

$$\sup_{\partial\Omega} |g(x,y)| < h(\Omega) \tag{1.17}$$

and the following tangential Lipschitz condition on the concave boundary $\partial^-\Omega$:

$$|g(\xi) - g(\xi_0)| \le l(\Omega) |(\xi - \xi_0, \nu_0)|$$
$$\text{for all } \xi_0 \in \partial^-\Omega \text{ and every } \xi \in \omega_0 = \omega(\xi_0). \tag{1.18}$$

Here we use the notations from Definition 2, and $(.,.)$ means the determinant of the plane vectors inscribed.

III. *Let* $\Gamma = \{(x, y, g(x, y)) \in \mathbb{R}^3 \mid (x, y) \in \partial\Omega\}$ *be the corresponding Jordan contour.*

Statement. *Then there exists exactly one solution* $\mathbf{x} \in \mathbf{P}(\Gamma)$ *of Plateau's problem. This solution* $\mathbf{x}$ *can be represented as a minimal graph* $z = \varsigma(x, y)$ *solving the Dirichlet–Problem* $\mathbf{P}(\Omega, g)$ *for the minimal surface equation.*

Remarks.

1.) On account of (1.18) the height representation of Γ projected onto the tangential plane of the contour orthogonal to ν_0 satisfies a Lipschitz condition.

2.) The Dirichlet–Problem $\mathbf{P}(\Omega, g)$ has been solved by C.P. Lau [7], F. Schulz [11], and G. Williams [12] by nonparametric methods; they also showed the Lipschitz condition (1.18) to be necessary.

3.) We shall give a parametric approach to this Dirichlet–Problem using the classical Dirichlet principle. Furthermore, we can explicitly determine the bound $h = h(\Omega)$.

4.) The existence of further (stable on unstable) parametric solutions of Plateau's problem is excluded by our Theorem.

The proof of our result will be given in the next section. It decisively depends on Proposition 1, where a parametric minimal surface is compared with a minimal graph. Utilizing Scherk's minimal surface as a comparison surface, we can proceed as in [10]. In Proposition 2 we prove an inclusion principle. Then we establish the transversality to the boundary for C^2–contours in Proposition 3. We use the stability of the solution $\mathbf{x}$ of Plateau's problem to see that $\mathbf{x}$ is a graph. By an approximation argument we then solve $\mathbf{P}(\Omega, g)$ for continuous data in Proposition 4. Invoking Proposition 1 again we complete the proof of the Theorem.

Finally, I would like to refer the reader to my joint work with Professor Stefan Hildebrandt [5] concerning partially free boundary value problems for minimal surfaces.

§ 2. The Proof.

We begin with the following

Proposition 1 (Parametric comparison principle).
Assumptions.

I. *On a domain* $\Omega \subset \mathbb{R}^2$ *let* $z = \varsigma(x,y) \in C^2(\Omega)$ *be a solution of the minimal surface equation* $L(\varsigma) = 0$ *in* Ω*. The corresponding minimal graph*

$$\mathbf{y}(x,y) := (x, y, \varsigma(x,y)), \qquad (x,y) \in \Omega$$

has the unit normal

$$\mathbf{Y}(x,y) := (1 + |\nabla\varsigma(x,y)|^2)^{-1/2}(-\varsigma_x, -\varsigma_y, 1), \qquad (x,y) \in \Omega\ .$$

Furthermore,we define the vector $\mathbf{e} := (0,0,1) \in \mathbb{R}^3$.

II. *On a domain* $B \subset \mathbb{C}$ *let*

$$\mathbf{x}(u,v) = (x(u,v), y(u,v), z(u,v)) : B \to \mathbb{R}^3 \in C^2(B)$$

be a nonconstant, parametric minimal surface satisfying (1.3) with possible branch points. In the usual way we define its unit normal $\mathbf{X}(u,v) : B \to \mathbb{R}^3$ *. For the associated plane map*

$$f(u,v) = (x(u,v), y(u,v)) : B \to \mathbb{R}^2$$

we assume $f(B) \subset \Omega$*. Finally, we define the functions*

$$\mathbf{y}(u,v) := \mathbf{y}(x(u,v), y(u,v)), \qquad (u,v) \in B$$

and

$$\mathbf{Y}(u,v) := \mathbf{Y}(x(u,v), y(u,v)), \qquad (u,v) \in B\ .$$

III. *Now we consider the auxiliary function*

$$\phi(u,v) := z(u,v) - \varsigma(x(u,v), y(u,v)), \qquad (u,v) \in B\ . \tag{2.1}$$

Statement. *Then this function ϕ satisfies the elliptic partial differential equation*

$$\begin{aligned}\frac{\partial}{\partial u}\{(\mathbf{Y}(u,v)\cdot\mathbf{e})\phi_u\}+\frac{\partial}{\partial v}\{(\mathbf{Y}(u,v)\cdot\mathbf{e})\phi_v\}&\\ -(\mathbf{e},\mathbf{Y}_v,\mathbf{X})\phi_u(u,v)-(\mathbf{Y}_u,\mathbf{e},\mathbf{X})\phi_v(u,v)&=0 \qquad in\ B. \qquad (2.2)\end{aligned}$$

Here $(.,.,.)$ denotes the triple product.

Proof.

1.) For the parametric minimal surface we have

$$\Delta\mathbf{x}=0,\ \mathbf{X}\wedge\mathbf{x}_u=\mathbf{x}_v,\ \mathbf{X}\wedge\mathbf{x}_v=-\mathbf{x}_u \qquad in\ B. \qquad (2.3)$$

The minimal graph in the parameters $(u,v)\in B$, which are neither regular nor conformal in general, fulfils the parameterinvariant differential equation

$$\mathbf{y}_u\wedge\mathbf{Y}_v+\mathbf{Y}_u\wedge\ \mathbf{y}_v=0 \qquad in\ B. \qquad (2.4)$$

2.) Abbreviating $\surd:=\sqrt{1+|\nabla\varsigma(x,y)|^2}\Big|_{\substack{x=x(u,v)\\ y=y(u,v)}}$ in B, we derive

$$\begin{aligned}\phi_u&=-\varsigma_x x_u-\varsigma_y y_u+z_u=\surd\,(\mathbf{Y}\cdot\mathbf{x}_u),\\ \phi_v&=-\varsigma_x x_v-\varsigma_y y_v+z_v=\surd\,(\mathbf{Y}\cdot\mathbf{x}_v).\end{aligned} \qquad (2.5)$$

Differentiating $\mathbf{y}(u,v)=(x(u,v),y(u,v),\varsigma(x(u,v),y(u,v)))$ in B we obtain

$$\mathbf{y}_u=\mathbf{x}_u+(\varsigma_x x_u+\varsigma_y y_u-z_u)\mathbf{e}=\ \mathbf{x}_u-\phi_u\mathbf{e}$$

and similarly $\mathbf{y}_v=\mathbf{x}_v-\phi_v\mathbf{e}$. This implies

$$\mathbf{x}_u=\mathbf{y}_u+\phi_u\,\mathbf{e},\mathbf{x}_v=\mathbf{y}_v+\phi_v\,\mathbf{e}\ . \qquad (2.6)$$

3.) Now we calculate with the aid of (2.5), (2.3), (2.4) and (2.6):

$$\begin{aligned}
\left(\frac{\phi_u}{\surd}\right)_u + \left(\frac{\phi_v}{\surd}\right)_v &= (\mathbf{Y}\cdot\mathbf{x}_u)_u + (\mathbf{Y}\cdot\mathbf{x}_v)_v \\
&= \mathbf{Y}\cdot\Delta\mathbf{x} + (\mathbf{Y}_u\cdot\mathbf{x}_u) + (\mathbf{Y}_v\cdot\mathbf{x}_v) \\
&= (\mathbf{x}_u, \mathbf{Y}_v, \mathbf{X}) + (\mathbf{Y}_u, \mathbf{x}_v, \mathbf{X}) \\
&= (\mathbf{y}_u + \phi_u\mathbf{e}, \mathbf{Y}_v, \mathbf{X}) + (\mathbf{Y}_u, \mathbf{y}_v + \phi_v\mathbf{e}, \mathbf{X}) \\
&= (\mathbf{y}_u \wedge \mathbf{Y}_v + \mathbf{Y}_u \wedge \mathbf{y}_v)\cdot\mathbf{X} \\
&\quad + (\mathbf{e}, \mathbf{Y}_v, \mathbf{X})\phi_u + (\mathbf{Y}_u, \mathbf{e}, \mathbf{X})\phi_v \\
&= (\mathbf{e}, \mathbf{Y}_v, \mathbf{X})\phi_u + (\mathbf{Y}_u, \mathbf{e}, \mathbf{X})\phi_v \qquad \text{in } B.
\end{aligned}$$

Observing $\frac{1}{\surd} = \mathbf{Y}(u,v)\cdot\mathbf{e}$ we arrive at the partial differential equation (2.2). □

Remark. This parametric comparison principle might be of independent interest. On account of the partial differential equation (2.2) the auxiliary function $\phi(u,v)$ measuring the deviation of $\mathbf{x}$ from the minimal graph $\mathbf{y}$ obeys the maximum principle, is accessible to the Hopf boundary point lemma, and allows asymptotic expansions of Hartman–Wintner–type at the zeroes of $\nabla\phi$.

In the following investigations we shall utilize the one–parametric family of minimal graphs $S(\beta), \beta \in (0,+\infty)$, discovered by H. F. Scherk in 1835.

Analytically, these surfaces $S(\beta)$ are given by

$$\begin{aligned}
&y = \sigma(x,z) := \tfrac{1}{\beta}(\log\cos(\beta z) - \log\cos(\beta x)) = \phi^{0,\beta}(x) - \phi^{0,\beta}(z) \\
&\text{for } (x,z) \in \Delta_\beta := \left\{(x,z) \in \mathbb{R}^2 \;\middle|\; |x| < \tfrac{\pi}{2\beta},\ |z| < \tfrac{\pi}{2\beta}\right\}; \qquad (2.7)
\end{aligned}$$

here we have $\beta \in (0,+\infty)$. Due to (1.14) the parameter $\beta \in (0,+\infty)$ gives the maximal sectional curvature of $S(\beta)$. Furthermore, we remark

$$\sigma(0,0) = 0,\ \sigma_x(0,0) = 0,\ \sigma_z(0,0) = 0\ . \qquad (2.8)$$

With the aid of (1.13) we easily verify the minimal surface equation $L(\sigma) = 0$ in Δ_β for Scherk's surface. On the set

$$C(\beta) := \left\{(x,y) \in \mathbb{R}^2 \;\middle|\; x \in (-\frac{\pi}{2\beta}, +\frac{\pi}{2\beta}),\ y \le -\frac{1}{\beta}\log\cos(\beta x)\right\}$$

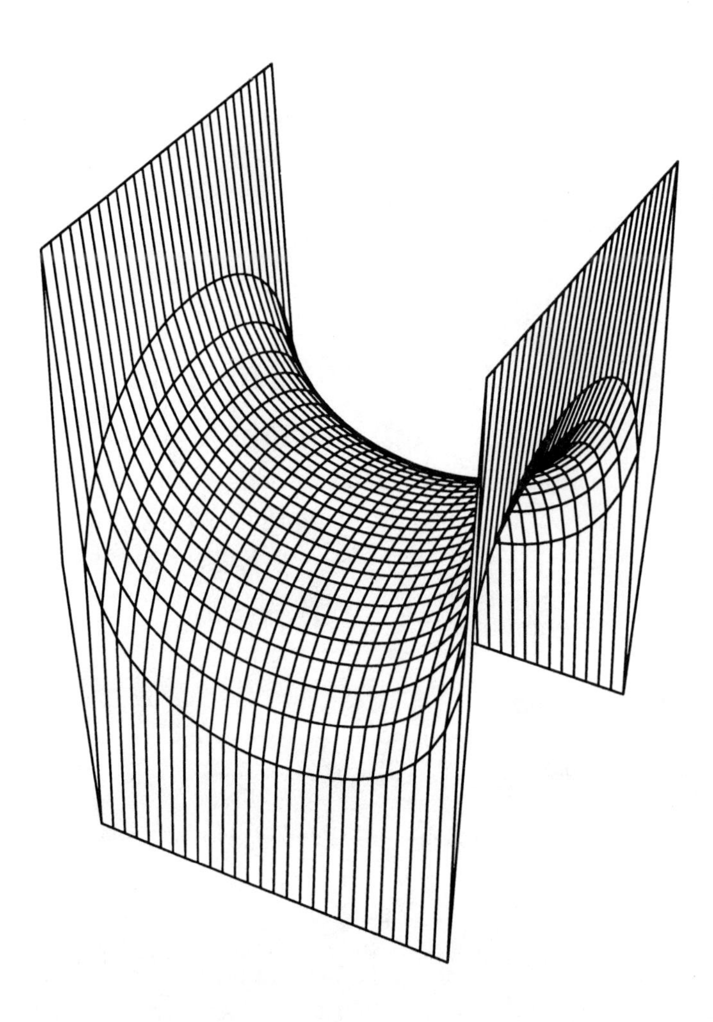

Figure 3: Scherk's minimal surface

we can represent $S(\beta)$ in the following form

$$z = \pm\frac{1}{\beta} \arccos\ \exp(\beta y + \log\cos(\beta x)), \qquad (x, y) \in C(\beta)\ . \tag{2.9}$$

Here $\arccos : [0, 1] \to [0, \frac{\pi}{2}]$ denotes the inverse function of $\cos : [0, \frac{\pi}{2}] \to [0, 1]$, which are both monotonically decreasing functions.

Proposition 2 (Inclusion principle). *Let the assumptions of the Theorem be satisfied and let* $\mathbf{x} \in \mathbf{P}(\Gamma)$ *be a solution of Plateau's problem with the plane map* f*. Then we have the inclusion*

$$f(B) \subset \Omega\ . \tag{2.10}$$

Proof.

1.) If $conv(\Omega)$ denotes the open convex hull of the set Ω, we immediately infer $f(B) \subset conv(\Omega)$ from the convex-hull-property for minimal surfaces. If (2.10) would be false, there exists a point $(u_1, v_1) \in B$ with the property

$$f(u_1, v_1) \in conv(\Omega)\backslash\Omega\ . \tag{2.11}$$

Then there exists a point $\xi_0 = (x_0, y_0) \in \partial^-\Omega$ with the normal $\nu_0 = \nu(x_0, y_0)$ such that the inclusion

$$f(u_1, y_1) \in \overline{D(\xi_0, \nu_0, 0, \beta)} \tag{2.12}$$

is correct. Let $(u_0, v_0) \in \partial B$ satisfy $f(u_0, v_0) = (x_0, y_0)$ and set $\mathbf{z}_0 = \mathbf{x}(u_0, v_0)$, $\mathbf{z}_0 \in \Gamma$. By a translation in the space $\mathbb{R}^3$ and rotation about the z-axis we can achieve

$$\mathbf{z}_0 = 0,\ \ \nu_0 = (0, 1)\ . \tag{2.13}$$

2.) Now we show that Scherk's surface $S(\beta)$ is a support surface to the contour Γ at the origin. By assumption the set

$$\omega_0 := \partial\Omega \cap D(\xi_0 - \gamma\nu_0, \nu_0, 0, \beta)$$

is a graph over the x–axis. Therefore the arc

$$\Gamma^* := \{(x, y, g(x, y)) \mid (x, y) \in \omega_0\} \subset \Gamma$$

above ω_0 can be parametrically represented in the following way:

$$\mathbf{y}(t) = (t, y(t), z(t)), \qquad t \in (t_-, t_+), \tag{2.14}$$

with $t_- = t_-(\xi_0) \in (-\frac{\pi}{2\beta}, 0)$ and $t_+ = t_+(\xi_0) \in (0, +\frac{\pi}{2\beta})$. For the C^0–functions y and z we have the conditions

$$y(0) = z(0) = 0 \ , \qquad |z(t)| \le \alpha|t| \quad \text{for all } t \in (t_-, t_+) \tag{2.15}$$

using the estimate (1.18). The auxiliary function

$$\theta(t) := y(t) - \sigma(t, z(t)), \qquad t \in (t_-, t_+)$$

satisfies the inequality

$$\begin{aligned} \theta(t) &= y(t) - \phi^{0,\beta}(t) + \phi^{0,\beta}(z(t)) \\ &\le y(t) - \phi^{0,\beta}(t) + \phi^{0,\beta}(\alpha|t|) \\ &\le y(t) - \phi^{0,\beta}(t) + \alpha\phi^{0,\beta}(t) \\ &= y(t) - \phi^{\alpha,\beta}(t) \le 0 \qquad \text{for all } t \in (t_-, t_+). \end{aligned} \tag{2.16}$$

Here we have used that $\phi^{0,\beta}$ is a symmetric, convex function and that

$$\Omega \cap D(\xi_0, \nu_0, \alpha, \beta) = \emptyset$$

has been assumed. Therefore Γ^* does not lie above the surface $S(\beta)$ w.r.t. the y–axis. From the assumption (1.17) we deduce

$$\sup_{\partial\Omega} |g(x, y)| < 2h = \frac{1}{\beta} \arccos \exp(-\beta\gamma) \ . \tag{2.17}$$

Now we consider the arc

$$\Gamma^{**} := \{(x, y, g(x, y)) \,|\, (x, y) \in \partial\Omega \backslash \omega_0\} \subset \Gamma \ ,$$

which does not intersect $S(\beta)$. This can be seen with the aid of (2.17), property ii) of Definition 2, and the representation (2.9) for Scherk's surface over the domain $C(\beta)$. Introducing the exterior set

$$E(\beta) := \big\{(x, y, z) \in \mathbb{R}^3 \mid (x, z) \in \Delta_\beta, \ y > \sigma(x, z)\big\}$$

we obtain

$$\Gamma \cap E(\beta) = \emptyset \ . \tag{2.18}$$

3.) We shall now show

$$x(\overline{B}) \cap E(\beta) = \emptyset \ . \tag{2.19}$$

From the boundary condition (2.17) we immediately infer

$$\sup_{\overline{B}} |z(u,v)| < \frac{\pi}{2\beta} \tag{2.20}$$

for the third component of $\mathbf{x}$. If (2.19) would be violated, we consider the nonvoid, open set

$$\Theta = \{(u,v) \in B \,|\, \mathbf{x}(u,v) \in E(\beta)\} \ .$$

With the aid of (2.20) we easily deduce

$$\mathbf{x}(\partial\Theta) \subset S(\beta) \ . \tag{2.21}$$

Then we consider the auxiliary function

$$\psi(u,v) := y(u,v) - \sigma(x(u,v), z(u,v)), \qquad (u,v) \in \overline{\Theta}$$

satisfying

$$\psi > 0 \text{ in } \Theta, \qquad \psi = 0 \text{ on } \partial\Theta \ . \tag{2.22}$$

Since ψ is subject to an elliptic partial differential equation established in Proposition 1, we arrive at a contradiction. Consequently (2.19) is correct.

4.) On account of (2.12) and (2.19) the auxiliary function ψ satisfies

$$\psi(u,v) \leq 0 \ \text{ in b}, \qquad \psi(u_1, v_1) = 0 \tag{2.23}$$

in a sufficiently small neighbourhood b of (u_1, v_1). This implies

$$\psi(u,v) \equiv 0$$

in b, which leads to an evident contradiction.

Therefore the stated inclusion (2.10) is valid. □

Proposition 3 (Transversality to the boundary).
Assumptions. *Let the assumptions of the Theorem be satisfied and inclusively* $g : \partial\Omega \to \mathbb{R} \in C^2(\partial\Omega)$. *Let* $\mathbf{x} \in \mathbf{P}(\Gamma)$ *be a solution of Plateau's problem with the unit normal* $\mathbf{X}$.

Statement. *Then we have*

$$\mathbf{x} \in C^1(\overline{B}), \qquad E(u,v) := |\mathbf{x}_u|^2 = |\mathbf{x}_v|^2 > 0 \ \text{ on } \partial B, \qquad \mathbf{X}\cdot\mathbf{e} > 0 \ \text{ on } \partial B\ .$$

Remark. This means that $\mathbf{x}$ has no branch points on the boundary, and $\mathbf{x}$ is a graph above the x,y–plane near the contour Γ.

Proof. Since Γ is a regular C^2–contour, we have $\mathbf{x} \in C^1(\overline{B})$ due to the investigations [3] and [4]. Let $(u_0, v_0) \in \partial B$ be given; we can achieve

$$\mathbf{x}(u_0, v_0) = 0, \qquad \nu_0 = \nu(x(u_0, v_0), y(u_0, v_0)) = (0, 1) \tag{2.24}$$

by a translation and a rotation about the z–axis. Rotating the disc B about the origin we can additionally assume $(u_0, v_0) = (0, 1)$.

Case I. $f(u_0, v_0) \in \partial^-\Omega$.

For a sufficiently small $\epsilon > 0$ we define the domain

$$b := \{w \in B \mid |w - i| < \epsilon\}$$

and consider the auxiliary function

$$\psi(u,v) := y(u,v) - \sigma(x(u,v), z(u,v)), \qquad (u,v) \in b \tag{2.25}$$

as in the proof of Proposition 2. Since ψ attains its maximum at the boundary point $(u_0, v_0) = (0, 1) \in \partial B$ we infer

$$0 < \psi_v(0,1) = (y_v - \sigma_x x_v - \sigma_z z_v)\Big|_{\substack{u=0\\ v=1}} = y_v(0,1) \tag{2.26}$$

from the Hopf boundary point lemma. This immediately implies

$$E(0,1) = |\mathbf{x}_u(0,1)|^2 = |\mathbf{x}_v(0,1)|^2 > 0\ . \tag{2.27}$$

Using the conformality relations of $\mathbf{x}$ and the geometry of Γ as in the proof of [10, Satz 2], we obtain

$$y_u(0,1) = 0, \qquad x_u(0,1) > 0\ . \tag{2.28}$$

Finally we arrive at

$$\mathbf{X}(0,1)\cdot\mathbf{e} = (E(0,1))^{-1}\left.\frac{\partial(x,y)}{\partial(u,v)}\right|_{(0,1)} = (E(0,1))^{-1}\, x_u(0,1)\, y_v(0,1) > 0\,.$$

Case II. $f(u_0,v_0)\in\partial^+\Omega$.

Then we take the auxiliary function $\psi(u,v) := y(u,v)$ and proceed as in Case I. □

With the aid of methods from [10] we shall now give a parametric approach to the nonparametric Dirichlet–Problem $\mathbf{P}(\Omega,g)$ for the minimal surface equation.

Proposition 4 (Existence result for the Dirichlet–Problem).
Let the assumptions of the Theorem be satisfied. Then there exists a solution $z=\varsigma(x,y)$ *of the Dirichlet–Problem* $\mathbf{P}(\Omega,g)$.

Proof.

1.) Let $g:\partial\Omega\to\mathbb{R}\in C^2(\partial\Omega)$ be satisfied. Then there exists a parametric solution $\mathbf{x}=\mathbf{x}(u,v)\in\mathbf{P}(\Gamma)\cap C^1(\overline{B})$, which is transversal to the boundary. More precisely

$$\mathbf{X}(u,v)\cdot\mathbf{e}>0 \quad \text{on } \partial B \tag{2.29}$$

holds true, due to Proposition 3. Now this area minimizing solution $\mathbf{x}$ is stable in the sense that the second variation of the area functional is nonnegative (cp. [10, Definition 6] and [10, Hilfssatz 8]). We can apply [10, Satz 3] and obtain

$$\mathbf{X}(u,v)\cdot\mathbf{e}>0 \quad \text{for all } (u,v)\in\overline{B}\,. \tag{2.30}$$

According to [10, Hilfssatz 7], the plane map

$$f(u,v) = (x(u,v),y(u,v)) : \overline{B}\to\overline{\Omega}\in C^1(\overline{B})$$

is a diffeomorphism. The height representation $\varsigma(x,y) := z\circ f^{-1}(x,y)$, $(x,y)\in\overline{\Omega}$, then solves the Dirichlet–Problem $\mathbf{P}(\Omega,g)$.

2.) Let us now consider $g : \partial\Omega \to \mathbb{R} \in C^0(\partial\Omega)$. We approximate g by functions $g_n : \partial\Omega \to \mathbb{R} \in C^2(\partial\Omega)$, $(n = 1,2,3,...)$ uniformly, respecting the Lipschitz condition. The contours

$$\Gamma_n := \{(x, y, g_n(x, y)) \mid (x, y) \in \partial\Omega\}, n = 1, 2, 3, ...$$

each bound a minimal graph $\mathbf{x}_n \in \mathbf{P}(\Gamma_n)$. As in [10, §4] we can extract a subsequence from the sequence $\{\mathbf{x}_n\}_{n=1,2,3,...}$ converging to a solution $\mathbf{x} \in \mathbf{P}(\Gamma)$. That $\mathbf{x}$ is again a graph solving $\mathbf{P}(\Omega, g)$, is guaranteed by [10, Hilfssatz 12]. This lemma shows that the class of minimal graphs is closed.

□

We are now prepared to give the complete

Proof of the Theorem. For the given contour

$$\Gamma := \left\{(x, y, g(x, y)) \in \mathbb{R}^3 \,|(x, y) \in \partial\Omega\right\}$$

there exists a solution $\mathbf{x} \in \mathbf{P}(\Gamma)$ which can be represented as a graph $z = \varsigma(x, y)$ solving $\mathbf{P}(\Omega, g)$, due to Proposition 4. Let now

$$\widetilde{\mathbf{x}} = \widetilde{\mathbf{x}}(u, v) = (\widetilde{x}(u, v), \widetilde{y}(u, v), \widetilde{z}(u, v)) \in \mathbf{P}(\Gamma)$$

be an arbitrary solution of Plateau's problem for this contour Γ. Then its plane map $\widetilde{f}(u, v) = (\widetilde{x}(u, v), \widetilde{y}(u, v))$, $(u, v) \in B$, satisfies the inclusion

$$\widetilde{f}(B) \subset \Omega, \tag{2.31}$$

due to Proposition 2. Now we consider the auxiliary function

$$\phi(u, v) := \widetilde{z}(u, v) - \varsigma(\widetilde{x}(u, v), \widetilde{y}(u, v)),\ (u, v) \in \overline{B}\ ,$$

and we observe

$$\phi \in C^2(B) \cap C^0(\overline{B})\ , \qquad \phi = 0 \ \text{ on } \partial B\ .$$

Since ϕ is subject to the elliptic partial differential equation given in Proposition 1, we conclude

$$0 \equiv \phi(u, v) = \widetilde{z}(u, v) - \varsigma(\widetilde{x}(u, v), \widetilde{y}(u, v)) \ \text{ in } \overline{B}\ . \tag{2.32}$$

From (2.32) we deduce the condition

$$\widetilde{\mathbf{X}}(u,v)\cdot\mathbf{e}\neq 0 \quad\text{for all } (u,v)\in B \tag{2.33}$$

for the normal of $\widetilde{\mathbf{x}}$. Since $\widetilde{f}:\partial B\to\partial\Omega$ is positively oriented we have

$$\widetilde{\mathbf{X}}(u,v)\cdot\mathbf{e}> 0 \quad\text{for all } (u,v)\in B\ . \tag{2.34}$$

Application of [10, Hilfssatz 7] shows that $\widetilde{f}:\overline{B}\to\overline{\Omega}$ is a diffeomorphism. Therefore the height representation $\widetilde{\varsigma}(x,y):=\widetilde{z}\circ\widetilde{f}^{-1}(x,y),\,(x,y)\in\overline{\Omega}$, solves $\mathbf{P}(\Omega,g)$. Since $\mathbf{P}(\Omega,g)$ has at most one solution due to the maximum principle, we condude

$$\varsigma(x,y)\equiv\widetilde{\varsigma}(x,y) \quad\text{in } \overline{\Omega}\ . \tag{2.35}$$

Therefore two solutions of $\mathbf{P}(\Gamma)$ coincide geometrically. They could only differ by a conformal parameter transformation keeping three points fixed. Consequently they are identical. □

References.

[1] Courant, R., *Dirichlet's principle, conformal mapping, and minimal surfaces*, Interscience: New York 1950.

[2] Dierkes, U.; Hildebrandt, S.; Küster, A.; Wohlrab, O., *Minimal surfaces I, II*, Grundlehren der mathematischen Wissenschaften 295/296. Springer: Berlin, Heidelberg, New York 1992.

[3] Heinz, E., *Über das Randverhalten quasilinearer elliptischer Systeme mit isothermen Parametern*, Math. Zeitschrift **113**, 99–105 (1970).

[4] Hildebrandt, S., *Über das Randverhalten von Minimalflächen*, Math. Annalen **165**, 1–18 (1966).

[5] Hildebrandt, S.; Sauvigny, F., *Uniqueness of stable minimal surfaces with partially free boundaries*, Journal of the Math. Society of Japan **47**, No. 3, 1995.

[6] Kneser, H., *Lösung der Aufgabe 41*, Jahresber. der dt. Math. Vereinigung **35**, 123–124 (1926).

[7] Lau, C.P., *The existence and nonexistence of nonparametric solutions to equations of minimal surface type*, Analysis **4**, 177–196 (1984).

[8] Nitsche, J.C.C., *Vorlesungen über Minimalflächen*, Grundlehren der mathematischen Wissenschaften **199**. Springer: Berlin, Heidelberg, New York 1975.

[9] Rado, T., Aufgabe 41, Jahresber der dt. Math. Vereinigung **35**, 49 (1926).

[10] Sauvigny, F., *Flächen vorgeschriebener mittlerer Krümmung mit eineindeutiger Projektion auf eine Ebene*, Math. Zeitschrift **180**, 41–67 (1982).

[11] Schulz, F.; Williams, G., *Barriers and existence results for a class of equations of mean curvature type*, Analysis **7**, 359–374 (1987).

[12] Williams, G., *The Dirichlet problem for the minimal surface equation with Lipschitz continuous boundary data*, Journal für die reine und angewandte Math. **354**, 123–140 (1984).

Received November 7, 1995.

Institut für Mathematik
der Brandenburgischen Technischen Universität,
Postfach 101344, D–03013 Cottbus, Germany

Dirichlet's energy on Teichmüller's moduli space is strictly pluri-subharmonic

A.J. TROMBA

Dirichlet's energy function on Teichmüller space is shown to be strictly pluri-subharmonic. This gives the new proof that Teichmüller space is a complex Stein manifold.

§ 0. Introduction.

Teichmüller space $\mathcal{T}(M)$ for a closed oriented compact surface M can be defined as the space $\mathcal{M}_{-1}/\mathcal{D}_0$ where $\mathcal{M}_{-1}$ is the manifold of Riemannian metrics of negative scalar curvature -1 and $\mathcal{D}_0$ the group of diffeomorphisms homotopic to the identity.

In [8] the author and Arthur Fischer introduced a Dirichlet's energy on Teichmüller space via the use of harmonic maps. The point in [8] was to show that Teichmüller space is a cell.

The purpose of this paper is to show that Dirichlet's energy is strictly pluri-subharmonic on $\mathcal{T}(M)$. Since Dirichlet's energy is proper, this provides a natural and elegant pluri-subharmonic exhaustion function on $\mathcal{T}(M)$.

In this paper we use the conventions and techniques of Riemannian Geometry developed with Arthur Fischer [6], [7], [8]. With this approach one shows directly that the quotient space $\mathcal{M}_{-1}/\mathcal{D}_0$ is a manifold with tangent space at a metric equivalence class [9] naturally identifiable with the holomorphic quadratic differentials on the complex curve $(M, c[g])$, where $c[g]$ is the unique complex structure associated to $[g]$.

This result is, of course, classically known through the application of the standard techniques of Teichmüller Theory.

§ 1. Preliminaries.

Let $\mathcal{A}$ be the space of almost complex structures on M compatible with its orientation and let $\mathcal{D}_0$ be the group of all diffeomorphisms of M homotopic to the identity. Then in [5], [6], [7], [8] classical Teichmüller space is defined to be the quotient $\mathcal{A}/\mathcal{D}_0$, where $\mathcal{D}_0$ acts on $\mathcal{A}$ by pull back. In [4] it is shown that $\mathcal{T}(M)$ has the structure of a C^∞ smooth $6(\text{genus}\, M) - 6$ dimensional manifold. If $\mathcal{M}_{-1}$ denotes the infinite dimensional manifold of Riemannian metrics of constant curvature -1, the $\mathcal{D}_0$ acts naturally on $\mathcal{M}_{-1}$ and $\mathcal{T}(M)$ is diffeomorphic to $\mathcal{M}_{-1}/\mathcal{D}_0$.

This diffeomorphism is described as follows (for details see [5], [13], [15]). There is a natural $\mathcal{D}$-equivariant diffeomorphism $\Phi : \mathcal{M}_{-1} \to \mathcal{A}$ given by

$$\Phi(g) = -g^{-1}\mu_g$$

where μ_g is the volume element of g. Φ then passes to a diffeomorphism $\tilde{\Phi}$ from $\mathcal{M}_{-1}/\mathcal{D}_0$ to $\mathcal{A}/\mathcal{D}_0$. Let $\theta : \mathcal{A} \to \mathcal{M}_{-1}$ be the inverse of Φ. For $J \in \mathcal{A}$, $\theta(J)$ is the unique Poincaré metric associated to J. Denote by $\tilde{\theta}$ the induced diffeomorphism from $\mathcal{A}/\mathcal{D}_0$ to $\mathcal{M}_{-1}/\mathcal{D}_0$. We also have a natural $\mathcal{D}_0$ invariant L_2-metric on $\mathcal{A}$ given by

$$\langle\langle H, K\rangle\rangle = \frac{1}{2}\int_M \operatorname{tr}(HK)\, d\mu_{\theta(J)}$$

and a natural L_2 splitting [13] of $T_J\mathcal{A}$, namely each $H \in T_J\mathcal{A}$ can be uniquely decomposed as

$$H = H^{TT} + L_X J \tag{1.1}$$

where $L_X J$ is the Lie derivative of J w.r.t. the vector field X on M, and H^{TT} denotes a $(1,1)$ tensor which is trace free and divergence free w.r.t. $\theta(J)$. The decomposition (1.1) is L_2-orthogonal. Since $\mathcal{D}_0$ acts as a group of isometries $\langle\langle , \rangle\rangle$ passes to a metric $\langle , \rangle$ on $\mathcal{T}(M) = \mathcal{A}/\mathcal{D}_0$.

Let us now consider the model $\mathcal{M}_{-1}/\mathcal{D}_0$ of $\mathcal{T}(M)$. The tangent space of $\mathcal{M}_{-1}$ at a metric, $g \in T_g\mathcal{M}_{-1}$ consists of those $(0,2)$ tensors h on M satisfying the equation

$$-\Delta(\operatorname{tr}_g h) + \delta_g\delta_g h + \frac{1}{2}(\operatorname{tr}_g h) = 0 \tag{1.2}$$

where $\operatorname{tr}_g h = g^{ij}h_{ij}$ is the trace of h w.r.t. the metric tensor g_{ij}, $\delta_g\delta_g h$ is the double covariant divergence of h w.r.t. g and Δ is the Laplace – Beltrami operator on functions. For example see [5], [13], [15] for details. The L_2-metric on $\mathcal{M}_{-1}$ is given by the inner product

$$\langle\langle h, k\rangle\rangle_g = \frac{1}{2}\int_M \operatorname{trace}(HK)\, d\mu_g \tag{1.3}$$

where $H = g^{-1}h$, $K = g^{-1}k$ are the $(1,1)$ tensors on M obtained from h and k via the metric g, or "by raising an index", i.e.

$$H^i_j = g^{ik} h_{kj}$$

and similarly for K.

The inner product (1.3) is $\mathcal{D}_0$ invariant. Thus $\mathcal{D}_0$ acts smoothly on $\mathcal{M}_{-1}$ as a group of isometries with respect to this metric, and consequently, as in the model $\mathcal{A}/\mathcal{D}_0$ for $\mathcal{T}(M)$, we have an induced metric on $\mathcal{T}(M)$ in such a way that the projection map $\pi : \mathcal{M}_{-1} \to \mathcal{M}_{-1}/\mathcal{D}_0$ becomes a Riemannian submersion [5]. In [5] it is shown that this induced metric is precisely the metric originally introduced by Weil, now called the Weil – Petersson metric, and that $\tilde{\theta} : \mathcal{A}/\mathcal{D}_0 \to \mathcal{M}_{-1}/\mathcal{D}$ is an isometry with respect to the two metrics just introduced.

We also denote by $\langle , \rangle$ this induced metric on $\mathcal{T}(M)$. We can characterize $\langle , \rangle$ as follows. From [5] one can show that given $g \in \mathcal{M}_{-1}$ every

$$h = h^{TT} + L_X g \tag{1.4}$$

where $L_X g$ is the Lie derivative of g w.r.t. some (unique) X and h^{TT} is a trace free, divergence free, symmetric tensor. Moreover the decomposition (1.4) is L_2-orthogonal. Recall that a conformal coordinate system (where $g_{ij} = \lambda \delta_{ij}$, λ some smooth positive function) is also a complex holomorphic coordinate system. In this system

$$h^{TT} = \mathrm{Re}(\xi(z)\, dz^2)$$

where Re is "real part" and $\xi(z)\, dz^2$ is a holomorphic quadratic differential. In fact, trace free, divergence free symmetric two tensors are precisely the real parts of holomorphic quadratic differentials.

Now $L_X g$ is always tangent to the orbit of $\mathcal{D}_0$ through g. As before we say that $L_X g$ is the *vertical* part of h in decomposition (1.4) and that h^{TT} represents the *horizontal* part of h. Given $h, k \in T_{[g]}\mathcal{T}(M)$ there are unique horizontal vectors $\tilde{h}, \tilde{k} \in T_g\mathcal{M}_{-1}$ such that $D\pi(g)\tilde{h} = h$ and $D\pi(g)\tilde{k} = k$. Then

$$\langle h, k\rangle_{[g]} = \langle\!\langle \tilde{h}, \tilde{k} \rangle\!\rangle_g .$$

The space of almost complex structures $\mathcal{A}$ has itself an almost complex structure. Recall

$$\mathcal{A} = \left\{ \mathcal{T} \in C^\infty(T(M)) | J^2 = -\,\mathrm{id} \right\},$$

i.e. each $J \in \mathcal{A}$ is a C^∞ $(1,1)$ tensor whose square is minus the identity on each tangent space T_xM. The tangent space to the infinite dimensional manifold $\mathcal{A}$ at J is the linear space of $(1,1)$ tensors H characterized by the equation $HJ = -JH$. $\mathcal{A}$ itself is a complex manifold [5] whose $\mathcal{D}$-invariant almost complex structure Ψ is characterized by

$$\begin{aligned} &\Psi_J : T_J\mathcal{A} \hookleftarrow \quad \text{with} \\ &\Psi_J(H) = JH. \end{aligned} \tag{1.5}$$

This almost complex structure passes naturally to a complex structure on $\mathcal{T}(M)$ which corresponds to multiplication by $i = \sqrt{-1}$ on the level of holomorphic quadratic differentials.

§ 2. Dirichlet's energy.

Suppose now that (N, g_0) is a Riemannian manifold of negative sectional curvature where the metric g_0 is fixed and that $s : (M,g) \to (N, g_0)$ is a smooth C^1 map viewed as a map from M, with some arbitrary metric $g \in \mathcal{M}_{-1}$ to N with its g_0 metric.

Define the Dirichlet's energy of s by the formula

$$E_g(s) = \frac{1}{2} \int_M |ds|^2 \, d\mu_g \tag{2.1}$$

where $ds^2 = \operatorname{trace} ds \otimes ds$ depends on both g and g_0.

By the embedding theorem of Nash – Moser we may assume that (N, g_0) is isometrically embedded in some Euclidean $\mathbf{R}^p$. Thus we can think of $s : (M,g) \to (N, g_0)$ as a map into $\mathbf{R}^p$ and Dirichlet's functional takes the equivalent form

$$E_g(s) = \frac{1}{2} \sum_{i=1}^{p} \int g(x) \left(\nabla_g s^i(x) \cdot \nabla_g s^i(x) \right) d\mu_g. \tag{2.2}$$

If $M = N$ there is another, equivalent, and useful way to express (2.1) and (2.2) using local conformal coordinate systems $g_{ij} = \lambda \delta_{ij}$ and $(g_0)_{ij} = \rho \delta_{ij}$ on (M,g) and (M, g_0) respectively, namely

$$E_g(s) = \frac{1}{4} \int_M \left[\rho(s(z)) |s_z|^2 + \rho(s(z)) |s_{\bar{z}}|^2 \right] dz d\bar{z}. \tag{2.3}$$

For fixed g, the critical points of E_g are then said to be *harmonic maps*. The following result is due to Eels – Sampson [4], Hartman [9] and Schoen –

Yau [17]. Eells – Sampson proved existence, Alber in 1964 and Hartman and Sampson [12] proved uniqueness and Schoen – Yau showed the harmonic map is a diffeomorphism, as stated below.

Theorem 2.1. *Given metrics g and g_0, with g_0 having negative sectional curvature then in every non-trivial homotopy class there exists a harmonic map $s(g) : (M, g) \to (N, g_0)$ which is the absolute minimum for E_g in this class. If the image of each such $s(g)$ does not collapse onto a geodesic arc, then $s(g)$ is unique in its homotopy class, and moreover $s(g)$ depends differentially on g in any H^r-topology, $r > 2$. If $M = N$ and $g_0 \in \mathcal{M}_{-1}$ $s(g)$ is a C^∞ diffeomorphism.*

Fix a homotopy class and assume that $s(g)$ depends differentially on g and is unique. Consider now the function

$$g \to E_g(s(g)).$$

This function on $\mathcal{M}_{-1}$ is $\mathcal{D}_0$-invariant and thus can be viewed as a function on Teichmüller space. To see this one must show that

$$E_{f^*g}(s(f^*(g))) = E_g(s(g)).$$

Let $c(g)$ be the complex structure associated to g, and induced by a conformal coordinate system for g. For $f \in \mathcal{D}_0$, $f : (M, f^*c(g)) \to (M, c(g))$ is holomorphic and consequently, since the composition of harmonic maps and holomorphic maps is still harmonic, we may conclude, by uniqueness in the given homotopy class, that

$$s(f^*g) = s(g) \circ f.$$

Since Dirichlet's functional is invariant under complex holomorphic changes of coordinates it follows immediately that

$$E_{f^*(g)}(s(g) \circ f) = E_g(s(g)).$$

Consequently for $[g] \in \mathcal{M}_{-1}/\mathcal{D}_0$ define the C^∞ smooth function

$$\tilde{E} : \mathcal{M}_{-1}/\mathcal{D}_0 \to \mathbf{R}$$

by

$$\tilde{E}[g] = E_g(s(g)).$$

In [14], [15] we prove the following

Theorem 2.2. *If* $s : (M,g) \to (N,g_0)$ *is harmonic the form* $\xi(z)\,dz^2 = \sum_{i=1}^{p} \left[\frac{\partial s^i}{\partial z}\right]^2 dz^2$ *is a holomorphic quadratic differential on the complex curve* $(M,c(g))$, *and thus* $\operatorname{Re}\xi(z)\,dz^2$ *represents a trace free, divergence free symmetric two tensor on* (M,g). *Therefore* $\operatorname{Re}\xi(z)\,dz^2$ *is a horizontal tangent vector to* $\mathcal{M}_{-1}$ *at* g, *and*

$$D\tilde{E}[g]h = -\frac{1}{2}\left\langle\!\left\langle \operatorname{Re}\xi(z)\,dz^2, \tilde{h}\right\rangle\!\right\rangle_g = -\frac{1}{2}\sum_l \int_M g(x)\left(\tilde{H}\nabla_g s^l, \nabla_g s^l\right) d\mu_g \tag{2.4}$$

where $\tilde{h}$ *is the horizontal lift of* $h = T_{[g]}\mathcal{T}(M)$ *and* $\tilde{H}$ *is obtained from* $\tilde{h}$ *by raising an index via* g.

Finally, if $N = M$, $g_0 \in \mathcal{M}_{-1}$ *and* $s(g)$ *is homotopic to the identity,* $[g_0]$ *is the only critical point of* $\tilde{E}$. *The Hessian of* $\tilde{E}$ *at* $[g_0]$ *is then given by*

$$D^2\tilde{E}[g_0](h,k) = \langle h,k\rangle$$

$h,k \in T_{[g_0]}\mathcal{T}(M)$. *That is, the second variation of this Dirichlet's energy function is the Weil – Petersson metric.*

In general, the second derivative of $\tilde{E}$ at an arbitrary $[g]$ will not be intrinsic. However for arbitrary metrics g we can ask for the second derivative of the function $g \to E_g(s(g)) = \hat{E}(g)$. For $g \in \mathcal{M}$, the space of all Riemannian metrics, we assume that E_g has a unique minimum $s(g)$ which depends differentially on g, which certainly holds in the case $N = M$, $g_0 \in \mathcal{M}_{-1}$. Thus we have the following two formulas (where we use the Einstein-summation convention).

Theorem 2.3. *For arbitrary* k

$$D\hat{E}(g)k = -\frac{1}{2}\int_M g(x)\left(K_T\nabla_g s^l, \nabla_g s^l\right) d\mu_g = -\frac{1}{2}\left\langle\!\left\langle \operatorname{Re}\xi(z)\,dz^2, k_T\right\rangle\!\right\rangle_g \tag{2.5}$$

where $K = (k)^\#$ *is the* $(1,1)$ *tensor obtained from* k *by raising an index via* g, *and* K_T *and* k_T *are the trace free parts of* K *and* k *respectively. For* k *trace free we have*

$$D^2\hat{E}(g)(k,k) = \frac{1}{4}\int_M \{k\cdot k\} g(x)\left(\nabla_g s^l, \nabla_g s^l\right) d\mu_g - \int_M g(x)\left(K\cdot\nabla_g s^l, \nabla w^l(k)\right) d\mu_g \tag{2.6}$$

where $h\cdot k = g^{ab}g^{cd}h_{ac}k_{bd} = \operatorname{tr}(HK)$, *and* $w^l = Ds^l(g)k$, *the derivative of* $s(g)$ *in the direction* k.

Proof.

$$\hat{E}(g) = \frac{1}{2}\int_M g^{ij}\frac{\partial s^\alpha}{\partial x^i}\cdot\frac{\partial s^\alpha}{\partial x^j}\,d\mu_g.$$

Thus

$$\begin{aligned} D\hat{E}(g)h &= -\frac{1}{2}\int_M h^{ij}\frac{\partial s^\alpha}{\partial x^i}\cdot\frac{\partial s^\alpha}{\partial x^j}\,d\mu_g \\ &+ \int_M g^{ij}\frac{\partial s^\alpha}{\partial x^i}\cdot\frac{\partial w^\alpha}{\partial x^j}\,d\mu_g + \frac{1}{4}\int_M g^{ij}\frac{\partial s^\alpha}{\partial x^i}\cdot\frac{\partial s^\alpha}{\partial x^j}(\operatorname{tr}_g h)\,d\mu_g \\ &= -\frac{1}{2}\int_M h^{ij}\frac{\partial s^\alpha}{\partial x^i}\cdot\frac{\partial s^\alpha}{\partial x^j}\,d\mu_g + \frac{1}{4}\int_M g^{ij}\frac{\partial s^\alpha}{\partial x^i}\cdot\frac{\partial s^\alpha}{\partial x^j}(\operatorname{tr}_g h)\,d\mu_g \end{aligned} \tag{2.7}$$

since the derivative of $g \to g^{ij}$ is $h \to -h^{ij}$ and the derivative of $g \to \mu_g$ is $h \to \frac{1}{2}(\operatorname{tr}_g h)\mu_g$. Also the term $\int\limits_M g^{ij}\frac{\partial s^\alpha}{\partial x^i}\cdot\frac{\partial w^\alpha}{\partial x^j}\,d\mu_g$ vanishes since $s(g)$ is a critical point of E_g.

Continuing we see that this is equal to

$$-\frac{1}{2}\int_M g^{il}h^i_l\frac{\partial s^\alpha}{\partial x^i}\cdot\frac{\partial s^\alpha}{\partial x^j}\,d\mu_g + \int_M(\frac{1}{4}\operatorname{tr}_g h)g^{ij}\frac{\partial s^\alpha}{\partial x_i}\cdot\frac{\partial s^\alpha}{\partial x^j}\,d\mu_g$$

which immediately yields the first equality in equation (2.5). The second follows directly from Theorem 2.2.

For the second derivative we note that $h^{ij} = g^{im}g^{lj}h_{mj}$ and so the derivative of $h \to h^{ij}$ w.r.t. g again in the direction h is $h \to -h^{im}g^{lj}h_{mj} - g^{im}h^{lj}h_{ml}$. In conformal coordinates $g_{ij} = \lambda\delta_{ij}$ it is easy to see that for h trace free this is equal to $-\frac{1}{\lambda}(h\cdot h)I$, $I =$ identity transformation.

This immediately implies that the derivative of

$$g \to -\frac{1}{2}\int_M h^{ij}\frac{\partial s^\alpha}{\partial x_i}\cdot\frac{\partial s^\alpha}{\partial x^j}\,d\mu_g$$

again in the direction of a trace free h is

$$h \to \frac{1}{2}\int_M (h\cdot h)g(x)\left(\nabla_g s^l, \nabla_g s^l\right)\,d\mu_g - \int h^{ij}\frac{\partial w^\alpha}{\partial x^i}\cdot\frac{\partial s^\alpha}{\partial x^j}\,d\mu_g.$$

In order to compute the derivative of

$$g \to \frac{1}{4}\int_M h^{ij}\frac{\partial s^\alpha}{\partial x_i}\cdot\frac{\partial s^\alpha}{\partial x^j}(\operatorname{tr}_g h)\,d\mu_g$$

again in the direction of a trace free k we must only compute the derivative of the term $g \to \operatorname{tr}_g h$ (not assuming h is trace free). But

$$\operatorname{tr}_g h = g^{ij}h_{ij}$$

and so the derivative is $k \to -k^{ij}h_{ij}$ or if $k = h$, $h^{ij}h_{ij} = h \cdot h$.

Thus the derivative of this second term w.r.t. g in the direction of (a trace free) h is

$$-\frac{1}{4}\int_M (h \cdot h)g(x)\left(\nabla_g s^l, \nabla_g s^l\right).$$

Noting that

$$\int_M h^{ij}\frac{\partial s^\alpha}{\partial x_i} \cdot \frac{\partial w^\alpha}{\partial x^j}\, d\mu_g = \int_M g(x)\left(h^\# \nabla_g s^l, \nabla_g w^l\right) d\mu_g$$

and adding the terms gives the desired result.

If the third term of (2.6) were positive it would immediately imply the pluri-subharmonicity of Dirichlet's energy. The next lemma shows that this is not the case.

Lemma 2.1.

$$-\int_M g(x)\left(\tilde{k}^\# \nabla_g s^l, \nabla w^l(\tilde{k})\right) d\mu_g \le 0. \tag{2.8}$$

Proof. Consider the map $g \to E_g(s(g))$. Since $s(g)$ is a critical point of E_g we have the relation

$$\frac{\partial E_g}{\partial s} \circ Ds(g)\tilde{k} \equiv 0$$

where $w(\tilde{k}) = Ds(g)(\tilde{k})$, for all g. Therefore the derivative of the expression directly above with respect to g, must be identically zero, or consequently we see that

$$0 \equiv \frac{\partial^2 E_g}{\partial g \partial s}(\tilde{k}, Ds(g)\tilde{k}) + \frac{\partial}{\partial s}\left[\frac{\partial E_g}{\partial s} \circ Ds(g)\tilde{k}\right] \circ Ds(g)\tilde{k}.$$

The first term on the right is expression (2.8). The second term is precisely the second variation of Dirichlet's energy E_g, $D^2E_g(w,w)$, at the critical point $s(g)$ in the direction w, where $w(\tilde{k}) = Ds(g)(\tilde{k})$. Since $s(g)$ is an absolute minimum it follows that $D^2E_g(w,w) \ge 0$.

$$\frac{\partial^2 E_g}{\partial g \partial s}(\tilde{k}, w) = -D^2E_g(w,w) = -\int_M g(x)\left(\tilde{k}^\# \nabla_g s^l, \nabla w^l(\tilde{k})\right) d\mu_g \le 0$$

which completes Lemma 1.

As a direct corollary we obtain the following formula for $D^2\hat{E}(g)$,

Lemma 2.2.

$$D^2\hat{E}(g)(h,h) = \frac{1}{4}\int_M \{h\cdot h\}g(x)\left(\nabla_g s^l, \nabla_g s^l\right) d\mu_g - D^2E_g(s)(w,w) \tag{2.9}$$

where $w = Ds(g)h$. □

As we shall later see, formula (2.9) illustrates the critical estimate necessary to prove that the Dirichlet's energy on $\mathcal{T}(M)$ is either pluri-subharmonic or convex along Weil – Petersson geodesics.

The necessary estimate is naturally that

$$I = D^2E_g(s)(w,w) < \frac{1}{4}\int_M \{h\cdot h\}g(x)\left(\nabla_g s^l, \nabla_g s^l\right) d\mu_g.$$

The proof of the next theorem shows that for general Riemannian manifolds (N, g_0) of negative sectional curvature the "best possible" estimate for the second variation of the associated Dirichlet's energy in directions $w = Ds(g)h$ is likely

$$I = D^2E_g(s)(w,w) < \frac{1}{2}\int_M \{h\cdot h\}g(x)\left(\nabla_g s^l, \nabla_g s^l\right) d\mu_g.$$

Theorem 2.4. *Let* (N, g_0) *be a Riemannian manifold of negative sectional curvature with* $s(g)$ *the unique harmonic map depending smoothly on* g *and let*

$$(g,s) \to E_g(s) \quad \text{and} \quad g \to \hat{E}(g)$$

the associated Dirichlet's energy. Then

$$I = D^2E_g(s)(w,w) < \frac{1}{2}\int_M \{\tilde{k}\cdot\tilde{k}\}g(x)\left(\nabla_g s^l, \nabla_g s^l\right) d\mu_g. \tag{2.10}$$

Proof. Fortunately, we have an explicit formula for the second variation of E_g at a minimum s, namely [4, p. 139] in conformal coordinates $g_{ij} = \lambda\delta_{ij}$ with local coordinates $(x,y) = (x^1, x^2)$ we have

$$\begin{aligned} I = D^2E_g(s)(w,w) &= \int_M \left\{\left\langle \nabla_{\frac{\partial}{\partial x}} w, \nabla_{\frac{\partial}{\partial x}} w\right\rangle + \left\langle \nabla_{\frac{\partial}{\partial y}} w, \nabla_{\frac{\partial}{\partial y}} w\right\rangle\right\} dx\wedge dy \\ &- \int_M \left\{\left\langle \mathcal{R}(w, \frac{\partial s}{\partial x})\frac{\partial s}{\partial x}, w\right\rangle + \left\langle \mathcal{R}(w, \frac{\partial s}{\partial y})\frac{\partial s}{\partial y}, w\right\rangle\right\} dx\wedge dy \end{aligned} \tag{2.11}$$

where $\langle , \rangle : \mathbf{R}^p \times \mathbf{R}^p \to \mathbf{R}$ is the Euclidean inner product and $\mathcal{R}$ is the curvature tensor of $(N, g_0) \subset \mathbf{R}^p$.

Since the curvature of (N, g_0) is negative we see that

$$I \geq \int_M \left\{ \left\langle \nabla_{\frac{\partial}{\partial x}}, w, \nabla_{\frac{\partial}{\partial x}} w \right\rangle + \left\langle \nabla_{\frac{\partial}{\partial y}}, w, \nabla_{\frac{\partial}{\partial y}} w \right\rangle \right\} dx \wedge dy$$

where strict inequality holds if $w \neq 0$. We are assuming that (N, g_0) is isometrically embedded in $\mathbf{R}^p$. For $x \in (N, g_0) \subset \mathbf{R}^p$ let $\Pi(x) : \mathbf{R}^p \to T_x M$ be the orthogonal projection of $\mathbf{R}^p$ onto the tangent space to M at x. Then the condition that s, $s : (M, g) \to (N, g_0)$, be harmonic can be written (in conformal coordinates) as

$$\Pi(s)\Delta s = \Pi(s) \left[\frac{\partial^2 s}{\partial x^2} + \frac{\partial^2 s}{\partial y^2} \right] = 0$$

$s = (s^1, \dots, s^k)$.

This can be written in terms of the metric g as

$$\Pi(s)(\sqrt{g}\Delta_g s) = 0$$

Δ_g = Laplace – Beltrami on the coordinate functions $(s^1, \dots, s^k)$. We know that the unique harmonic map s depends on g, so let us write this as

$$\Pi(s(g))(\sqrt{g}\Delta_g(s(g)) = 0 \tag{2.12}$$

and this holds for all g.

Differentiating (3.12) w.r.t. g in the direction of a trace free (w.r.t. g) tensor h we obtain

$$D\Pi(s)(w)(\sqrt{g}\Delta_g s(g)) + \Pi(s(g)) \left\{ \frac{\partial}{\partial g} [\sqrt{g}\Delta_g](h) \right\} s + \Pi(s(g))(\sqrt{g}\Delta_g w) \equiv 0 \tag{2.13}$$

or in short

$$D\Pi(s)w(\Delta s) + \Pi(s) \left\{ \frac{\partial}{\partial g} [\sqrt{g}\Delta_g](h) \right\} s + \Pi(s)\Delta w \equiv 0$$

where $w = Ds(g)h$.

Now

$$\sqrt{g}\Delta_g s = \frac{\partial}{\partial x^j} \left[\sqrt{g} g^{ij} \frac{\partial s}{\partial x^i} \right].$$

Therefore, using the fact that $\operatorname{tr}_g h = 0$,

$$\frac{\partial}{\partial g}\left\{\sqrt{g}\Delta_g(h)\right\}(s) = -\frac{\partial}{\partial x^j}\left[\sqrt{g}h^{ij}\frac{\partial s}{\partial x^i}\right].$$

But necessarily the second variation $I = D^2E_g(s)(w,w)$ equals

$$I = -\int_M \left\{ \langle D\Pi(s)(w)(\Delta s) + \Pi(S)(\Delta w), w\rangle \right\} dx \wedge dy.$$

Integrating by parts we get

$$I = \int_M h^{11}\frac{\partial s}{\partial x}\cdot\frac{\partial w}{\partial x}\sqrt{g}\,dx\wedge dy + \int_M h^{22}\frac{\partial s}{\partial y}\cdot\frac{\partial w}{\partial y}\sqrt{g}\,dx\wedge dy$$
$$+ \int_M h^{12}\frac{\partial s}{\partial x}\cdot\frac{\partial w}{\partial y}\sqrt{g}\,dx\wedge dy + \int_M h^{21}\frac{\partial s}{\partial y}\cdot\frac{\partial w}{\partial x}\sqrt{g}\,dx\wedge dy$$

where $\cdot$ denotes the $\mathbf{R}^p$ inner product. Thus

$$I = \int_M \left[h^{21}\frac{\partial s}{\partial y} + h^{11}\frac{\partial s}{\partial x}\right]\cdot\frac{\partial w}{\partial x}\sqrt{g}\,dx\wedge dy$$
$$+ \int_M \left[h^{22}\frac{\partial s}{\partial y} + h^{12}\frac{\partial s}{\partial x}\right]\cdot\frac{\partial w}{\partial y}\sqrt{g}\,dx\wedge dy. \quad (2.14)$$

Since

$$\nabla_{\frac{\partial}{\partial y}} w = \Pi(s)\frac{\partial w}{\partial y}$$
$$\nabla_{\frac{\partial}{\partial x}} w = \Pi(s)\frac{\partial w}{\partial x}$$

we see that this is equal to

$$\int_M \left[h^{21}\frac{\partial s}{\partial y} + h^{11}\frac{\partial s}{\partial x}\right]\cdot\left[\nabla_{\frac{\partial}{\partial x}} w\right]\sqrt{g}\,dx\wedge dy$$
$$+ \int_M \left[h^{22}\frac{\partial s}{\partial y} + h^{12}\frac{\partial s}{\partial x}\right]\cdot\left[\nabla_{\frac{\partial}{\partial y}} w\right]\sqrt{g}\,dx\wedge dy.$$

Applying the Schwartz inequality and using the fact that $g_{ij} = \lambda\delta_{ij}$, $\sqrt{g} = \lambda$ we obtain

$$I \leq \left\{\sqrt{\int_M \left\|h^{22}\frac{\partial s}{\partial y} + h^{12}\frac{\partial s}{\partial x}\right\|^2 \lambda^2\,dx\wedge dy}\right\}\left\{\sqrt{\int_M \left\|\nabla_{\frac{\partial}{\partial y}} w\right\|^2 dx\wedge dy}\right\}$$
$$+ \left\{\sqrt{\int_M \left\|h^{11}\frac{\partial s}{\partial x} + h^{21}\frac{\partial s}{\partial y}\right\|^2 \lambda^2\,dx\wedge dy}\right\}\left\{\sqrt{\int_M \left\|\nabla_{\frac{\partial}{\partial x}} w\right\|^2 dx\wedge dy}\right\} \quad (2.15)$$

where

$$\left\| \nabla_{\frac{\partial}{\partial x}} w \right\|^2 = \Pi(s)\frac{\partial w}{\partial x} \cdot \Pi(s)\frac{\partial w}{\partial x}.$$

Now the right hand side of (3.15) can be written as

$$\sqrt{A_1}\sqrt{B_1} + \sqrt{A_2}\sqrt{B_2} \le \sqrt{(A_1 + A_2)(B_1 + B_2)}$$

and since $h^{11} = -h^{22}$, we see that I is less than or equal to

$$\left\{ \sqrt{\int_M \{(h^{11})^2 + (h^{12})^2\} \left\{ \|\frac{\partial s}{\partial x}\|^2 + \|\frac{\partial s}{\partial y}\|^2 \right\} \lambda^2 \, dx \wedge dy} \right\}$$
$$\left\{ \sqrt{\int_M \left\{ \left\| \nabla_{\frac{\partial}{\partial x}} w \right\|^2 + \left\| \nabla_{\frac{\partial}{\partial y}} w \right\|^2 \right\} dx \wedge dy} \right\}.$$

Furthermore, since $h^{ij} = \frac{1}{\lambda^2} h_{ij}$ this is equal to

$$\left\{ \sqrt{\frac{1}{2} \int_M (h \cdot h) g(s) \left[\nabla_g s^l, \nabla_g s^l\right] d\mu_g} \right\}$$
$$\sqrt{\int_M \left\{ \left\| \nabla_{\frac{\partial}{\partial x}} w \right\|^2 + \left\| \nabla_{\frac{\partial}{\partial y}} w \right\|^2 \right\} dx \wedge dy}.$$

If $w \neq 0$ this is strictly less than

$$\sqrt{\frac{1}{2} \int_M (h \cdot h) g(s) \left[\nabla_g s^l, \nabla_g s^l\right] d\mu_g}$$
$$\sqrt{I}$$

or

$$\sqrt{I} < \sqrt{\frac{1}{2} \int_M (h \cdot h) g(s) \left[\nabla_g s^l, \nabla_g s^l\right] d\mu_g}$$

whence

$$I < \frac{1}{2} \int_M (h \cdot h) g(s) \left[\nabla_g s^l, \nabla_g s^l\right] d\mu_g. \tag{2.16}$$

If $w = 0$, the inequality (2.16) clearly holds. This establishes Theorem 3.4.

For Dirichlet's energy to be convex along Weil – Petersson geodesics we need the inequality

$$I < \frac{1}{4}\int_M (h\cdot h)g(s)\left[\nabla_g s^l, \nabla_g s^l\right]\, d\mu_g. \tag{2.17}$$

In order to see the necessity of (2.17) for Weil – Petersson convexity we proceed as follows.

Let $\sigma(t)$ be a geodesic on Teichmüller space $\mathcal{T}(M)$. We can lift $\sigma(t)$ to a smooth path $\tilde{\sigma}(t)$ in $\mathcal{M}_{-1}$ with the property that for each t, $\tilde{\sigma}'(t)$ is horizontal.

We know that $\mathcal{M}_{-1} \subset \mathcal{M}$ the space of all metrics which itself is an open subset of the space of all symmetric tensors S_2. Thus the second derivative $\sigma''(t)$ can be thought of as an element of S_2. Let $S_2^{TT}(\sigma)$ be the space of trace free divergence free symmetric two tensors with the metric σ and let

$$\Pi_{\tilde{\sigma}} : S_2 \to S_2^{TT}(\tilde{\sigma})$$

be the L_2-orthogonal projection. Then σ being a geodesic implies that

$$\Pi_{\tilde{\sigma}}\tilde{\sigma}'' = 0. \tag{2.18}$$

Consider the map $t \to \tilde{E}(\sigma(t))$. The second derivative of this map is by Theorem 2.3 and Lemma 2.2

$$D\hat{E}(\tilde{\sigma}(t))(\tilde{\sigma}''(t)) + \frac{1}{4}\int \{h\cdot h\}g(x)\left(\nabla_g s^l, \nabla_g s^l\right)\, d\mu_g - D^2E_g(s)(w,w) \tag{2.19}$$

where $g = \tilde{\sigma}(t)$, $h = \tilde{\sigma}'(t)$, and $w = \frac{d}{dt}s(g)$. If inequality (2.17) holds then the difference of the second two terms above is positive. We claim the first term is zero. By Theorem 2.3 the first term is equal to

$$-\frac{1}{2}\left\langle\!\left\langle \operatorname{Re}\xi(z)\,dz^2, \{\tilde{\sigma}''(t)\}_T\right\rangle\!\right\rangle_g$$

since $\operatorname{Re}\xi(z)\,dz^2$ is already trace free this is equal to

$$\begin{aligned} &-\frac{1}{2}\left\langle\!\left\langle \operatorname{Re}\xi(z)\,dz^2, \sigma''(t)\right\rangle\!\right\rangle_g \\ =&-\frac{1}{2}\left\langle\!\left\langle \operatorname{Re}\xi(z)\,dz^2, \Pi_\sigma\sigma''\right\rangle\!\right\rangle_g \\ &-\frac{1}{2}\left\langle\!\left\langle \operatorname{Re}\xi(z)\,dz^2, \Pi_\sigma^c\sigma''\right\rangle\!\right\rangle_g \end{aligned} \tag{2.20}$$

where $\Pi_\sigma^c = \mathrm{id} - \Pi_\sigma$. Since $\Pi_\sigma \tilde{\sigma}'' = 0$ the first term in (2.20) vanishes. But Π_σ^c projects onto all $(0,2)$ tensors of the form $\rho \cdot g$ which is pointwise L_2-orthogonal to all trace free tensors, in particular to $\mathrm{Re}\,\xi(z)\,dz^2$ and so the second term in (2.20) also vanishes.

Therefore we have

$$\frac{d^2\tilde{E}}{dt^2}(\sigma(t)) = \frac{1}{4}\int \{h \cdot h\} g(x) \left[\nabla_g s^l, \nabla_g s^l\right] d\mu_g - D^2 E_g(s)(w,w). \tag{2.21}$$

Clearly if (2.17) holds then $\tilde{E}$ is Weil – Petersson convex. However the author sees no way of improving estimate (2.16) to estimate (2.17) even if the range $N \equiv M$.

Fortunately when $N = M$ we can show that the Levi-form $\partial\bar{\partial}\tilde{E} = \sum_{\alpha,\beta} \frac{\partial^2 \tilde{E}}{\partial z^\alpha \partial \bar{z}^\beta} \xi_\alpha \bar{\xi}_\beta > 0$ which we now proceed to do.

§ 3. Dirichlet's energy is pluri-subharmonic.

In the last sections, for the sake of generality we have considered various possible Dirichlet's energies on Teichmüller's space. These energies depended on the choice of a range manifold, metric and specific homotopy class.

In the remaining two sections, when we refer to *the Dirichlet's energy* $\tilde{E}$ we mean that the range "manifold" N is identical to the domain M, that the range metric g_0 has curvature -1 and that the unique harmonic map $s(g)$ is chosen homotopic to the identity.

What estimate is needed to prove that $\tilde{E} : \mathcal{T}(M) \to \mathbb{R}$ is pluri-subharmonic? Let

$$h = h_{11}\,dx^2 + 2h_{12}\,dxdy + h_{22}\,dy^2$$

be a trace free and divergence free symmetric $(0,2)$ tensor. Then in conformal coordinates $h_{22} = -h_{11}$ and $h_{11} + ih_{12}$ is anti-holomorphic.

Remark. Since orientation preserving conformal coordinates are also complex coordinates, multiplication by i in these coordinates is well defined and we shall exploit this fact.

Let H denote $h_1^1 + ih_2^1$ and $iH = -h_2^1 + ih_1^1$ (this corresponds to multiplying the associated $(1,1)$ tensor by J) and $ih = -h_{12}\,dx^2 + 2h_{11}\,dxdy + h_{12}\,dy^2$ (this corresponds to multiplying the associated holomorphic quadratic differential $\{(h_{11} - ih_{12})\,dz^2\}$ by $-i$).

For $s(g) : (M, g) \to (N, g_0)$ the unique harmonic map, we now denote by w^h the derivative $D_g s(g)h$ for h a trace free and divergence free $(0,2)$ tensor and

$$w^{ih} = Ds(g)(ih).$$

Let $\sigma_1(t)$, $\sigma_2(t)$ be Weil – Petersson geodesics with $\sigma_1(0) = \sigma_2(0) = g$ and whose tangents $\sigma_1^1(t)|_{t=0} = h$ and $\sigma_2^1(t)|_{t=0} = ih$.

Then since the Weil – Petersson metric is Kähler it follows that the Levi-form $\partial\bar{\partial}\tilde{E} > 0$ iff

$$\frac{d\tilde{E}}{dt^2}(\sigma_1(t))|_{t=0} + \frac{d\tilde{E}}{dt^2}(\sigma_2(t))|_{t=0} > 0.$$

This is a consequence of the fact that about each point in $\mathcal{T}(M)$ there is a holomorphic coordinate system with $\sigma_j''(t) \equiv 0$ in this coordinate system. For example see [2].

By (2.21) this means we must show that

$$D^2 E_g(s)(w^h, w^h) + D^2 E_g(s)(w^{ih}, w^{ih}) < \frac{1}{2}\int_M \{h \cdot h\} g(x) \left(\nabla_g s^l, \nabla_g s^l\right) d\mu_g \tag{3.1}$$

since $\{ih \cdot ih\} = \{h \cdot h\}$.

In [14] we show that in complex coordinates on Teichmüller space, if $\tilde{H} = \sum \gamma_\alpha \tilde{H}_\alpha + \rho_\alpha J\tilde{H}_\alpha = \sum \xi_\alpha \tilde{H}_\alpha$ and $\xi_\alpha = \gamma_\alpha + i\rho_\alpha$, where $\tilde{H}_\alpha$ the horizontal lift to $\mathcal{A}$ of a basic H_α over $\mathbb{C}$ of $T_{[J]}\mathcal{T}(M)$, $H = \sum \xi_\alpha H_\alpha$, $[g] = \tilde{\theta}[J]$ then considering $\tilde{E} : \mathcal{A}/\mathcal{D}_0 \to \mathbf{R}$ we have

$$\begin{aligned}\sum_{\alpha,\beta} \frac{\partial^2 \tilde{E}}{\partial z^\alpha \partial \bar{z}^\beta}\xi_\alpha\bar{\xi}_\beta &= D^2\tilde{E}[J](H,H) + D^2\tilde{E}[J](\hat{J}H, \hat{J}H) \\ &= \frac{1}{2}\int_M \{h \cdot h\} g(x) \left(\nabla_g s^l, \nabla_g s^l\right) d\mu_g - D^2 E_g(s)(w^h, w^h) \\ &\qquad - D^2 E_g(s)(w^{ih}, w^{ih}) \end{aligned} \tag{3.2}$$

where h is the $(0,2)$ tensor obtained from $\tilde{H}$ by lowering an index (via g) and where $iH = D\pi(J\tilde{H})$ is the almost complex structure on $\mathcal{T}(M)$ induced by the almost complex structure (1.5) on $\mathcal{A}$.

This again shows that the pluri-subharmonicity of $\tilde{E} : \mathcal{T}(M) \to \mathbb{R}$ depends on establishing inequality (3.1).

From this point on we shall assume that $M = N$. The main idea here is to establish a formula for two-dimensional $N = M$ analogous to what we obtained for arbitrary dimensional N in (2.15).

First let us establish some convention. Write conformal coordinates on (M, g) as $g_{ij} = \lambda\delta_{ij}$ and on (N, g_0) as $(g_0)_{ij} = \rho\delta_{ij}$. Let $s = s(g)$ be the unique harmonic diffeomorphism homotopic to the identity. We introduce two covariant differentiation operators "along s".

Let

$$\nabla_{\frac{\partial}{\partial x}} w = \frac{\partial w^\alpha}{\partial x} + \Gamma^\alpha_{ij} \frac{\partial s^i}{\partial x} w^j$$

$$\nabla_{\frac{\partial}{\partial y}} w = \frac{\partial w^\alpha}{\partial y} + \Gamma^\alpha_{ij} \frac{\partial s^i}{\partial y} w^j$$

where Γ^α_{ij} are the Christofel symbols of g_0. Define

$$\nabla_{\frac{\partial}{\partial z}} = \frac{1}{2}\left[\nabla_{\frac{\partial}{\partial x}} - i\nabla_{\frac{\partial}{\partial y}}\right]$$

$$\nabla_{\frac{\partial}{\partial \bar{z}}} = \frac{1}{2}\left[\nabla_{\frac{\partial}{\partial x}} + i\nabla_{\frac{\partial}{\partial y}}\right].$$

A simple computation yields the following

Theorem 3.1. *In conformal coordinates*

$$\nabla_{\frac{\partial}{\partial z}} w = \frac{\partial w}{\partial z} + (\log\rho)_s s_z w$$

$$\nabla_{\frac{\partial}{\partial \bar{z}}} w = \frac{\partial w}{\partial \bar{z}} + (\log\rho)_s s_{\bar{z}} w$$

where

$$2\rho(\log\rho)_s = \rho_{s_1} - i\rho_{s_2}, \quad s = s_1 + is_2.$$

Remark. The harmonic map equation can be written either as

$$\nabla_{\frac{\partial}{\partial z}} s_{\bar{z}} = 0 \quad \text{or} \quad \nabla_{\frac{\partial}{\partial \bar{z}}} s_z = 0.$$

We are now in a position to state our main computational result:

Theorem 3.2. *Let $\nu^h = D_g s(g)h$ and let w be any vector field over s; i.e. $w(x) \in T_{s(x)}M$ for all $x \in M$. Then the second variation of Dirichlet's energy can be expressed as*

$$D^2E_g(s)(\nu^h, w) = 2\,\mathrm{re}\int \rho\left\{s_z \overline{\nabla_{\frac{\partial}{\partial \bar{z}}} w} + \overline{s_{\bar{z}}}\nabla_{\frac{\partial}{\partial z}} w\right\} H\, dx \wedge dy$$

$H = h^1_1 + ih^1_2.$

Note: This is the complex two dimensional analog to (2.14).

Proof. The general equation of a harmonic map is

$$\frac{1}{\sqrt{g}}\frac{\partial}{\partial x_j}\left[g^{ij}\sqrt{g}\frac{\partial}{\partial x_j}s^\alpha\right]+\Gamma^\alpha_{\gamma\beta}\frac{\partial s^\alpha}{\partial x_i}\frac{\partial s^\beta}{\partial x_j}g^{ij}=0 \tag{3.3}$$

where $\Gamma^\alpha_{\gamma\beta}(s(g))$ depends on s and therefore on g. The idea is to differentiate (3.3) w.r.t. g and then integrate this against a vector field w over g. *The derivatives of all those terms containing $s(g)$ when differentiated and integrated against ν will yield the second variation of Dirichlet's energy.* It is the remaining terms we must calculate.

To begin, we recall the formulas for the Christofel symbols in conformal coordinates, namely

$$\Gamma^1_{12}=\frac{1}{2\rho}\frac{\partial\rho}{\partial y}=\Gamma^1_{21}=\Gamma^2_{22}=-\Gamma^2_{11}$$

and

$$\Gamma^1_{11}=\frac{1}{2\rho}\frac{\partial\rho}{\partial x}=\Gamma^2_{21}=\Gamma^2_{12}=-\Gamma^1_{22}$$

where we write the local coordinates as (x,y).

We first concentrate on the second term of (3.3) namely $\Gamma^\alpha_{\gamma\beta}\frac{\partial s^\gamma}{\partial x_i}\frac{\partial s^\beta}{\partial x_j}g^{ij}$.

Taking the derivative of the g^{ij} term w.r.t. g and then take the inner product with w we obtain the expression

$$-\rho\Gamma^\alpha_{\gamma\beta}\frac{\partial s^\gamma}{\partial x_i}\frac{\partial s^\beta}{\partial x_j}h^{ij}w^\alpha$$

which is equal to

$$-\rho\Gamma^\alpha_{11}\frac{\partial s^1}{\partial x_i}\frac{\partial s^1}{\partial x_j}w^\alpha-\rho\Gamma^\alpha_{22}\frac{\partial s^2}{\partial x_i}\frac{\partial s^2}{\partial x_j}h^{ij}w^\alpha-2\rho\Gamma^2_{12}\frac{\partial s^1}{\partial x_i}\frac{\partial s^2}{\partial x_j}w^\alpha h^{ij}.$$

We deal with this term by term. Writing

$$2\rho_s=\frac{\partial\rho}{\partial s_1}-i\frac{\partial\rho}{\partial s_2},\qquad 2\rho_{\bar{s}}=\frac{\partial\rho}{\partial s_1}+i\frac{\partial\rho}{\partial s_2}$$

we find that

$$-\rho\Gamma^\alpha_{11}\frac{\partial s^1}{\partial x_i}\frac{\partial s^1}{\partial x_j}w^\alpha h^{ij}=-\operatorname{re}(\rho_{\bar{s}}w)\left\{\frac{\partial s^1}{\partial x_i}\frac{\partial s^1}{\partial x_j}h^i_j\right\}$$

$$-\rho\Gamma^\alpha_{22}\frac{\partial s^2}{\partial x_i}\frac{\partial s^2}{\partial x_j}h^{ij}w^\alpha=-\operatorname{re}(\rho_{\bar{s}}w)\left\{\frac{\partial s^2}{\partial x_i}\frac{\partial s^2}{\partial x_j}h^i_j\right\}$$

and

$$-2\rho\Gamma^{\alpha}_{12}\frac{\partial s^1}{\partial x^i}\frac{\partial s^2}{\partial x^j}w^\alpha h^{ij} = -\operatorname{im}(\rho_{\overline{s}}w)\left\{2\frac{\partial s^1}{\partial x^i}\frac{\partial s^2}{\partial x^j}h^i_j\right\}$$

The sum of these three terms can be written as (using that h is trace free, and lowering an index on h)

$$\begin{aligned}
&-\lambda\operatorname{re}(\rho_{\overline{s}}w)\left\{(\frac{\partial s^1}{\partial x})^2+(\frac{\partial s^2}{\partial y})^2-(\frac{\partial s^1}{\partial y})^2-(\frac{\partial s^2}{\partial x})^2\right\}h^1_1\\
&-\lambda\operatorname{re}(\rho_{\overline{s}}w)\left\{2\left[\frac{\partial s^1}{\partial x}\cdot\frac{\partial s^1}{\partial y}-\frac{\partial s^2}{\partial x}\cdot\frac{\partial s^2}{\partial y}\right]h^1_2\right\}-\lambda\operatorname{im}(\rho_{\overline{s}}w)\\
&\cdot\left\{2\left[\frac{\partial s^1}{\partial x}\cdot\frac{\partial s^2}{\partial y}-\frac{\partial s^1}{\partial y}\cdot\frac{\partial s^2}{\partial y}\right]h^1_1+2\left[\frac{\partial s^1}{\partial x}\cdot\frac{\partial s^2}{\partial y}+\frac{\partial s^1}{\partial y}\cdot\frac{\partial s^2}{\partial x}\right]h^1_2\right\}.
\end{aligned}$$

Using the expressions for

$$4[s_z]^2=\left\{\left[\frac{\partial}{\partial x}-i\frac{\partial}{\partial y}\right]s\right\}^2,$$
$$4[s_{\overline{z}}]^2=\left\{\left[\frac{\partial}{\partial x}-i\frac{\partial}{\partial y}\right]s\right\}^2,$$

and $\overline{s_{\overline{z}}}s_z$ the above sum can be written simply as

$$\begin{aligned}
&-\frac{2}{\lambda}\operatorname{re}(\rho_{\overline{s}}w)\Big\{\operatorname{re}\left[[s_z]^2+[s_{\overline{z}}]^2\right]h^1_1+\operatorname{im}\left[[s_{\overline{z}}]^2-[s_z]^2\right]h^1_2\Big\}\\
&\qquad\qquad-\frac{2}{\lambda}\operatorname{re}(\rho_{\overline{s}}w)\Big\{\operatorname{im}\left[[s_{\overline{z}}]^2+[s_z]^2\right]h^1_1+\operatorname{re}\left[[s_z]^2-[s_{\overline{z}}]^2\right]h^1_2\Big\}.
\end{aligned}$$

Setting $H=h^1_1+ih^1_2$ we may simplify this sum to equal

$$-\frac{2}{\lambda}\operatorname{re}\Big\{(\rho_s\overline{w})\left[[s_z]^2H+[s_{\overline{z}}]^2\overline{H}\right]\Big\} \tag{3.4}$$

which gives us an expression for the derivative of the second term of (3.3) omitting the derivative w.r.t. $s(g)$.

Integrating this over M we obtain

$$-2\operatorname{re}\int(\rho_s\overline{w})\left[[s_z]^2H+[s_{\overline{z}}]^2\overline{H}\right]. \tag{3.5}$$

We now consider the term $\frac{1}{\sqrt{g}}\frac{\partial}{\partial x_j}\left\{g^{ij}\sqrt{g}\frac{\partial}{\partial x_i}s^\alpha\right\}$.

Differentiating w.r.t. g in the direction of a trace free h, omitting again the derivative of $s(g)$ w.r.t. g and taking the inner product with w and integrating by parts we get

$$-\int \rho \frac{\partial}{\partial x_j}\left\{h^i_j \frac{\partial}{\partial x^i} s^\alpha\right\} w^\alpha$$
$$= \int \left[\frac{\partial \rho}{\partial s^1}\frac{\partial s^1}{\partial x^i} + \frac{\partial \rho}{\partial s^2}\frac{\partial s^2}{\partial x^j}\right]\frac{\partial s^\alpha}{\partial x^i} w^\alpha h^i_j + \int \rho \frac{\partial s^\alpha}{\partial x^i}\frac{\partial w^\alpha}{\partial x^j} h^i_j. \quad (3.6)$$

Let us concentrate on the first term of (3.6). Setting $\alpha = 1$ we obtain the expression

$$\int_M \left\{\frac{\partial \rho}{\partial s^1}\left[[\frac{\partial s^1}{\partial x}]^2 - [\frac{\partial s^1}{\partial y}]^2\right] w^1 h^1_1\right\}$$
$$+\int \left\{\frac{\partial \rho}{\partial s^2}\left[\frac{\partial s^2}{\partial x}\cdot\frac{\partial s^1}{\partial x} - \frac{\partial s^2}{\partial y}\cdot\frac{\partial s^1}{\partial y}\right] w^1 h^1_1\right\} + 2\int \left\{\frac{\partial \rho}{\partial s^1}\left[\frac{\partial s^1}{\partial x}\cdot\frac{\partial s^1}{\partial y}\right] w^1 h^1_2\right\}$$
$$+\int \left\{\frac{\partial \rho}{\partial s^2}\left[\frac{\partial s^2}{\partial y}\cdot\frac{\partial s^1}{\partial x} + \frac{\partial s^2}{\partial x}\cdot\frac{\partial s^1}{\partial y}\right] w^1 h^1_2\right\}$$

which can be written as

$$\int_M \left\{\frac{\partial \rho}{\partial s^1} \operatorname{re}\left[[s_z]^2 + [s_{\overline{z}}]^2 + 2\overline{s}_{\overline{z}} s_z\right] w^1 h^1_1\right\}$$
$$+\int \left\{\frac{\partial \rho}{\partial s^2} \operatorname{im}\left[[s_z]^2 + [s_{\overline{z}}]^2\right] w^1 h^1_1\right\}$$
$$+\int \left\{\frac{\partial \rho}{\partial s^1} \operatorname{im}\left[[s_{\overline{z}}]^2 - [s_z]^2 - 2\overline{s}_{\overline{z}} s_z\right] w^1 h^1_2\right\} +$$
$$\int \left\{\frac{\partial \rho}{\partial s^2} \operatorname{re}\left[[s_z]^2 - [s_{\overline{z}}]^2\right] w^1 h^1_2\right\}.$$

Letting $\alpha = 2$ we obtain the sum

$$\int \left\{\frac{\partial \rho}{\partial s^1} \operatorname{im}\left[[s_z]^2 + [s_{\overline{z}}]^2\right] w^2 h^1_1\right\}$$
$$-\int \left\{\frac{\partial \rho}{\partial s^2} \operatorname{re}\left[[s_{\overline{z}}]^2 + [s_z]^2 - 2\overline{s}_{\overline{z}} s_z\right] w^2 h^1_1\right\}$$
$$+\int \left\{\frac{\partial \rho}{\partial s^1} \operatorname{re}\left[[s_z]^2 - [s_{\overline{z}}]^2\right] w^2 h^1_2\right\} +$$
$$\int \left\{\frac{\partial \rho}{\partial s^2} \operatorname{im}\left[[s_z]^2 - [s_{\overline{z}}]^2 - 2\overline{s}_{\overline{z}} s_z\right] w^2 h^1_2\right\}.$$

Let us now sum the terms containing $\overline{s}_{\overline{z}}s_z$ and we obtain

$$2\int \frac{\partial\rho}{\partial s^1}(\mathrm{re}\,\overline{s}_{\overline{z}}s_z)w^1h_1^1 - 2\int \frac{\partial\rho}{\partial s^1}(\mathrm{im}\,\overline{s}_{\overline{z}}s_z)w^1h_2^1$$
$$+ 2\int \frac{\partial\rho}{\partial s^2}(\mathrm{re}\,\overline{s}_{\overline{z}}s_z)w^2h_1^1 - 2\int \frac{\partial\rho}{\partial s^2}(\mathrm{im}\,\overline{s}_{\overline{z}}s_z)w^2h_2^1$$

which can be further simplified to

$$4\int [\mathrm{re}(\rho_s w)\,\mathrm{re}(H\overline{s}_{\overline{z}}s_z)]\,. \tag{3.7}$$

The sum of the remaining terms can be written as

$$\int \left\{\frac{\partial\rho}{\partial s^1}\,\mathrm{re}\left[[s_z]^2 + [s_{\overline{z}}]^2\right]w^1h_1^1\right\} + \int \left\{\frac{\partial\rho}{\partial s^2}\,\mathrm{im}\left[[s_z]^2 + [s_{\overline{z}}]^2\right]w^1h_1^1\right\}$$
$$+ \int \left\{\frac{\partial\rho}{\partial s^1}\,\mathrm{im}\left[[s_{\overline{z}}]^2 - [s_z]^2\right]w^1h_2^1\right\} + \int \left\{\frac{\partial\rho}{\partial s^2}\,\mathrm{re}\left[[s_z]^2 - [s_{\overline{z}}]^2\right]w^1h_2^1\right\}$$
$$+ \int \left\{\frac{\partial\rho}{\partial s^1}\,\mathrm{im}\left[[s_z]^2 + [s_{\overline{z}}]^2\right]w^2h_1^1\right\} - \int \left\{\frac{\partial\rho}{\partial s^2}\,\mathrm{re}\left[[s_{\overline{z}}]^2 + [s_z]^2\right]w^2h_1^1\right\}$$
$$+ \int \left\{\frac{\partial\rho}{\partial s^1}\,\mathrm{re}\left[[s_z]^2 - [s_{\overline{z}}]^2\right]w^2h_2^1\right\} + \int \left\{\frac{\partial\rho}{\partial s^2}\,\mathrm{im}\left[[s_z]^2 - [s_{\overline{z}}]^2\right]w^2h_2^1\right\}.$$

Adding the terms containing $[s_z]^2$ we obtain

$$2\int \mathrm{re}\left\{\rho_{\overline{s}}[s_z]^2\right\}w^1h_1^1 - 2\int \mathrm{im}\left\{\rho_{\overline{s}}[s_z]^2\right\}w^1h_2^1$$
$$+ 2\int \mathrm{im}\left\{\rho_{\overline{s}}[s_z]^2\right\}w^2h_1^1 + 2\int \mathrm{re}\left\{\rho_{\overline{s}}[s_z]^2\right\}w^2h_1^1$$

which can be simplified to

$$2\int \mathrm{re}\left\{\rho_{\overline{s}}\overline{w}[s_z]^2H\right\}. \tag{3.8}$$

Adding the terms containing $[s_{\overline{z}}]^2$ we obtain

$$2\int \mathrm{re}\left\{\rho_{\overline{s}}[s_{\overline{z}}]^2\right\}w^1h_1^1 + 2\int \mathrm{im}\left\{\rho_{\overline{s}}[s_{\overline{z}}]^2\right\}w^1h_2^1$$
$$+ 2\int \mathrm{im}\left\{\rho_{\overline{s}}[s_{\overline{z}}]^2\right\}w^2h_1^1 - 2\int \mathrm{re}\left\{\rho_{\overline{s}}[s_{\overline{z}}]^2\right\}w^2h_2^1$$
$$= 2\int \mathrm{re}\left\{\rho_{\overline{s}}[s_{\overline{z}}]^2\right\}\mathrm{re}\{wH\} + \int \mathrm{im}\left\{\rho_{\overline{s}}[s_{\overline{z}}]^2\right\}\mathrm{im}\{wH\}$$
$$= 2\int \mathrm{re}\left\{\rho_{\overline{s}}\overline{w}[s_{\overline{z}}]^2\overline{H}\right\}.$$

Adding this to (3.8) we obtain

$$2\,\mathrm{re}\int \rho_{\overline{s}}\overline{w}\left\{[s_z]^2 H + [s_{\overline{z}}]^2 \overline{H}\right\}. \tag{3.9}$$

Therefore taking the derivative of (3.3), w.r.t. g excepting the terms containing $s(g)$ and integrating against w we obtain the sum of (3.5) and (3.9) (which cancel each other) and (3.7) and the second term of (3.6). This leaves us with

$$\int \rho \frac{\partial s^\alpha}{\partial x^i}\frac{\partial w^\alpha}{\partial x^j} h^i_j + 4\int (\mathrm{re}\,\rho_s w)\,\mathrm{re}(H \overline{s}_{\overline{z}} s_z). \tag{3.10}$$

Let us write out the expression from Theorem 3.2, namely

$$2\,\mathrm{re}\int \rho\left\{s_z \overline{\nabla_{\frac{\partial}{\partial \overline{z}}} w} + \overline{s}_{\overline{z}} \nabla_{\frac{\partial}{\partial z}} w\right\} H\, dx \wedge dy. \tag{3.11}$$

Using the formulas for $\nabla_{\frac{\partial}{\partial z}}$ and $\nabla_{\frac{\partial}{\partial \overline{z}}}$ we see that this is equal to

$$\mathrm{re}\,2\int \rho\left\{s_z \overline{w}_{\overline{z}} + \overline{s}_{\overline{z}} w_z\right\} H + 4\int \mathrm{re}(s_z \overline{s}_{\overline{z}})\,\mathrm{re}(\rho_s w). \tag{3.12}$$

Now the first term of (3.12) is the first term of (3.10) in complex notation. This establishes the equality of (3.10) and (3.11).

Finally if we take the derivative of (3.3) w.r.t. g for all those terms depending on $s(g)$ and then integrate by parts we obtain the negative of the second variation of Dirichlet's energy E_g. Since the total derivative of (3.3) w.r.t. g must be zero we obtain the equality

$$0 = -D^2 E_g(s)(\nu^h, w) + 2\,\mathrm{re}\int \rho\left\{s_z \overline{\nabla_{\frac{\partial}{\partial \overline{z}}} w} + \overline{s}_{\overline{z}} \nabla_{\frac{\partial}{\partial z}} w\right\} H\, dx \wedge dy$$

which finally establishes Theorem 3.2.

We are now ready to use this formula for $D^2 E_g(s)(\nu^h, w)$ to establish the fact that Dirichlet's energy $\tilde{E} : \mathcal{T}(M) \to \mathbb{R}$ is strictly pluri-subharmonic. For simplicity denote the second variation of Dirichlet's energy E_g, $D^2 E_g(s)$ by Q.

Our computational trick will be to write

$$2\left\{Q(w^{ih}, w^{ih}) + Q(w^h, w^h)\right\} \tag{3.13}$$

as

$$Q\left[w^{ih}, w^{ih} + i w^h\right] + Q\left[w^h, w^h + i w^{ih}\right] + Q\left[w^{ih}, w^{ih} - i w^h\right] + Q\left[w^h, w^h + i w^{ih}\right].$$

Now

$$Q\left[w^{ih}, w^{ih} + iw^h\right] + Q\left[w^h, w^h - iw^{ih}\right]$$

is of the form

$$Q(w^{ih}, iw) + Q(w^h, w) \tag{3.14}$$

where $w = w^h + iw^{ih}$.

Since

$$Q(w^h, \nu) = 2\,\mathrm{re}\int \rho\left\{s_z\overline{\nabla_{\frac{\partial}{\partial \bar{z}}}\nu} + \overline{s_{\bar{z}}}\nabla_{\frac{\partial}{\partial z}}\nu\right\} H$$

expression (3.14) can be expressed as

$$4\,\mathrm{re}\int \rho\left\{s_z\overline{\nabla_{\frac{\partial}{\partial \bar{z}}}w}\right\} H.$$

Moreover

$$Q\left[w^{ih}, w^{ih} - iw^h\right] + Q\left[w^h, w^h + iw^{ih}\right]$$

is of the form

$$Q(w^{ih}, -i\nu) + Q(w^h, \nu)$$

where $\nu = w^h + iw^{ih}$.

Thus

$$\begin{aligned}
&2\left\{Q(w^{ih}, w^{ih}) + Q(w^h, w^h)\right\}\\
&= \mathrm{re}\,4\int \rho\left\{s_z\overline{\nabla_{\frac{\partial}{\partial \bar{z}}}(w^h - iw^{ih})}\right\} H + \mathrm{re}\,4\int \rho\left\{\overline{s_{\bar{z}}}\nabla_{\frac{\partial}{\partial z}}(w^h - iw^{ih})\right\} H.
\end{aligned} \tag{3.15}$$

Applying Schwarz's inequality we see that

$$\begin{aligned}
&2\left\{Q(w^{ih}, w^{ih}) + Q(w^h, w^h)\right\} \qquad (3.16)\\
&\le 2\int \rho\left[|s_z|^2 + |s_{\bar{z}}|^2\right]|H|^2 + 2\int \rho\left|\nabla_{\frac{\partial}{\partial \bar{z}}}(w^h - iw^{ih})\right|^2\\
&+ 2\int \rho\left|\nabla_{\frac{\partial}{\partial z}}(w^h + iw^{ih})\right|^2\\
&= \frac{1}{2}\int \{h\cdot h\}g(x)\left(\nabla_g s^l, \nabla_g s^l\right)\,d\mu_g + 2\int \rho\left\{\left|\nabla_{\frac{\partial}{\partial \bar{z}}}w^h\right|^2 + \left|\nabla_{\frac{\partial}{\partial z}}w^h\right|^2\right\}\\
&+ 2\int \rho\left\{\left|\nabla_{\frac{\partial}{\partial \bar{z}}}w^{ih}\right|^2 + \left|\nabla_{\frac{\partial}{\partial z}}w^{ih}\right|^2\right\} + 4\int\left\langle\nabla_{\frac{\partial}{\partial \bar{z}}}iw^h, \nabla_{\frac{\partial}{\partial \bar{z}}}w^{ih}\right\rangle\\
&- 4\int\left\langle\nabla_{\frac{\partial}{\partial z}}iw^h, \nabla_{\frac{\partial}{\partial z}}w^{ih}\right\rangle
\end{aligned}$$

where $\langle , \rangle$ is the inner product on (M, g_0).

The last terms arise from the fact that in conformal coordinates

$$\left\langle \nabla_{\frac{\partial}{\partial \bar z}} iw^h, \nabla_{\frac{\partial}{\partial \bar z}} w^{ih} \right\rangle = \frac{1}{2}\rho \left\{ \overline{\nabla_{\frac{\partial}{\partial \bar z}} iw^h} \nabla_{\frac{\partial}{\partial \bar z}} w^{ih} + \nabla_{\frac{\partial}{\partial \bar z}} iw^h \overline{\nabla_{\frac{\partial}{\partial \bar z}} w^{ih}} \right\}$$

and similarly for $\left\langle \nabla_{\frac{\partial}{\partial z}} w^{ih}, \nabla_{\frac{\partial}{\partial z}} iw^h \right\rangle$.

Let us concentrate on these last two terms. Integrating by parts we see that their sum equals

$$-4\left\{ \int \left\langle i\left[\nabla_{\frac{\partial}{\partial z}} \nabla_{\frac{\partial}{\partial \bar z}} w^h - \nabla_{\frac{\partial}{\partial \bar z}} \nabla_{\frac{\partial}{\partial z}} w^h \right], w^{ih} \right\rangle \right\}$$
$$= 2\int \left\langle \nabla_{\frac{\partial}{\partial x}} \nabla_{\frac{\partial}{\partial y}} w^h - \nabla_{\frac{\partial}{\partial y}} \nabla_{\frac{\partial}{\partial x}} w^h, w^{ih} \right\rangle .$$

But the last term is an integral of a curvature term, namely

$$2\int \left\langle \mathcal{R}\left[\frac{\partial s}{\partial x}, \frac{\partial s}{\partial y} \right] w^h, w^{ih} \right\rangle \, dx \wedge dy$$

where $\mathcal{R}$ is the curvature tensor of (M, g_0). Thus we arrive at the inequality

$$2\left\{ Q(w^{ih}, w^{ih}) + Q(w^h, w^h) \right\} \le \frac{1}{2}\int \{h \cdot h\} g(x) \left(\nabla_g s^l, \nabla_g s^l \right) \, d\mu_g$$
$$+ 2\int \rho \left\{ \left| \nabla_{\frac{\partial}{\partial \bar z}} w^h \right|^2 + \left| \nabla_{\frac{\partial}{\partial z}} w^h \right|^2 \right\} + 2\int \rho \left\{ \left| \nabla_{\frac{\partial}{\partial \bar z}} w^{ih} \right|^2 + \left| \nabla_{\frac{\partial}{\partial z}} w^{ih} \right|^2 \right\}$$
$$+ 2\int \left\langle \mathcal{R}\left[\frac{\partial s}{\partial x}, \frac{\partial s}{\partial y} \right] w^h, w^{ih} \right\rangle \, dx \wedge dy$$
$$= \int \rho \left\{ \left| \nabla_{\frac{\partial}{\partial x}} w^h \right|^2 + \left| \nabla_{\frac{\partial}{\partial y}} w^h \right|^2 \right\} + \int \rho \left\{ \left| \nabla_{\frac{\partial}{\partial x}} w^{ih} \right|^2 + \left| \nabla_{\frac{\partial}{\partial y}} w^{ih} \right|^2 \right\}$$
$$+ 2\int \left\langle \mathcal{R}\left[\frac{\partial s}{\partial x}, \frac{\partial s}{\partial y} \right] w^h, w^{ih} \right\rangle \, dx \wedge dy + \frac{1}{2}\int \{h \cdot h\} g(x) \left(\nabla_g s^l, \nabla_g s^l \right) \, d\mu_g .$$

Taking into account the formula (2.11) for the second variation of Dirichlet's energy E_g we may bring the two integrals on the right to the left side of inequality (3.16) to conclude the inequality

$$Q(w^{ih}, w^{ih}) + Q(w^h, w^h) - \int \left\langle \mathcal{R}\left[w^{ih}, \frac{\partial s}{\partial x_l} \right] \frac{\partial s}{\partial x_l}, w^{ih} \right\rangle \, dx \wedge dy$$
$$- \int \left\langle \mathcal{R}\left[w^h, \frac{\partial s}{\partial x_l} \right] \frac{\partial s}{\partial x_l}, w^h \right\rangle \, dx \wedge dy$$
$$\le \frac{1}{2}\int \{h \cdot h\} g(x) \left(\nabla_g s^l, \nabla_g s^l \right) \, d\mu_g + 2\int \left\langle \mathcal{R}\left[\frac{\partial s}{\partial x}, \frac{\partial s}{\partial y} \right] w^h, w^{ih} \right\rangle \, dx \wedge dy. \tag{3.17}$$

Since (M, g_0) has negative curvature -1 and s is a diffeomorphism the sum of curvature terms on the left side of inequality (3.17) is strictly positive.

Our next step is to show that the curvature term on the left of (3.17) can be used to "absorb" the curvature term on the right hand side. In this case we shall again explicitly use the fact that the range is two dimensional and that s is a diffeomorphism.

With these facts in mind we may write

$$w^{ih} = c\frac{\partial s}{\partial x} + d\frac{\partial s}{\partial y}$$

and

$$w^{h} = a\frac{\partial s}{\partial x} + b\frac{\partial s}{\partial y}.$$

Then it follows that

$$-\int \left\langle \mathcal{R}\left[w^{ih}, \frac{\partial s}{\partial x_l}\right]\frac{\partial s}{\partial x_l}, w^{ih}\right\rangle - \int \left\langle \mathcal{R}\left[w^{h}, \frac{\partial s}{\partial x_l}\right]\frac{\partial s}{\partial x_l}, w^{h}\right\rangle$$
$$= -\int (a^2+b^2+c^2+d^2)\left\langle \mathcal{R}\left[\frac{\partial s}{\partial x}, \frac{\partial s}{\partial y}\right]\frac{\partial s}{\partial y}, \frac{\partial s}{\partial x}\right\rangle \quad (3.18)$$

and that

$$\int \left\langle \mathcal{R}\left[\frac{\partial s}{\partial x}, \frac{\partial s}{\partial y}\right] w^h, w^{ih}\right\rangle = \int (ad-bc)\left\langle \mathcal{R}\left[\frac{\partial s}{\partial x}, \frac{\partial s}{\partial y}\right]\frac{\partial s}{\partial y}, \frac{\partial s}{\partial x}\right\rangle. \quad (3.19)$$

Clearly

$$2\left|\int \left\langle \mathcal{R}\left[\frac{\partial s}{\partial x}, \frac{\partial s}{\partial y}\right] w^h, w^{ih}\right\rangle\right|$$
$$\leq -\int (a^2+b^2+c^2+d^2)\left\langle \mathcal{R}\left[\frac{\partial s}{\partial x}, \frac{\partial s}{\partial y}\right]\frac{\partial s}{\partial y}, \frac{\partial s}{\partial x}\right\rangle. \quad (3.20)$$

We therefore arrive at the final inequality

$$Q(w^{ih}, w^{ih}) + Q(w^h, w^h) \leq \frac{1}{2}\int \{h \cdot h\} g(x) \left(\nabla_g s^l, \nabla_g s^l\right) d\mu_g. \quad (3.21)$$

We would like to show that inequality (3.21) is strict. Our strategy will be as follows: The proof of inequality (3.21) depended on two inequalities, namely (3.16) and (3.20). If either of these were strict, our proof would show that inequality (4.20) is also strict. Our goal is two show that if both inequalities, (3.16) and (3.20) are equalities, then w^h and w^{ih} are both zero. In this case strict inequality in (3.21) is trivially satisfied. We begin with inequality (3.20). It follows at once that equality holds iff $a = d$ and $b = -c$. Thus since s is a diffeomorphism we may write:

$$w^{ih} = -b\frac{\partial s}{\partial x} + a\frac{\partial s}{\partial y}$$
$$w^h = a\frac{\partial s}{\partial x} + b\frac{\partial s}{\partial y}.$$

Now a simple computation shows that

$$\begin{aligned} w^h + iw^{ih} &= (a - ib)s_{\overline{z}} = 2\tau(z)s_{\overline{z}} \\ w^h - iw^{ih} &= (a + ib)s_z = 2\tau(z)s_z \end{aligned} \tag{3.22}$$

where $\tau(z) := a+ib$. We now observe that τ is actually a negative differential (i.e. a differential of order -1, $\tau(z)\,dz^{-1}$. The easiest way to see this is to consider (M, g_0) as isometrically embedded in $\mathbb{R}^p$ with $J_0(x) : T_xM \to T_xM$ the almost complex structure on (M, g_0). Then

$$w^h - iw^{ih} = 2\tau(z)s_z.$$

So

$$\tau(z) = \frac{w^h - iw^{ih}}{2s_z}.$$

But $s_z\,dz$ is a differential, and therefore $\tau(z)\,dz^{-1}$ is a negative differential on (M, g). However such a negative differential is a vector field on (M, g) (one just checks how $\tau(z)$ transforms under conformal changes of coordinates). From this point on we shall simply write $\tau(z)$ for $\tau(z)\,dz^{-1}$, but will keep in mind that τ is a vector field. Now let us return to inequality (3.16). First observe that since s is harmonic,

$$\nabla_{\frac{\partial}{\partial z}} s_{\overline{z}} = \nabla_{\frac{\partial}{\partial \overline{z}}} s_z = 0.$$

Therefore

$$\frac{1}{2}\nabla_{\frac{\partial}{\partial \overline{z}}}\{w^h - iw^{ih}\} = \nabla_{\frac{\partial}{\partial \overline{z}}}(\tau s_z) = \tau_{\overline{z}} s_z + \tau(z)\nabla_{\frac{\partial}{\partial \overline{z}}} s_z = \tau_{\overline{z}} s_z.$$

Rewrite (3.15) as

$$2\{Q(w^{ih}, w^{ih}) + Q(w^h, w^h)\}$$
$$= 8\int_M \langle Hs_z, \tau_{\overline{z}} s_z\rangle\,dxdy + 8\int_M \langle Hs_{\overline{z}}, \overline{\tau_{\overline{z}} s_{\overline{z}}}\rangle\,dxdy.$$

If $H = 0$ then $h = 0$ and then w^h and w^{ih} are 0. In this case, strict inequality in (3.21) is satisfied. Now assume that $H \neq 0$. In this case H is a 1–1 tensor corresponding to a holomorphic quadratic differential by raising an index. This implies that both $\operatorname{Re} H$ and $\operatorname{Im} H$ vanish on at most a finite set on M. Consequently in going from (3.15) to (3.16) we have equality iff

$$\tau_{\overline{z}} = \kappa(z)H, \qquad \tau_{\overline{z}} = \mu(z)\overline{H}$$

where κ and μ are real valued functions defined everywhere on M except possibly on a finite set. Since $\operatorname{Re} H \neq 0$, $\operatorname{Im} H \neq 0$ on a dense set, $\tau_{\overline{z}} = 0$ everywhere. Thus τ is a holomorphic vector field on (M, g). Since the only holomorphic vector field is zero, $\tau \equiv 0$. Since $\tau \equiv 0$, (3.22) implies that both w^h and w^{ih} must be zero. Therefore we have strict inequality in (3.21). Putting all these fact together we arrive at

Theorem 3.3. *Dirichlet's energy $\tilde{E} : \mathcal{T}(M) \to \mathbf{R}$ on Teichmüller's moduli space is strictly pluri-subharmonic.*

References.

[1] L. Bers and L. Ehrenpreis, *Holomorphic convexity of Teichmüller Space.* Bull AMS **70** (1964), 761–764.

[2] P. Deligne, P. Griffiths, J. Morgan and D. Sullivan, *Real homotopy theory of Kähler manifolds.* Inv. math. **29** (1975), 245–274.

[3] C. Earle and J. Eells, *Deformations of Riemannian surfaces.* Lecture Notes in Mathematics **103**, Springer – Verlag

[4] J. Eells and J.H. Sampson, *Harmonic mappings of Riemannian manifolds.* Amer. J. Math. **86** (1964), 109–160.

[5] A.E. Fischer and A.J. Tromba, *On a purely Riemannian proof of the structure and dimension of the unramified moduli space of a compact Riemannian surface.* Math. Ann. **267** (1984), 311–345.

[6] A.E. Fischer and A.J. Tromba, *On the Weil – Petersson metric on Teichmüller space.* Trans. AMS **284** (1984), 319–335.

[7] A.E. Fischer and A.J. Tromba, *Almost complex principle bundles and the complex structure on Teichmüller space.* Crelles J. **252**, 151–160.

[8] A.E. Fischer and A.J. Tromba, *A new proof that Teichmüller space is a cell.* Trans. AMS **303**, No. 1 Sept. (1987), 257–262.

[9] Hartmann, *On homotopic harmonic maps.* Can. J. Math. **19** (1967), 673–687.

[10] S.P. Kerckhoff, *The Nielsen Realization Problem.* Bull Amer. Math. Soc. **2** (1980), 452–454.

[11] J. Nielsen, *Abbildungsklassen endlicher Ordnung.* Acta Math. **75** (1942), 23–115.

[12] J.H. Sampson, *Some properties and applications of harmonic mappings.* Ann. Sci. Ecole Normale Sup. **11** (1978), 211–228.

[13] A.J. Tromba, *On a natural algebraic affine connection on the space of almost complex structures and the curvature of Teichmüller space with respect to its Weil - Petersson metric.* Manuscripta Math. **56**, Fas. 4 (1986), 475–497.

[14] A.J. Tromba, *On an energy function for the Weil - Petersson metric.* Manuscripta Math. **59** (1987), 249–266.

[15] A.J. Tromba, *Teichmüller Theory in Riemannian Geometry.* Lecture Notes in Mathematics, ETH Zürich, Birkhäuser, 1993.

[16] A.J. Tromba, *On the Levi-form for Dirichlet's energy on Teichmüller space.* SFB 256 preprint, Bonn.

[17] R. Shoen and S.T. Yau, *On univalent harmonic maps between surfaces.* Inventiones mathematics **44** (1978), 265–278.

[18] S. Wolpert, *Geodesic length functionals and the Nielsen problem.* J. Diff. Geometry **25** (1987), 275–295.

[19] H. Zieschang, *Finite Groups of Mapping Classes of Surfaces.* Lecture Notes in Mathematics **875**, Springer - Verlag 1981.

RECEIVED NOVEMBER 20, 1995.

UNIVERSITY OF CALIFORNIA
SANTA CRUZ, CALIFORNIA
USA 95065

The Plateau Problem for Boundary Curves with Connectors

HENRY C. WENTE

I. Introduction.

Let γ_1, γ_2 be a pair of disjoint rectifiable Jordan curves in $\mathbb{R}^3$. By a connector we mean a rectifiable arc σ joining a point S on γ_1 with a point T on γ_2 where these are the only intersection points of σ with γ_1, γ_2. We shall be interested in oriented rectifiable Jordan curves γ_n constructed by removing a small segment around S on γ_1, a small segment around T on γ_2 and joining what's left of γ_1, γ_2 by two arcs σ_n^-, σ_n^+ parallel to σ to form γ_n which resembles $\gamma_1 \vee \gamma_2$ joined by a narrow bridge.

Our first result concerns minimal surfaces. Let M_i be a simply connected minimal surface of least area spanning γ_i with area a_i and set $A_d = a_1 + a_2$. We also consider genus 0 surfaces of annular type whose oriented boundary is $\gamma_1 \vee \gamma_2$. Let A_0 be the infimum of area among all such connected surfaces with boundary $\gamma_1 \vee \gamma_2$. If the Douglas condition $A_0 < A_d$ is satisfied then we know that there exists an annular genus 0 minimal surface with boundary $\gamma_1 \vee \gamma_2$ and area A_0.

Suppose now that $A_0 < A_d$ and let σ be a connector of γ_1 and γ_2 with the property that σ is not contained in any annular minimal surface solving the genus 0 Douglas Problem for $\gamma_1 \vee \gamma_2$ with area A_0. Let γ_n be a sequence of rectifiable Jordan curves based on γ_1, γ_2 with bridge connector approximating σ as described above. For any such γ_n let $B_0(\gamma_n)$ be the infimum of area of disk type surfaces spanning γ_n and denote by $B_1(\gamma_n)$ the infinum of area for genus one surfaces spanning γ_n. We show that for narrow enough bridges (n large) we will have $B_1(\gamma_n) < B_0(\gamma_n)$. The genus one Douglas condition is satisfied and there exists a genus one minimizer for γ_n.

The classic example is where γ_1, γ_2 are circles of equal radii lying in parallel planes and whose centers lie on the same normal line, the axis. Suppose that the distance between the two circles is such that the minimal surface of least area is a section of a catenoid. In this case we have $A_0 < A_d$

where the degenerate minimizer is two flat disks. Let σ be a connector not lying on the catenoid. For example, we could choose σ to be a straight line segment parallel to the axis connecting γ_1 to γ_2. Form the corresponding bridge in the obvious manner to produce γ_n. There are two genus 0 minimal surfaces often produced in texts but the real minimizer (for narrow necks) must have genus one. See R. Courant [1] for example [see Figure 1].

Theorem 2.1:

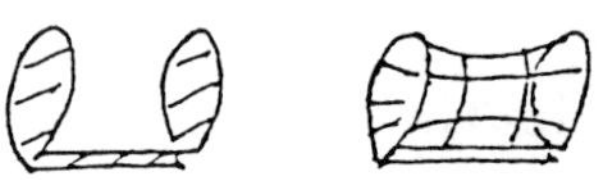

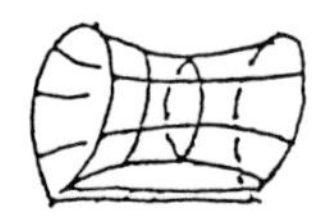

Two Genus 0 Minimal Surfaces

The Genus 1 Minimizer

Figure 1.

A similar but more interesting result occurs when we consider the same type of problem but with a volume constraint. The boundary curves γ_1, γ_2 with connector σ and curves γ_n are as above. Now we study solutions to the volume constrained Plateau Problem whose minimizers are surfaces of constant mean curvature (cmc surfaces). Let Σ_1 be a solution to the volume constrained Plateau Problem (VCPP) so that Σ_1 is a simply connected cmc surface with boundary γ_1 which minimizes area and encloses volume V_1 and let Σ_2 be a similar solution for boundary γ_2 with volume V_2, see [6]. Let their areas be a_1, a_2, respectively. Now let K be a prescribed constant and consider the paired problem to find Σ_1, Σ_2 so that the total area $a_1 + a_2$ is minimized where $V_1 + V_2 = K$. The solution is a pair of surfaces of the same mean curvature enclosing total volume K and with total area $A_d = a_1 + a_2$. We also let A_0 be the infimum of areas of annular type genus 0 surfaces spanning $\gamma_1 \vee \gamma_2$ and which enclose the oriented volume $V = K$. Again suppose K is chosen so that the Douglas condition $A_0 < A_d$ is satisfied. This is sufficient to guarantee that there exists an annular genus 0 minimizer, see [4]. Now let σ be a connector with the property that it is not contained in any minimizer of annular type. Let γ_n be a sequence of bridge type Jordan curves based on $(\gamma_1, \gamma_2, \sigma)$ and consider the volume constrained Plateau Problem on γ_n. We set $B_0(\gamma_n, K)$ to be the infimum of area among simply connected surfaces spanning γ_n and enclosing oriented volume K, while we let $B_1(\gamma_n, K)$ be the infimum of area among all oriented surfaces of genus one spanning γ_n and enclosing oriented volume $V = K$. Our second theorem is that the condition $A_0 < A_d$ for the boundary $\gamma_1 \vee \gamma_2$ implies that $B_1(\gamma_n, K) < B_0(\gamma_n, K)$ for n large (narrow necks). The appropriate Douglas condition is satisfied for γ_n and there exists a genus one minimizer for the volume constrained Plateau Problem with boundary γ_n.

The picture for the volume constrained Plateau Problem is less obvious than for the standard Plateau Problem. As a prototype for this situation consider the case where γ_1 and γ_2 are circles of unit radius lying in the same plane which are exterior to each other at a distance d apart. For large volumes K it is clear that the condition $A_0 < A_d$ is satisfied. Now let σ be the straight line segment connecting γ_1 to γ_2 whose length is precisely the distance between γ_1 and γ_2 and let γ_n be the planar Jordan curve whose connecting segments σ_n^-, σ_n^+ are both parallel to σ at a distance ϵ_n apart, forming a dumbell-shaped planar Jordan curve. Theorem 2 asserts that if n is large enough then the minimizer for the volume constrained problem will have genus one. It will resemble a complete sphere with two disks removed whose boundaries are γ_1 and γ_2 to which a narrow bridge spanning σ_n^+, σ_n^- is attached. Note that if the two circles were removed and the boundary curves γ_n were just narrow rectangles then for large volumes the minimizers Σ_n to the volume constrained Plateau Problem for γ_n would resemble large spheres with a narrow rectangle excised. As the width of γ_n shrinks to zero one expects the spherically shaped surface Σ_n to become detached from the narrow rectangle which is itself converging to a degenerate limit of a single wire. The presence of the two circles γ_1, γ_2 serves to pin the sphere down, preventing it from detaching [see Figure 2].

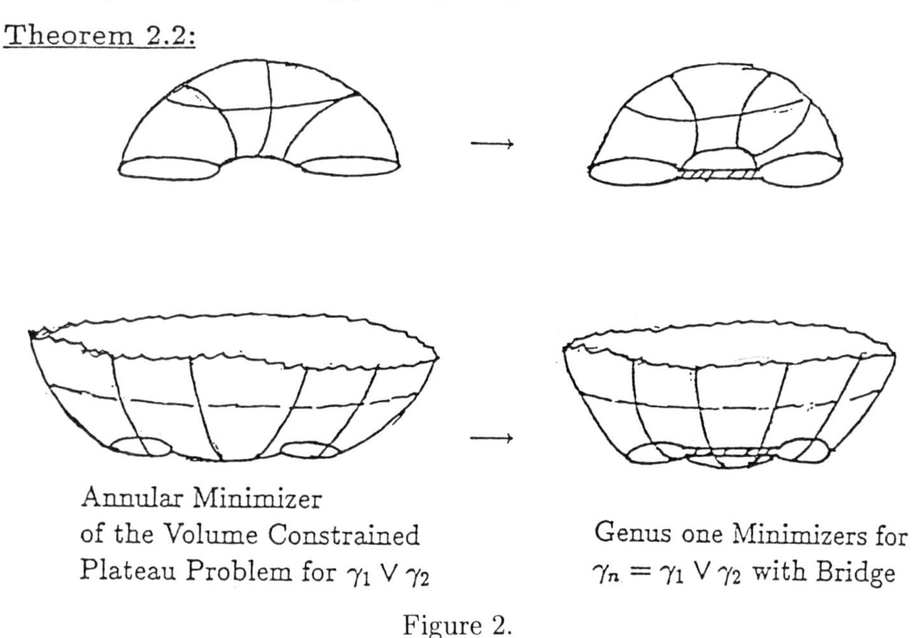

Annular Minimizer of the Volume Constrained Plateau Problem for $\gamma_1 \vee \gamma_2$

Genus one Minimizers for $\gamma_n = \gamma_1 \vee \gamma_2$ with Bridge

Figure 2.

Finally our last theorem considers the case where γ_1, γ_2 are disjoint Jordan curves symmetric to each other upon reflection about a plane of symmetry Π which separates them. For example γ_1, γ_2 might be (as above) disjoint circles lying in the same plane. Let K be chosen so large that the minimizer to the degenerate genus zero volume constrained Plateau Problem is not symmetric with respect to the plane Π. For large volumes one expects a large spherical cup spanning one curve (say γ_1) and a small spherical cup spanning γ_2. Now suppose that $A_d \leq A_0$ (the Douglas condition is not satisfied) and that there is no connected annular type minimizer for this problem. This can be guaranteed by placing γ_1, γ_2 sufficiently far apart. Now let γ_n be a symmetrically constructed dumbell-shaped Jordan curve obtained by connecting γ_1, γ_2 by a narrow bridge. We claim that for sufficiently narrow bridges the solution to the genus zero volume constrained Plateau Problem is not symmetric about Π even though γ_n is [see Figure 3].

Theorem 2.3:

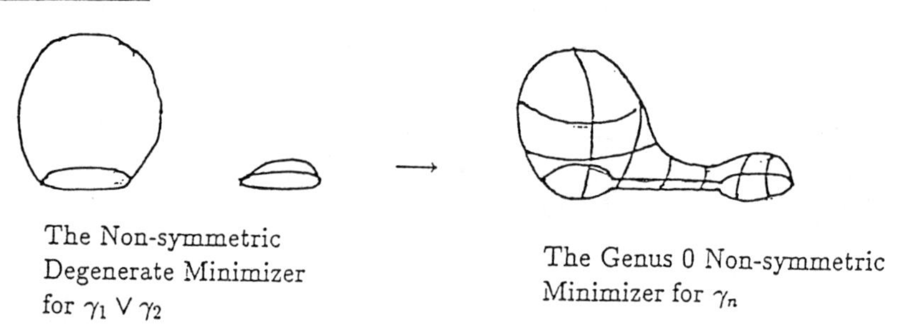

Figure 3.

Our approach is the same in all three theorems. We solve the appropriate genus zero problem for γ_n and consider what happens as the size of the bridge's neck goes to zero. In Section II we set up precisely the conditions on $(\gamma_1, \gamma_2, \sigma)$ which are needed, state the theorems and prove them.

II. Precise Statement and Proofs.

Definition 2.1. Construction of curves γ_n based on γ_1, γ_2 and connector σ.

Let γ_1, γ_2 be disjoint oriented rectifiable Jordan curves and let σ be a connector joining S in γ_1 to T in γ_2 which is otherwwise disjoint from $\gamma_1 \cup \gamma_2$. Let $S_n^- and S_n^+$ be located on γ_1 so that S lies inside the oriented segment (S_n^-, S_n^+) where the length of this segment is less than ϵ_n (where $\epsilon_n \to 0$, say

$\epsilon_n \leq 1/n$). Similarly let T_n^-, T_n^+ be picked from γ_2 so that T lies inside the oriented segment (T_n^+, T_n^-) with the length of this segment also less than ϵ_n. Now let σ_n^- be an oriented segment going from S_n^- to T_n^- which is otherwise disjoint from $\gamma_1 \cup \gamma_2$, let σ_n^+ be an oriented segment going from T_n^+ to S_n^+ also disjoint from $\gamma_1 \cup \gamma_2$. Let $\gamma_1^{(n)}$ be γ_1 with the arc $(S_n^- S_n^+)$ deleted and $\gamma_2^{(n)}$ be γ_2 with (T_n^+, T_n^-) deleted.

The oriented rectifiable Jordan curve γ_n is the union of $\gamma_1^{(n)}, \sigma_n^-, \gamma_2^{(n)}$ and σ_n^+. Finally we impose the following conditions.

1. The curves σ_n^-, σ_n^+ have length less than $2l + 1$ where l is the length of σ and both curves lie within a tubular neighborhood around σ of width ϵ_n.

2. There is a simply connected surface Σ_n with boundary consisting of the four arcs, $\sigma_n^-, (T_n^- T_n^+), \sigma_n^+$ and $(S_n^+ S_n^-)$. We suppose that $|\Sigma_n|$ is bounded by $K\epsilon_n \cong (2l + 1)\epsilon_n$.

3. There is a constant C greater than one such that for any two points $p\epsilon\sigma_n^-, q\epsilon\sigma_n^+$ there is an arc Γ_{pq} from p to q lying on Σ_n such that the length of Γ_{pq} divided by the distance from p to q is less than C.

We think of Σ_n as the ribbon shaped surface bounded by σ_n^+, σ_n^- forming a bridge from γ_1 to γ_2. One might choose Σ_n as a minimal surface, for example.

Definition 2.2. For γ an oriented rectifiable Jordan Curve in R^3 we let $\mathcal{S}_0(\gamma)$ be the set of all vector functions $x(w)\epsilon C^0(\bar{B}) \cap C^1(B)$ where B is the open unit disk in R^2 with finite Dirichlet Integral.

$$D(x) = \int\int_B (|x_u|^2 + |x_v|^2)dudv \tag{2.3}$$

whose boundary values are a weakly monotonic representation of the Jordan Curve γ. The closure of $\mathcal{S}_0(\gamma)$ in the Sobolev space $W^{1,2}$ we denote by $\mathcal{S}(\gamma)$.

$$m(\gamma) = \inf\{D(x)/2 \text{ for } x\epsilon\mathcal{S}(\gamma)\} \tag{2.3}$$

The minimum value $m(\gamma)$ is taken on by a conformally parameterized minimal surface $x_0(w)$ of least area in $\mathcal{S}(\gamma)$ so that $D(x_0) = 2A(x_0) = 2m(\gamma)$ where

$$\Delta x_0 \equiv 0 \text{ on } B \tag{2.3a}$$

$$|x_u| = |x_v| \ , \ (x_u \cdot x_v) = 0 \text{ on } B \tag{2.3b}$$

$$x|\delta B \text{ is a monotonic representation of } \gamma. \tag{2.3c}$$

Definition 2.3. Let γ_1, γ_2 be two disjoint Jordan Curves. The solution to degenerate Plateau Problem for the pair $\gamma_1 \vee \gamma_2$ is a pair of minimal surfaces $x_i \epsilon \mathcal{S}(\gamma_i)$ of least area where we set

$$A_d = A_d(\gamma_1, \gamma_2) = m(\gamma_1) + m(\gamma_2). \tag{2.4}$$

For $0 < r < 1$ let $\mathcal{A}_r = \{w | r < |w| < 1\}$ be an annular domain. We set $\mathcal{S}_0(\gamma_1, \gamma_2) = \{x(w) | x : \bar{\mathcal{A}}_r \to R^3$ for some $\mathcal{A}_r\}$ where $x(w) \epsilon C^0(\bar{\mathcal{A}}_r) \cap C^1(\mathcal{A}_r)$ with finite Dirichlet Integral $D(x)$ such that $x(w)$ on the unit circle is a weakly monotonic representation of γ_2 while on the inner circle $x(w)$ describes γ_1. $\mathcal{S}(\gamma_1, \gamma_2)$ is then the closure of $\mathcal{S}_0(\gamma_1, \gamma_2)$ in the Sobolev space $W^{1,2}$. We now set

$$A_0 = A_0(\gamma_1, \gamma_2) = \inf\{D(x)/2 \text{ for } x \epsilon \mathcal{S}_0(\gamma_1, \gamma_2)\}. \tag{2.5}$$

If the Douglas condition $A_0 < A_d$ is satisfied then there exists an annular type minimizer $x_0 \epsilon \mathcal{S}(\gamma_1, \gamma_2)$ which is a conformally parameterized minimal surface of least area A_0 and satisfying

$$\Delta x = 0 \tag{2.6a}$$

$$|x_u| = |x_v|, \ (x_u \cdot x_v) = 0 \tag{2.6b}$$

The boundary values of $x(w)$ are monotonic representations of γ_1, γ_2 (see Courant [1], for example). (2.6c)

Theorem 2.1. *Let γ_1, γ_2 be a pair of rectifiable oriented Jordan curves and let σ be a connector joining $S \epsilon \gamma_1$ to $T \epsilon \gamma_2$. We consider γ_n, a sequence of rectifiable Jordan curves based on $(\gamma_1, \gamma_2, \sigma)$ satisfying the conditions of Definition 2.1. Let A_d be the infimum of area for the degenerate genus zero Plateau Problem and A_0 be the infimum of area among annular type surfaces with boundary $\gamma_1 \vee \gamma_2$. Suppose that the Douglas Condition $A_0 < A_d$ is satisfied so that both problems have minimizers where σ is not contained in any annular minimizer.*

Let $B_0(\gamma_n)$ be the minimum of area for genus 0 surfaces spanning γ_n and $B_1(\gamma_n)$ be the infimum of area among all genus one surfaces spanning γ_n. We assert that for n sufficiently large $B_1(\gamma_n) < B_0(\gamma_n)$. This implies that the Douglas Condition is satisfied for the genus one problem for γ_n and so there exists a genus one minimizer.

Proof. First we establish the following inequalities.

$$\limsup B_0(\gamma_n) \leq A_d, \quad \limsup B_1(\gamma_n) \leq A_0. \tag{2.7}$$

By the conditions of Definiton 2.1 there is a simply connected ribbon surface $\sum_n$ "filling in" the bridge joining γ_1 to γ_2 with area bounded by $K\epsilon_n \to 0$ where ϵ_n is the width of the bridge. Choose $\sum_1, \sum_2$ as least area simply connected minimizers for the Plateau Problem with boundaries γ_1, γ_2. Now form a simply connected surface with boundary γ_n by joining these surfaces $\Omega_n = \sum_n \cup \sum_1 \cup \sum_2$. The area of Ω_n satisfies

$$|\Omega_n| = |\Sigma_n| + |\Sigma_1| + |\Sigma_2| \leq A_d + K\epsilon_n$$

By Morrey's ϵ-conformal representation theorem, there is a parametric representation $y_n \epsilon \mathcal{S}(\gamma_n)$ of Ω_n so that

$$D(y_n)/2 \leq A_d + (K+1)\epsilon_n.$$

This shows that $B_0(\gamma_n) \leq A_d + (K+1)\epsilon_n$ and thus $\limsup B_0(\gamma_n) \leq A_d$.

Similarly, let $y \epsilon \mathcal{S}(\gamma_1, \gamma_2)$ be an annular minimizer with $D(y)/2 = A_0$. Such y exists by the Douglas condition. Call the surface determined by $y, \sum_0$. Now let $\Omega_n = \sum_0 \cup \sum_n$ be the genus one surface spanning γ_n. Again one can find a parametric representation for Ω_n, say $y_n \epsilon \mathcal{S}(\gamma_1 \vee \gamma_2, G_1)$ where G_1 is a domain of genus one with simply connected boundary such that

$$D(y_n)/2 \leq A_0 + (K+1)\epsilon_n.$$

It follows that $B_1(\gamma_n) \leq A_0 + (K+1)\epsilon_n$ and thus $\limsup B_1(\gamma_n) \leq A_0$.

Now let $x_n \epsilon \mathcal{S}(\gamma_n)$ be a minimizer of the Dirichlet Integral so that $D(x_n) = 2B_0(\gamma_n)$. Furthermore, by choosing a subsequence if necessary we may suppose $\text{limit}_{n\to\infty} D(x_n)/2 = \liminf B_0(\gamma_n)$.

We impose a three point condition on the mappings $x_n(w)$ as follows. Pick Q_1, Q_2, Q_3, Q_4 such that

a) $Q_1, Q_2 \epsilon \gamma_1$ with (Q_1, Q_2, S) following the orientation of γ_1.

b) $Q_3, Q_4 \epsilon \gamma_2$ with (T, Q_3, Q_4) following the orientation of γ_2.

Each Q_i is at a finite distance from the connector σ and it follows that each Q_i lies on the Jordan curve γ_n. Now pick three points q_1, q_2, q_3 on the unit circle (say $q_1 = e^{3\pi i/4}, q_2 = e^{5\pi i/4}, q_3 = e^{-\pi i/4}$) and impose the three point condition $x_n(q_i) = Q_i (i = 1, 2, 3)$ on the mappings $x_n(w)$. Since $x_n|\delta B$ is a strictly monotonic representation of the Jordan curve γ_n there is a unique $q_4^{(n)}$ on δB mapped onto Q_4 by the function $x_n(w)$. On the unit circle the point $q_4^{(n)}$ lies between q_3 and q_1. There are three possibilities.

Possibility 1. For some subsequence (relabeled) we have $q_4^{(n)} \to q_1$. We now apply the Courant-Lebesgue Lemma [1, p. 102] which asserts that given a point $p \epsilon \bar{B}$ and $\delta > 0$ there is a number $\bar{r}, \delta < \bar{r} < \sqrt{\delta}$ so that the length, L of the curve $x_n(C_{\bar{r}})$ where $C_{\bar{r}}$ is the circular arc in B with center p and radius $\bar{r}$ satisfies $L^2 < 8\pi M / \log(1/\delta)$ where M is the value of the Dirichlet integral. Since $q_4^{(n)} \to q_1$ there is a circular arc in $\bar{B}$ connecting a point of (q_1, q_2) to $(q_3, q_4^{(n)})$ of total length approaching zero. This means that the distance from the arc $(Q_1, Q_2) \subset \gamma_1$ is arbitrarily close to the arc $(Q_3, Q_4) \subset \gamma_2$, a contradiction.

Possibility 2. For some subsequence $q_4^{(n)}$ (relabeled) we have $q_4^{(n)} \to q_3$. We apply the Courant-Lebesgue Lemma again. There is a (small) circular arc C_n in $\bar{B}$ connecting a point on (q_2, q_3) to a point on $(q_4^{(n)}, q_1)$ such that the length L_n of $x_n(C_n)$ shrinks to zero. This image curve $x_n(C_n)$ connects a point on the path $(Q_2, Q_3) \subset \gamma_n$ containing the connector σ_n^- to a point on (Q_4, Q_1) containing the connector σ_n^+. We are going to use the arc $x_n(C_n)$ to disconnect the surface $x_n(\bar{B})$ into two pieces. It will allow us to conclude that

$$A_d \leq \liminf B_0(\gamma_n). \tag{2.8}$$

This along with (2.7) and the assumption $A_0 < A_d$ lets us conclude $B_1(\gamma_n) < B_0(\gamma_n)$.

(i) Suppose that $x_n(C_n)$ connects a point P_n^- on σ_n^- to a point P_n^+ on σ_n^+ where $L_n = \text{length}[X_n(C_n)] \to 0$. By the conditions of Definition 2.1 there is an arc Γ_n lying on the bridge $\sum_n$ joining P_n^- to P_n^+ with length $l(\Gamma_n) \leq CL_n$ approaching zero also. Thus there is a simply connected surface G_n with boundary $[X_n(C_n)] \vee \Gamma_n$ such that $\text{Area}(G_n) = |G_n|$ converges to zero. The curve Γ_n cuts $\sum_n$ into two pieces. One piece $\sum_{n,1}$ touches the curve γ_1 while the other piece $\sum_{n,2}$ touches γ_2. By assumption $|\Sigma_{n,i}| \leq |\sum_n| \leq K \cdot \epsilon_n$. There is one simply connected surface with boundary γ_1 consisting of $X_n(B_{n,1}) \vee \sum_{n,1} \vee G_n$ where $B_{n,1}$ is one of the two subregions of B divided by C_n which contains (q_1, q_2, C_n). This surface can be represented by a map $y_n \epsilon S(\gamma_1)$ which is ϵ-conformal and satisfying $D(y_n) < 2A(y_n) + \epsilon$ where

$$A(y_n) \leq [D(x_n, B_{n,1})/2] + |\Sigma_{n,1}| + |G_n|.$$

Similarly there is a map $z_n \epsilon S(\gamma_2)$ with $D(x_n) < A(z_n) + \epsilon$ and

$$A(z_n) \leq [D(x_n, B_{n,2})/2] + |\Sigma_{n,2}| + |G_n|$$

This gives

$$\frac{D(y_n) + D(z_n)}{2} \leq \frac{D(x_n)}{2} + |\Sigma_n| + 2|G_n| + \epsilon$$

where $|\Sigma_n| \to 0$ and $|G_n| \to 0$. This leads to (2.8).

(ii) Suppose that $x_n(C_n)$ connects P_n^- on σ_n^- to P_n^+ on γ_2 so that P_n^+ is on the subarc (Q_4, T_n^+) of γ_2. In this case we can assert that the length of the oriented subarc (P_n^+, T_n^+) of γ_2 has length approaching zero. We attach to $x_n(C_n)$ this sub arc, giving us an arc connecting P_n^- on σ_n^- to T_n^+ on σ_n^+. The length of this enlarged arc still becomes small and we can repeat our previous construction. All other possibilities can be handled in similar fashion. The inequality (2.8) remains true in all cases. When Possibility 2 occurs the theorem follows.

Possibility 3. There is a point q_4 on the arc (q_3, q_1) such that for some subsequence (relabeled) we have $q_4^{(n)} \to q_4$.

We claim that the boundary functions $x_n|\partial B$ are equicontinuous despite the fact that the limit curve of the sequence γ_n is not a rectifiable Jordan curve since both σ_n^+ and σ_n^- converge to σ (although with reverse orientation). First we can reparameterize the x_n so that $x_n(q_i) = Q_i$ $(i = 1, 2, 3, 4)$ without significantly disturbing the Dirichlet Integrals.

Consider $x_n(w)$ restricted to the closed arc $[q_1, q_2, q_3, q_4]$ where we delete the open subarc (q_4, q_1) of ∂B. The image of this arc under the map x_n is the Jordan arc $[Q_4, Q_1, Q_2, Q_3]$ which contains σ_n^- and parts of γ_1, γ_2. This is a Jordan arc which converges in the sense of Frechet to the Jordan arc $(Q_1, S) \vee \sigma \vee (T, Q_4)$ where $(Q_1, S) \subset \gamma_1$ and $(T, Q_4) \subset \gamma_2$. It follows from the Courant-Lebesgue Lemma that the functions $x_n(w)$ converge uniformly when restricted to any closed subarc of (q_1, q_2, q_3, q_4). By a similar argument we conclude that the boundary functions $x_n(w)$ also converge uniformly on any closed subarc of (q_3, q_4, q_1, q_2). It follows that the harmonic functions $x_n(w)$ converge uniformly to a limit harmonic function $x_0(w)$ whose boundary values are a representation of the rectifiable curve consisting of $\gamma_1 \vee \gamma_2$ and the connector σ traversed twice with opposite orientations. The topological boundary of $x_0(w)$ is $\gamma_1 \vee \gamma_2$. Call this boundary $\gamma_0 = \gamma_1 \vee \sigma^- \vee \gamma_2 \vee \sigma^+$.

The surface determined by $x_0(w)$ is actually an annular surface with boundary $\gamma_1 \vee \gamma_2$ which has a crease along the connector σ. It follows that there is an annular mapping $y_0 \epsilon \mathcal{S}(\gamma_1 \vee \gamma_2, \mathcal{A}_r)$ where $\mathcal{A}_r$ is an annular domain so that $D(y_0) = D(x_0)$. But

$$D(x_0) \leq \liminf D(x_n) = \liminf[2B_0(\gamma_n)]$$

This implies that $A_0 \le D(x_0)/2 \le \liminf B_0(\gamma_n)$. However we must have $A_0 < D(x_0)/2$ since otherwise $x_0(w)$ would be an annular minimizer for boundary $\gamma_1 \vee \gamma_2$. Our conditions on σ (that it not lie on any mimizer for this problem) prevent $x_0(w)$ from being a minimizer. Therefore we have

$$A_0 < \liminf B_0(\gamma_n). \tag{2.0}$$

This inequality along with (2.7) shows that $B_1(\gamma_n) < B_0(\gamma_n)$ for n sufficiently large. □

Our next theorem shows that the same result holds if we impose a volume constraint into the problem. If $x(w)$ is a member of $\mathcal{S}_0(\gamma)$ then the oriented volume functional is

$$V(x) = (1/3)\int\int[x \cdot (x_u \wedge x_v)]dudv.$$

This functional extends continuously to $\mathcal{S}(\gamma)$ although it is not weakly continuous there. Properties of this functional and relevant existence theorems can be found in several sources [5, 6]. In particular we use the following result.

For any constant K there is an $x_0 \epsilon \mathcal{S}(\gamma)$ which minimizes the Dirichlet Integral $D(x)$ subject to the constraint $V(x) = K$. $x_0(w) \epsilon C^0(\bar{B}) \cap C^\infty(B)$ and satisfies

$$\Delta x = 2H(x_u \wedge x_v) \text{ for some constant H.} \tag{2.10a}$$

$$x(w) \text{ is a conformal immersion on } B. \tag{2.10b}$$

$$x|\partial B \text{ is a monotonic representation of } \gamma. \tag{2.10c}$$

Thus $x(w)$ is a conformal immersion of a surface with constant mean curvature H.

Theorem 2.2. *Let γ_1, γ_2 be a pair of rectifiable oriented Jordan curves with connector σ joining $S \epsilon \gamma_1$ to $T \epsilon \gamma_2$. Let γ_n be a sequence of rectificable Jordan curves based on $(\gamma_1, \gamma_2, \sigma)$ satisfying the conditions of Definition of 2.1. For any constant K let A_d be the infimum of area for the degenerate volume constrained Plateau Problem for volume $V = K$ and let A_0 be the infimum of area among annular surfaces spanning $\gamma_1 \vee \gamma_2$ enclosing volume $V = K$. Suppose the Douglas condition $A_0 < A_d$ is satisfied so that both problems have minimizers where σ is not contained in any annular minimizer.*

Let $B_0(\gamma_n)$ be the infimum of area for genus zero surfaces spanning γ_n with volume constraint $V = K$ and $B_1(\gamma_n)$ be the infimum of area among all *genus one surfaces spanning γ_n with volume $V = K$. For sufficiently large n we will have $B_1(\gamma_n) < B_0(\gamma_n)$ and so the Douglas condition is satisfied for the genus one volume constrained Plateau Problem.*

Proof. The method of proof is parallel to that of Theorem 2.1. No new essential difficulties appear. For a given constant K we define A_d as in the minimal surface case (2.6) except we impose the constraint $V(x_1)+V(x_2) = K$ where $x_i \epsilon \mathcal{S}(\gamma_i)$. This degenerate problem has a minimizer, being a pair of CMC surfaces of the same mean curvature. Also $A_0 = A_0(\gamma_1, \gamma_2, K)$ is identical to the minimal surface case (2.5) except that we impose the constraint $V(x) = K$. The Douglas condition $A_0 < A_d$ guarantees that there is an annular type minimizer to the VCPP for γ_1, γ_2 with $V(x) = K$.

The quantities $B_0(\gamma_n), B_1(\gamma_n)$ parallel the minimal surface definition except for the volume constraint $V(x) = K$. There is a minimizer $x_n \epsilon \mathcal{S}(\gamma_n)$ where x_n satisfies (2.10) and $B_0(\gamma_n) = D(x_n)/2$.

The proof now parallels the proof of Theorem 2.1. The only change is making higher order adjustments to the volume functional as needed. We shall work out the case Possibility 3 as listed in Theorem 2.1. First we find as before that (2.7) remains true.

Possibility 1 of Theorem 2.1 cannot happen here for the same reasons. Possibility 2 is also similar. One can effect a disconnection leading to the inequality (2.8) which along with (2.7) leads to the desired result.

Now consider Possibility 3. We have $x_n \epsilon \mathcal{S}(\gamma_n)$ with $V(x_n) = K$ and $x_n(w)$ is a solution to the VCPP satisfying (2.10) with $D(x_n) = 2B_0(\gamma_n)$. The boundary functions are equicontinuous and there is a subsequence (relabeled) converging weakly to $x_0(w) \epsilon \mathcal{S}(\gamma_0)$ where γ_0 is just $\gamma_1 \vee \sigma^- \vee \gamma_2 \vee \sigma^+$ which topologically is $\gamma_1 \vee \gamma_2$. $x_0(w)$ is also a solution to (2.10). As in [6] there is a sphere S such that

$$V(x_0) + V(S) = K \tag{2.11a}$$
$$D(x_0) + 2A(S) \leq \liminf D(x_n). \tag{2.11b}$$

There are two possibilities. First suppose the sphere $S = 0$. We have $V(x_0) = K$ and $A(x_0) \leq \liminf B_0(\gamma_n)$. But x_0 is just an annular surface with boundary $\gamma_1 \vee \gamma_2$ with a crease σ. If $A(x_0) = A_0$ then x_0 would provide a minimizer for the annular volume constrained Plateau Problem. This contradicts our condition for choice of the bridge σ. Thus $A_0 < \liminf B_0(\gamma_n)$ and the theorem follows.

If $S \neq 0$ then there is a mapping $x^* \epsilon \mathcal{S}(\gamma_0)$ with $V(x^*) = K$ and $D(x^*) < D(x_0) + 2A(S)$. This is the sphere attaching Lemma [6]. This immediately implies $A_0 < \liminf B_0(\gamma_n)$. □

We go on to our third problem. For convenience we shall focus our discussion on the following specific curves.

Boundary Curves for Problem 3. Let γ_1 be the unit circle in the $x-y$ plane whose equation is $(x+a)^2 + y^2 = 1$ with center $(-a, 0)$ lying in the left half space $x < 0$ while its reflective image about the plane $x = 0$ is γ_2 given by $(x-a)^2 + y^2 = 1$ where $a > 1$. The connector σ is the line segment on the x-axis joining the two circles with length $l = 2(a-1)$ while γ_n will be the rectifiable Jordan curve based on $(\gamma_1, \gamma_2, \sigma)$ where σ_n^+ is a line segment on $y = \epsilon_n = 1/n$ and σ_n^- lies on $y = -\epsilon_n = -1/n$, as in Definition 2.1. We note that $\gamma_1 \vee \gamma_2$ and γ_n are reflectively symmetric about the vertical planes $x = 0$ and $y = 0$ in R^3.

We always assume that our surfaces are symmetric about the plane $y = 0$. Observe that if $\sum$ were an embedded cmc surface with boundary $\gamma_1 \vee \gamma_2$ or γ_n which is contained in the upper half space then it must have this symmetry via an Alexandrov reflection argument [8]. We want to demonstrate the existence of cmc surfaces spanning γ_n which will have this symmetry about $y = 0$ but not be symmetric about the plane $\Pi : \{x = 0\}$.

Definition 2.4 (The Degenerate Minimizers). As in the statement of Theorem 2.2, for a given constant K we denote by A_d the minimum area for the pair of spherical caps $\sum_1, \sum_2$ which solve the degenerate volume constrained Plateau problem where γ_1, γ_2 are unit circles, where $V_1 + V_2 = K$ and $|\sum_1| + |\sum_2| = A_d$. We denote by A_s the solution to the same VCPP except that we also impose symmetry about the plane Π. For any K the values A_d, A_s are independent of the distance between γ_1 and γ_2.

We *now assume* that the total volume $V = K$ is *greater than* $4\pi/3$ which implies that the minimizer giving A_d is not symmetric but consists of one large spherical cap (say at γ_1) and one small cap at γ_2. The minimizer for the symmetric problem will be two large spherical caps and so $A_d < A_s$. Note that in this case one might allow the symmetric minimizer to be two small symmetric caps spanning γ_1, γ_2 with a large complete sphere in the middle. This would be the symmetric minimizer when the exposed volume is large.

Even if we allow this configuration as a symmetric minimizer we still will have $A_d < A_s$.

We now consider annular type cmc surfaces which span $\gamma_1 \vee \gamma_2$.

Lemma 2.1. *Let M be any constant. If the distance between the Jordan curves γ_1, γ_2 is sufficiently large then the area of any annular type cmc surface spanning $\gamma_1 \vee \gamma_2$ has a value greater than M.*

Proof. Step 1: Let $x(w) : A \to R^3$ be a conformally parameterized cmc surface with mean curvature H where A is an annular domain such that the boundary values of $x(w)$ describe γ_1 and γ_2. Then H is bounded by

$$2|H| \leq C/A = 2 \text{ or } |H| \leq 1 \tag{2.12}$$

where $C = C(\gamma_1, \gamma_2)$ is the circumference of $\gamma_1 \vee \gamma_2$ and $A = A(\gamma_1, \gamma_2)$ is the area of the two regions in the $x - y$ plane bounded by γ_1, γ_2.

Proof of Step 1. $x(w)$ is a solution to (2.10). Therefore it's z-component function satisfies

$$\Delta z = 2HJ(x, y) = 2H(x_u y_v - x_v y_u). \tag{2.13}$$

Integrate this over the annular region $\mathcal{A}$ keeping in mind that

$$\int\int J(x, y)dudv = A.$$

We find

$$2HA = \int z_v ds.$$

Now we use the conformality of the mapping (2.10b) and inequality (2.12) follows.

Step 2: Let $l = 2a - 2 = 2n\delta$ with $\delta > 0$ and divide the region between γ_1 and γ_2 into n subregions bounded by vertical planes parallel to $x = 0$ of width 2δ each. Suppose $\sum$ is a connected cmc surface with boundary $\gamma_1 \vee \gamma_2$ and mean curvature $0 \leq H \leq 1$. Since $\sum$ must pass through each of the n strip regions of thickness 2δ one can pick a point on the surface which lies at the middle of each of these regions and estimate by the monotonicity formula the area $A(R)$ of that part of $\sum$ inside a ball of radius R centered at the selected midpoints. We find

$$A(R)e^{2RH} \geq \pi R^2 \quad 0 < R \leq \delta. \tag{2.14}$$

Since $0 \leq H \leq 1$ we have $A(\delta) \geq (\pi\delta^2)e^{-2\delta}$. We have n such estimates. This gives us

$$|\sum| \geq (\pi l/2)(\delta e^{-2\delta}).$$

Finally we choose $2\delta = 1$ and we get $|\sum| \geq (\pi l)/(4e)$. The Lemma now follows.

Suppose that the total volume $V > 4\pi$ so that $A_d < A_s$ as in Definition 2.4 and that $l = 2a - 2$ is sufficiently large so that the area of any connected cmc surface spanning $\gamma_1 \vee \gamma_2$ has area greater than A_s (and also A_d).

Theorem 2.3. *Let $\gamma_1, \gamma_2, \gamma_n$ be the set of boundary curves as stipulated for Problem 3. Suppose the volume $V = K > 4\pi$ is chosen so that $A_d < A_s$ and that the distance between γ_1, γ_2 is so large that any annular cmc surface with boundary $\gamma_1 \vee \gamma_2$ has area greater than A_s. Let $\mathcal{S}(\gamma_n)$ be those vector functions of finite Dirichlet integral spanning γ_n where we also impose the equivariance condition*

$$x(u, -v) = R_1 \cdot x(u, v)$$

where $R_1(x, y, z) \equiv (x, -y, z)$ so that $x(w)$ is symmetric with respect to reflection about the y-axis. Let $\mathcal{T}(\gamma_n)$ be those members of $\mathcal{S}(\gamma_n)$ which also satisfy

$$x(-u, v) = R_2 \cdot x(u, v)$$

where $R_2(x, y, z) = (-x, y, z)$ so that $x(w)$ is also symmetric about the plane Π.

For a given constant K let

$$B_0(\gamma_n) = \min\{D(x)/2 \textit{ for } x \epsilon \mathcal{S}(\gamma_n), V(x) = K\}$$
$$B_s(\gamma_n) = \min\{D(x)/2 \textit{ for } x \epsilon \mathcal{T}(\gamma_n), V(x) = K\}$$

We claim that $B_0(\gamma_n) < B_s(\gamma_n)$ for large n. The minimizer for the volume contained Plateau problem is not symmetric about Π.

Proof. As in the proof of the previous theorems one can easily show that

$$\limsup B_0(\gamma_n) \leq A_d, \quad \limsup B_s(\gamma_n) \leq A_s. \tag{2.15}$$

Suppose we have a sequence of minimizers $x_n(w)$ for the symmetric problem so that $B_s(\gamma_n) = D(x_n)/2$. Since $x_n(w)$ is equivariant with respect to both reflections we already have imposed a 4-point condition, $x(p_i) = P_i (i = 1, 2, 3, 4)$ in an obvious manner. Now let Q_1 be a point on γ_1 in the second quadrant, Q_2 its symmetric point in the third quadrant, Q_4 its symmetric point in the first qudrant, and $Q_3 = -Q_1$. Let $q_i^{(n)}$ on the unit circle be chosen as preimages under x_n of Q_i.

Due to the equivariance of the mappings $x_n(w)$ the $q_i^{(n)}$ share the same symmetry as Q_i. Suppose that we have chosen a subsequence (relabeled) so that

$$\text{limit} D(x_n)/2 = \liminf B_s(\gamma_n)$$

There are three possibilities as in the proof of the earlier theorems.

Possibility 1. $q_1^{(n)} \to (0,1)$. In this case $q_2^{(n)} \to (0,-1)$, $q_3^{(n)} \to (0,-1)$, $q_4^{(n)} \to (0,1)$ as well. By the Courant Lemma there is a circular arc C_n with center at $(0,1)$ whose image $\Gamma_n = x(C_n)$ has length approaching zero and which connects a point of γ_1 to a point of γ_2. This is not possible.

Possibility 2. There is a subsequence $q_1^{(n)} \to (-1,0)$ and thus $q_2^{(n)} \to (-1,0), q_3^{(n)} \to (1,0), q_4^{(n)} \to (1,0)$. In this case there are two circular arcs C_n, C_n' one centered at $(-1,0)$ and the other at $(1,0)$ symmetric with respect to the plane $u = 0$, such that the lengths of their image curves $x_n(C_n) = \Gamma_n$ are shrinking to zero. C_n will connect a point on the oriented subarc of $\gamma_n, [(0,1/n), Q_1]$ to its reflected point about the y-axis. This arc will lie in the plane $x < 0$. $\Gamma_n' = x(C_n')$ is a similar arc of short length in the right half plane. As in the proof of Theorem 2.2 these two arcs serve to disconnect the surface $x_n(w)$ into three pieces. One is a simply connected piece with boundary γ_1, the second is a symmetric piece spanning γ_2 while the middle piece will be a closed surface. Replace this middle piece by a round sphere of the same volume. Repeat the argument as presented in the proof of the earlier theorems and we get

$$A_s \leq \liminf B_s(\gamma_n). \tag{2.16}$$

This along with (2.15) and the inequality $A_d < A_s$ establishes the theorem in this case.

Possibility 3. $q_1^{(n)} \to q_1$ a point on the unit circle in the second quadrant.

In this case (as in Theorem 2.1) it follows that the boundary values are equicontinuous.

There is a limit function $x_0(w)$ which is equivariant with respect to both reflection maps, whose boundary values will describe the curve $\gamma_1 \vee \sigma^- \vee \gamma_2 \vee \sigma^+$ where the boundary σ is traversed twice so that its topological boundary is $\gamma_1 \vee \gamma_2$. We also get that there is a sphere S such that

$$V(x_0) + V(S) = K \tag{2.17a}$$

$$D(x_0) + 2A(s) \leq \liminf D(x_n). \tag{2.17b}$$

The function $x_0(w)$ is a conformally parameterized surface of constant mean curvature satisfying (2.10). We may regard $x_0(w)$ as an annular cmc surface symmetric with respect to both planes with boundary γ_1, γ_2 containing a crease along the connector σ.

Let W_s be the infimum of the areas of annular type surfaces with boundary $\gamma_1 \vee \gamma_2$ and symmetric in the planes $x = 0$ and $y = 0$.

Suppose that the sphere S is zero in (2.18). Then $D(x_0) \leq \text{limit} D(x_n) = \liminf 2B_s(\gamma_n)$. But $V(x_0) = K$ and we conclude that $W_s \leq \liminf B_s(\gamma_n)$. Now $W_s \leq A_s$. If $W_s = A_s$ then we conclude that $A_s = \text{limit} B_s(\gamma_n)$ and so $B_0(\gamma_n) < B_s(\gamma_n)$ for large n using (2.15). If $W_s < A_s$ then the Douglas condition is satisfied for the class of symmetric annular surfaces spanning $\gamma_1 \vee \gamma_2$. There will be an annular type minimizer $z_0(w)$, a cmc surface with boundary $\gamma_1 \vee \gamma_2$. By our choice of γ_n we know that any annular type cmc surface with boundary $\gamma_1 \vee \gamma_2$ has large Dirichlet Integral. Such a surface cannot produce a minimizer. We must have $W_s = A_s$.

Finally if $S \neq \Phi$ then there exist a $y_0(w)$ equivariant with respect to the full reflection group with the same boundary data as $x_0(w)$ satisfying

$$V(y_0) = K$$
$$D(y_0) < D(x_0) + 2A(S) \leq \liminf D(x_n)$$

This immediately implies that $D(y_0) < \liminf B_s(\gamma_n) \leq \limsup B_s(\gamma_n) \leq A_s$, leading to the Douglas condition $W_s < A_s$ again. We have a contradiction. $\square$

References.

[1] Courant, R. *Dirichlet's Principle, Conformal Mapping, and Minimal Surfaces*, Interscience, New York 1950.

[2] Douglas, J. *Minimal Surfaces of Higher Topological Structure.* Proc. Natl. Acad. Sci. USA 24.343-353 (1938).

[3] Fomenko, A.T. *The Plateau Problem, Historical Survey I.* Gordon & Breach (translated from Russian). 1989.

[4] Patnaik, Umadhar, *Volume Constrained Douglas Problem and the Stability of Liquid Bridges between Two Coaxial Tubes*, Dissertation, Univ. of Toledo (1994).

[5] Steffen, K. *Flachen konstanter mittlerer Krummung mit vorgegebenem Volumen oder Flacheninhalt.* Arch. Ration. Mech. Anal. 49, 99-128 (1972).

[6] Wente, H. *A General Existence Theorem for Surfaces of Constant Mean Curvature*, Math. Z. 120, 277-288 (1971).

[7] Wente, H. *The Plateau Problem for Symmetric Surfaces.* Arch. Ration. Mech. Anal. 60, 149-169 (1976).

[8] Wente, H. *The Symmetry of Sessile and Pendant Drops*, Pac. Jour. Math. Vol. 88, No. 2, 387-398 (1980).

RECEIVED NOVEMBER 8, 1995.

DEPARTMENT OF MATHEMATICS
UNIVERSITY OF TOLEDO
TOLEDO, OHIO 43606

Half of Enneper's Surface Minimizes Area

Brian White [1]

Let C be a smooth embedded closed curve in $\mathbf{R}^3$ (or more generally in a smooth riemannian 3-manifold) and let S be an area minimizing disk with boundary C. It has long been known (from the work of Douglas, Rado, Morrey [M2], Hildebrandt [H], and Heinz [HH]) that S must be a smooth immersion except, perhaps, at finitely many branch points. Around 1970, Osserman [O1], Gulliver [G], and Alt [A1, A2] proved that such branch points could not occur in the interior of S. It is still not known, however, whether such branch points can occur at the boundary.

(On the other hand, if S is a surface that minimizes area among all smooth oriented surfaces with boundary C, then S must be a smooth embedded surface with no singularities anywhere [F, HS].)

In 1973, Gulliver and Lesley [GL] proved that such boundary branch points do not occur if the boundary curve is analytic. Their result raises the question: can a nonplaner complete area minimizing surface of finite total curvature with a straight line boundary be area minimizing? For if the answer were no, one could show, using the Gulliver-Lesley theorem, that boundary branch points are impossible in area minimizing disks (see the discussion at the end of this paper).

This paper proves that there is such a complete area minimizing surface, namely half of Enneper's surface.

Theorem 1. *Half of Enneper's surface minimizes area.*

Proof. Let K and K' be smooth convex closed curves in the upper half plane $\{(x,y) : y \geq 0\}$ such that the intersection of K and K' is an interval I on the x-axis.

Let $\gamma : [0,4] \to \mathbf{R}^2$ be a smooth closed curve such that

(1) $\gamma|[0,1]$ traces out I,

[1] The author was partially funded by NSF grants DMS-9207704 and DMS 9504456

(2) $\gamma|(1,2)$ traces out $K \setminus I$,

(3) $\gamma|[2,3]$ traces out I again, and

(4) $\gamma|(3,4)$ traces out $K' \setminus I$.

Let π and π' be the orthogonal projections from $\mathbf{R}^3$ onto the xy-plane and the z-axis, respectively. For $n = 1, 2, 3, \ldots$, let $\gamma_n : [0,4] \to \mathbf{R}^3$ be a smooth closed curve such that

(1) $\pi \circ \gamma_n \equiv \gamma$,

(2) $\pi' \circ \gamma_n|[0,1] \equiv 0$,

(3) $\pi' \circ \gamma_n|[1,2]$ increases smoothly from 0 to $1/n$,

(4) $\pi' \circ \gamma_n|[2,3] \equiv 1/n$, and

(5) $\pi' \circ \gamma_n|[3,4]$ decreases smoothly from $1/n$ to 0.

Let D_n be an area-minimizing disk with boundary γ_n. Since the total curvature of γ_n is slightly less than 4π, D_n has no branch points. (This is by the Gauss-Bonnet theorem —any interior or boundary branch point contributes at least 2π to the total curvature $|\int_{D_n} K\,dA|$. See [N, §380] for details.)

Let K_n be the largest principal curvature on D_n, and let $p_n \in D_n$ be a point at which the principal curvature is K_n. If a subsequence of the K_n were bounded, then the corresponding subsequence of the D_n's (or a further subsequence thereof) would converge smoothly to an immersed minimal disk D with boundary γ [W1]. By the convex hull property, D would lie in the xy-plane. But by elementary topology, γ does not bound any immersed disk in the xy-plane. (One way to prove this is by the Gauss-Bonnet theorem: if it bounded an immersed disk Δ, then

$$2\pi = \int_{\partial\Delta} k + \int_{\Delta} K = 4\pi + 0,$$

a contradiction.) Thus $K_n \to \infty$.

Translate D_n by $-p_n$ and dilate by K_n to get a new sequence of area minimizing surfaces D'_n such that

(1) $0 \subset D'_n$,

(2) The principal curvatures of the D'_n are ≤ 1 everywhere and $= 1$ at the origin,

(3) The total curvature of D'_n is $< 2\pi$ (by Gauss-Bonnet).

It follows (see [W1]) that a subsequence of the D'_n (which for notational simplicity we will assume to be the original sequence) converges smoothly on compact subsets of $\mathbf{R}^3$ to a surface D' such that

(1) D' has total curvature $\leq 2\pi$,

(2) D' has nonzero curvature at the origin,

(3) D' is area minimizing,

(4) D' has quadratic area growth. That is, $\text{area}(D' \cap \mathbf{B}(x,r))/r^2$ is bounded. (This follows from the monotonicity formula applied to the D_n; see remark 4 at the end of this paper.)

Note since D_n is contained in a slab of width $1/n$, D'_n is contained in a slab of width K_n/n, so D' is contained in a slab of width $\delta = \liminf K_n/n$. This limit cannot be 0 since D' is not flat.

Note that the boundary of D' must be a limit of dilates of the γ_n. In particular, $\partial D'$ must be either

(1) the empty set,

(2) two parallel lines (δ apart), or

(3) one line L.

Case (1) is impossible because there are no complete stable orientable minimal surfaces without boundary in $\mathbf{R}^3$ [DP, FS].

Case (2) is also impossible. For suppose not. Then (after a rigid motion and dilation) we may assume D' lies in the region

$$W = \{(x,y,z) : x \geq 0, \quad |y| \leq 1\}$$

and that the boundary of D' consists of the two parallel lines L_1 and L_{-1} (oriented in the same way), where

$$L_i = \{(x,y,z) : x = 0, \quad y = i\}$$

Note that the function

$$p \mapsto \nu(p) \cdot \mathbf{e}_3 \tag{$*$}$$

is a bounded jacobi field on D' that vanishes on $\partial D'$. (Here $\nu(p)$ is the unit normal to D' at p.) Since D' is stable, the lowest eigenfunction of the jacobi operator does not change sign. (Since D' is non-compact, to prove this requires using a cut-off function; see [W2, §5.1].) Thus the function (*) does not vanish except at the boundary.

Define a function $\theta(z)$ by

$$\nu((0,1,z)) = (\cos\theta(z), \sin\theta(z), 0)$$

where ν is the unit normal to D'. Note that the range of $\theta(\cdot)$ is an interval (α, β) with $0 \le \alpha < \beta \le \pi/2$. Since the function (*) does not change sign on D', the function $\theta(\cdot)$ is monotonic, say increasing.

Let D'' be a subsequential limit (as $k \to \infty$) of the surfaces $D' + k\mathbf{e}_3$ obtained by translating D' by k in the positive z-direction. Note that D'' lies in W, has L_{-1} and L_1 as boundary, and that all along L_1, its unit normal is $(\cos\alpha, \sin\alpha, 0)$. It follows that D'' is flat. Since it lies in the half-slab W, it must either be a pair of half planes (parallel to the xz plane) or the strip $\{(0,y,z) : |y| \le 1\}$ joining L_{-1} and L_1. The latter is impossible, because L_1 and L_{-1} are oriented in the same direction. Thus D'' consists of half planes parallel to the xz plane. That means $\alpha = \pi/2$, which means (since $\alpha \le \theta(\cdot) \le \beta \le \pi/2$) that $\theta(\cdot)$ is identically $\pi/2$ along L_1, which means that D' is flat, a contradiction.

Thus the boundary of D' must be a straight line L. If we reflect D' throught L, we get a complete non-flat minimal surface of finite total curvature $\le 4\pi$. The only such surfaces are the catenoid and Enneper's surface [O2, 9.4]. Since the catenoid does not contain a straight line, D' must be half of Enneper's surface. □

Remarks.

1. If (contrary to the result in this paper) a half-plane were the only complete area minimizing surface with straight-line boundary and finite total curvature, then we would have a proof that an area minimizing disk D with smooth embedded boundary could not have boundary branch points. The argument is as follows. By pushing the boundary of D in along D along some portion of ∂D without boundary branch points, we may assume that D is the unique area minimizing disk with its boundary $\partial D = \Gamma$. Let Γ_i be a sequence of real analytic closed curves that converge smoothly to D, and for each i, let D_i be an area minimizing disk with boundary Γ_i. By the theorems of Gulliver, Alt, and Osserman, D_i has no interior branch points, and by the theorem of Gulliver-Lesley, it has no boundary branch points, either. If the principal curvatures of the D_i are uniformly bounded, then a subsequence of the D_i will converge smoothly to a smooth immersed area minimizing disk D^* with $\partial D^* = \Gamma = \partial D$. Since D is uniquely minimizing, $D^* = D$, so D has no boundary branch points.

Thus if D has boundary branch points, the principal curvatures of the D_i must be blowing up. Now by translating and dilating the D_i and passing to a limit as in the proof of theorem 1, we get a complete area minimizing surface with straight line boundary and finite total curvature.

Of course the fact that half of Enneper's surface minimizes area does not imply that boundary branch points do occur. But it does mean that no local curvature estimates will rule them out.

2. Let $M \subset \mathbf{R}^3$ be a complete area minimizing surface with straight line boundary. I conjecture that if M has finite total curvature or if it has quadratic area growth, then M must be half of Enneper's surface. This is not true without the hypothesis on total curvature or area growth, since the half helicoid is absolutely area minimizing (even as an integral current).

3. So far in this paper, "area minimizing" has been used in the classical sense. That is, the allowed comparison surfaces are those obtained by compactly supported deformations that vanish on the boundary. Does half of Enneper's surface also minimize area as an integral current? The following theorem gives some information.

Theorem 2. *There exists a non-planar area minimizing integral current in* $\mathbf{R}^3$ *with straight line boundary and quadratic area growth.*

The proof is as in the proof of theorem 1, except that one lets D_n be the area minimizing integral current with boundary γ_n. As before, one gets a limit surface D'. One does not know that D' has finite total curvature, but D' does have quadratic area growth In fact, if one applies the monotonicity formula carefully (see remark 4 below), one sees that

$$\lim_{R\to\infty} \frac{\text{area}(D' \cap \mathbf{B}(0,R))}{\pi R^2} \leq \frac{3}{2}.$$

Of course if the conjecture in remark 2 above is correct, then the surface whose existence is asserted in theorem 2 can only be half of Enneper's surface.

4. The well-known monotonicity formula for minimal surfaces is usually stated for balls that do not intersect the boundary curve. To conclude quadratic area growth in the proof of theorem 1, we need the monotonicity formula for arbitrary balls, namely:

Monotonicity Theorem. *Let D be an m-dimensional area minimizing surface with boundary Γ. Let E be the exterior cone over Γ:*

$$E = \{rx : r \geq 1, \quad x \in \Gamma\}$$

Then

$$\frac{\text{area}(D \cap \mathbf{B}(0,r)) + \text{area}(E \cap \mathbf{B}(0,r))}{r^m}$$

is an increasing function of r.

The proof of monotonicity via cone-comparison (cf. [M1, §9.3]) also establishes this more general monotonicity.

References.

[A1] H. W. Alt, *Verzweigungspunkte von H-Flächen, I*, Math. Z. **127** (1972), 333–362.

[A2] ____________, ____________-*II*, Math. Ann. **201** (1973), 33–55.

[DP] M. doCarmo and C.-K. Peng *Stable complete minimal surfaces in $\mathbf{R}^3$ are planes*, Bull. Amer. Math. Soc. **1** (1979), 903–906.

[F] W. Fleming *On the oriented plateau problem*, Rend. Circ. Mat. Palermo **11** (1962), 1–22.

[FS] D. Fischer-Colbrie and R. Schoen *The structure of complete stable minimal surfaces in , 3-manifolds*, Comm. Pure Appl. Math. **33** (1980), 199-211.

[G] R. Gulliver *Regularity of minimizing surfaces of prescribed mean curvature*, Ann. of Math. **97** (1973), 275–305.

[GL] R. Gulliver and F. Lesley *On the boundary branch points of minimizing surfaces*, Arch. Rat. Mech. Anal. **52** (1973), 20–25.

[HS] R. Hardt and L. Simon *Boundary regularity and embedded solutions for the oriented plateau problem*, Ann. of Math. **110** (1979), 439–486.

[H] S. Hildebrandt *Boundary behavior of minimal surfaces*, Arch. Rat. Mech. Anal. **35** (1969), 47–82.

[HH] E. Heinz and S. Hildebrandt *Some remarks on minimal surfaces in riemannian manifolds*, Comm. Pure Appl. Math. **23** (1970), 371–377.

[M1] F. Morgan *Geometric Measure Theory: A Beginner's Guide*, Academic Press 1988.

[M2] C. B. Morrey *The problem of plateau on a riemannian manifold*, Ann. of Math. **49** (1948), 807–851.

[N] J. C. C. Nitsche *Lectures on minimal surfaces, I*, Cambridge Univ. Press 1989.

[O1] R. Osserman *A proof of the regularity everywhere of the classical solution to Plateau's problem*, Ann. of Math. **91** (1970), 550–569.

[O2] ____________*A survey of minimal surfaces*, 2nd edition, Dover 1986.

[W1] B. White *Curvature estimates and compactness theorems in 3-manifolds for surfaces that are stationary for parametric elliptic functionals* , Invent. Math. **88** (1987), 243–256.

[W2] ____________*Existence of smooth embedded surfaces of prescribed genus that minimize parametric even elliptic functionals on three-manifolds*, J. Diff. Geom. **33** (1991), 413–443.

Received June 16, 1995.

Mathematics Department
Stanford University
Stanford CA 94305
E-mail address: white@math.stanford.edu

Foliation by constant mean curvature spheres on asymptotically flat manifolds

Rugang Ye

Dedicated to the 60th birthday of Prof. S. Hildebrandt

0. Introducton.

The main result of this paper is the following.

Main Theorem. *Let M^{n+1} be an asymptotically flat manifold, $n \geq 2$, and Ω an end of M having nonzero mass. Then there is on Ω a smooth codimension one foliation $\mathcal{F}_0$ by constant mean curvature spheres. $\mathcal{F}_0$ is balanced and regular at ∞. Moreover, it is the unique weakly balanced and regular C^2 foliation on Ω by closed hypersurfaces of constant mean curvature.*

"Balanced at ∞" means that near ∞ the leaves approach geodesic spheres of a fixed center. "Weakly balanced" roughly means that the "geodesic centers" of the leaves do not shift to ∞ as fast as the farthest points on the leaves. "Regular at ∞" means that the rescaled second fundamental form of the leaves is uniformly bounded. For the precise definitions we refer to the next section. Note that $\mathcal{F}_0$ actually foliates $\overline{\Omega \setminus K}$ for a compact region K. In the statement of the theorem the phrase "on Ω" is used in a more general way than standard. For the precise meaning of uniqueness we refer to the Uniqueness Theorem in §2.

We obtained the existence part of this result (and a somewhat weaker uniqueness result) at the end of 1988. (In [Ye1], we showed that around a nondegenerate critical point of the scalar curvature function in a Riemannian manifold, there exists a unique regular foliation by constant mean curvature spheres. We mentioned that the arguments extend to yield the existence result stated in the Main Theorem.) Then we obtained the uniqueness part of the Main Theorem. Recently there has been more interest in this problem and several colleagues have inquired about the details of the proof of this result. Also recently, we showed in [Ye2] that in dimension 3 all *diameter-pinched* (see Definition 5) C^2 foliations by constant mean curvature spheres are regular. Since "diameter-pinched" implies "weakly balanced", the Main Theorem implies the following

Strong Uniqueness Theorem. *Let M be a 3-dimensional asymptotically flat manifold, then on each end of nonzero mass of M there is a unique diameter-pinched C^2 foliation by constant mean curvature spheres.*

For details we refer to [Ye2]. We remark that the "diameter-pinched" condition is (much) weaker than the "balanced" condition.

Asymptotically flat manifolds arise in general relativity. The significance of the above results is to provide a canonical and regular geometric structure on asymptotically flat manifolds. Though the usual concept of asymptotical flatness is sufficient for applications, it is expressed in terms of coordinates and hence is not geometrically canonical. (This is a problem on S. T. Yau's list [Ya] of open problems in differential geometry. Note that in [B] a good understanding of the asymptotical coordinates is provided.) The canonical geometric structure provided by the above results serves to make the concept of asymptotical flatness more geometrical and canonical. In particular, it immediately makes the mass a geometric invariant. It also serves as a geometric linkage between different asymptotical coordinates by the way of its construction. Indeed, it should be possible to completely characterize asymptotically flat ends of nonzero mass in terms of balanced and regular foliations by constant mean curvature spheres. One also expects further applications in general relativity. A philosophical implication is a concept of "center of universe" which may be defined as the region surrounded by the foliations.

The program of constructing foliations by constant mean curvature spheres on asymptotically flat manifolds was initiated by S. T. Yau. He and G. Huisken have an independent proof of existence of foliations by constant mean curvature spheres on asymptotically flat ends of *positive* mass [HY1][HY2]. They apply the mean curvature flow to deform Euclidean spheres in asymptotical coordinates. Positive mass implies a stability estimate which is employed to show convergence of the flow. On the other hand, Yau and Huisken [HY2] showed that on a 3-dimensional asymptotically flat manifold of positive mass, the constructed foliations is the unique foliation by *stable* spheres of constant mean curvature. In comparison, our existence and uniqueness results hold for both positive and negative mass and in all dimensions. But we note that our conditions on foliations for the uniqueness are rather different from that of Yau–Huisken. Their condition is stability as just mentioned. Our conditions are weak balance and regularity in general dimensions, and diameter pinching in dimension 3. These have a different flavor than Yau and Huisken's stability uniqueness. It is unknown whether uniqueness holds without any condition. In the local Riemannian situation, we do have such a universal uniqueness result in dimension 3, see [Y2]. But the situation of asymptotically flat ends is more subtle, see the discussion below.

Now we would like to discuss the proof of the Main Theorem. There are several delicate aspects here. It is natural to attempt to perturb Euclidean spheres in asymptotical coordinates in order to construct constant mean curvature spheres. But a naive application of the implicit function theorem does not work, because the linearized operator of the constant mean curvature equation in the problem has a nontrivial kernel. To resolve this difficulty, we apply the crucial idea in [Ye1] of moving centers. Namely we first perturb the center of the asymptotical coordinates and then perform normal perturbation of the Euclidean spheres. Both in [Ye1] and here the effect of the center perturbations is asymptotically degenerate. We remove this degeneracy by carefully expanding the equation and balancing the center perturbations against the normal perturbations. On the other hand, it is important to control the magnitude of the center perturbations in order to retain the foliation property. The situation of asymptotically flat ends is more delicate than the local picture in [Ye1], because the asymptotical flat structure deteriorates when the center is shifted too far away. This problem is even more serious for uniqueness than for existence, and it is the reason for the requirement of weak balance. A priori, a foliation by constant mean curvature spheres can differ dramatically from the one we constructed. We have to obtain strong geometric control of the leaves in order to derive uniqueness. Note that weak balance is a fairly weak geometric condition. We think that it is necessary for uniqueness.

We acknowledge interesting discussions with G. Huisken.

1. Moving Centers and Perturbation.

Since we deal with ends of asymptotically flat manifolds, it is convenient to introduce the concept of asymptotically flat ends. An asymptotically flat manifold is then a complete Riemannian manifold which is the union of a compact region and finitely many asymptotically flat ends.

Definition 1. Let M be a complete Riemannian manifold of dimension $n+1$ with $n \geq 2$. Let g be the metric of M. A closed domain Ω of M is called an *asymptotically flat end* if there is a coordinate map from Ω to $\mathbb{R}^{n+1} \setminus \mathring{\mathbb{B}}_{R_0}(o)$ for some $R_0 > 0$ such that on this coordinate chart the metric g can be written

$$g_{ij}(x) = \left(1 + \frac{\sigma}{r^{n-1}}\right) \delta_{ij} + h_{ij}(x), \tag{1.1}$$

where $r = |x|$, σ is a constant and the h_{ij}'s satisfy

$$h_{ij} = O\left(\frac{1}{r^n}\right), \; h_{ij,k} = O\left(\frac{1}{r^{n+1}}\right), \; h_{ij,k\ell} = O\left(\frac{1}{r^{n+2}}\right),$$
$$h_{ij,k\ell m} = O\left(\frac{1}{r^{n+3}}\right), \; h_{ij,k\ell mm'} = O\left(\frac{1}{r^{n+4}}\right),$$

as $r \to \infty$ ($h_{ij,k}$ etc. denote partial derivatives, e.g. $h_{ij,k} = \frac{\partial}{\partial x^k} h_{ij}$). The constant σ is called the *mass* or *energy* of Ω. (This differs from the usual definition [LP] by a dimensional factor.)

Note that more general concept of asymptotically flat manifolds have been introduced in the literature, see e.g. [LP] and [Ye2]. But asymptotically flat manifolds as defined here are the most important. (See e.g. [SY]. For technical reasons we required decay of up to the fourth order derivatives of h_{ij}. This same condition is also assumed in [HY1] and [., HY2]) Hence we focus on them in this paper.

Definition 2. Let $\mathcal{F}$ be a foliation of codimension 1 on an asymptotically flat end Ω, whose leaves are all closed. We say that $\mathcal{F}$ is *balanced at* ∞ or *balanced*, if there is a point $p \in \Omega$ such that

$$\frac{\operatorname{dist}(p, S)}{\operatorname{diam}(p, S)} \to 1 \text{ for } S \in \mathcal{F} \text{ as } \operatorname{dist}(p, S) \to \infty,$$

where $\operatorname{diam}(p, S) = \max\limits_{q \in S} \operatorname{dist}(p, q)$.

Definition 3. Let $\mathcal{F}$ be as above. For $S \in \mathcal{F}$, let Ω_S be the open domain on the outside of S. Put

$$s(S) = \max_{p \in \Omega \setminus \Omega_S} \frac{\operatorname{dist}(p, S)}{\operatorname{diam}(p, S)}.$$

A point $p \in \Omega \setminus \Omega_S$ is called *geodesic center* of S, if the maximum $s(S)$ is achieved at p. We say that $\mathcal{F}$ is *weakly balanced*, if there are geodesic centers $p(S)$ of S such that

$$\limsup \frac{\operatorname{dist}(p_0, p(S))}{\operatorname{diam}(p(S), S)} < 1 \text{ for } S \in \mathcal{F} \text{ as } \operatorname{diam}(p_0, S) \to \infty,$$

where p_0 is a fixed point in Ω. We shall denote the above limit by $b(S)$.

Definition 4. Let $\mathcal{F}$ be as above. Furthermore, assume that $\mathcal{F}$ is of class C^2. We say that $\mathcal{F}$ is *regular at* ∞ or *regular*, if

$$\limsup_{\operatorname{diam}(S)\to\infty} \|A_S\|_{C^0(S)} \operatorname{diam}(S) < \infty \text{ for } S \in \mathcal{F},$$

where A_S denotes the second fundamental form of S.

Definition 5. Let $\mathcal{F}$ be as above. We say that $\mathcal{F}$ is *diameter-pinched at* ∞ or *diameter-pinched*, if

$$\limsup \frac{\operatorname{diam}(p_0, S)}{\operatorname{dist}(p_0, S)} < \infty \text{ for } S \in \mathcal{F} \text{ as } \operatorname{diam}(p_0, S) \to \infty,$$

where p_0 is a fixed point in Ω.

Let Ω be an asymptotically flat end of dimension $n+1$, $n \geq 2$, whose mass σ is nonzero. We can identify Ω with $\mathbb{R}^{n+1} \setminus \mathring{\mathbb{B}}_{R_0}$ for some $R_0 > 0$. Without loss of generality, we assume $R_0 = 1$. To construct a balanced and regular foliation by constant mean curvature spheres on Ω, we apply the moving center method in [Ye1]. Let ν denote the inward Euclidean unit normal of $S^n := \partial B_1(o)$ and α_r the dilation $x \mapsto rx$ for $r > 0$. For $\varphi \in C^2(S^n)$ and $\tau \in \mathbb{R}^{n+1}$ we define $S^n_\varphi = \{x + \varphi(x)\nu(x) : x \in S^n\}$ and $S_{r,\tau,\varphi} = \alpha_r(\tau(S^n_\varphi))$, where the action of τ is defined to be the translation by τ. Note that $S_{r,\tau,0} = \partial B_r(\tau)$ and S^n_φ is an embedded C^2 surface if only $\|\varphi\|_{C^1} \leq \varepsilon_0$ for some $\varepsilon_0 \in (0, \frac{1}{4})$. (We use the standard metric on S^n unless otherwise stated.) For $0 < r < \frac{1}{4}$, $\|\varphi\|_{C^1} \leq \varepsilon_0$, $\tau \in \mathbb{R}^{n+1}$ and $x \in S^n$ with $|\tau| + r + \|\varphi\|_{C^0} \leq 1$, we put

$$H(r,\tau,\varphi)(x) = \frac{1}{r} \text{ (the inward mean curvature of the surface } S_{r,\tau,\varphi}$$
$$\text{at } \frac{1}{r}(\tau + x + \varphi(x)\nu(x))),$$

where of course we use the metric g on Ω. By (1.1), it is easy to see that $H(r,\tau,\varphi)$ extends to $r = 0$ in a C^3 fashion. We are going to compute $H(r,\tau,0)$. Fix r_0, τ_0 and $x_0 \in S^n$. We need to compute the mean curvature of $\partial B_{1/r_0}(\tau_0/r_0)$ near $y_0 = (\tau_0 + x_0)/r_0$. Choose coordinate vector fields $X_1, \ldots, X_n$ on $\partial B_{1/r_0}(\tau_0/r_0)$ near y_0 such that they are orthonormal at y_0 (with respect to the metric g). We can extend them to coordinate vector fields around y_0 in the following way: $\widetilde{X}_i((\tau_0+x)/r) = \frac{r_0}{r}X_i((\tau_0+x)/r_0)$. We set $Y((\tau_0+x)/r) = -x$. Then we have for $y = (\tau_0+x)/r_0$

$$\begin{aligned}\sum_{i=1}^{n} g(\nabla_{X_i}Y, X_i)\Big|_y &= \frac{1}{2}\sum_{i=1} Yg(\widetilde{X}_i, \widetilde{X}_i)\Big|_y \\ &= nr_0 + \frac{1}{2}\sum_{i=1} Yg(X_i, X_i)\Big|_y ,\end{aligned} \tag{1.2}$$

where ∇ denotes the Levi–Civita connection and X_i is the euclidean parallel translation of the original X_i along the radial lines. But

$$\sum_{i=1}^{n} g(X_i, X_i)\Big|_y = g_{ii}(y) - g_{ij}(y)x^i x^j,$$

where the summation convention is used on the right hand side for $1 \le i, j \le n+1$. Hence

$$\sum_{i=1}^{n} g(\nabla_{X_i}Y, X_i)\Big|_y = nr_0 + \frac{1}{2}(g_{ii,k}(y)x^k - g_{ij,k}(y)x^i x^j x^k). \tag{1.3}$$

Next let Z denote the inward unit normal of $\partial B_{1/r_0}(\tau_0)$. Z can be constructed explicitly from Y by standard methods of linear algebra. Indeed, there are linearly independent vectors $v_1(y), \dots, v_n(y)$ which are linear functions of x and are orthogonal to x in the Euclidean metric. Applying the Gram–Schmidt procedure we obtain from the v_i's orthonormal tangent vectors $w_1(y), \dots, w_n(y)$ at y. Every w_i can be written in the form $F(g_{ij}(y), x)$ for a smooth function F of t_{ij}, $1 \le i, j \le n+1$ and x. Finally, set $Z_1 = \sum_{i=1}^{n} h(Y, w_i)w_i$, where h denotes the tensor h_{ij} at y. Then

$$Z = \frac{Y - Z_1}{g(Y - Z_1, Y - Z_1)^{1/2}}.$$

We compute

$$g(Y - Z_1, Y - Z_1) = 1 + \frac{\sigma}{|y|^{n-1}} + h_{ij}Y^iY^j - 2g_{ij}Z_1^iY^j + g_{ij}Z_1^iZ_1^j,$$

whence

$$\begin{aligned}Z - Y = -\frac{\sigma}{2|y|^{n-1}}Y + \frac{1}{|y|^{2n-2}}F_1(g_{ij}, h_{ij}, |y|^{-1}, x) \\ + h_{ij}F_{ij}(g_{k\ell}, h_{k\ell}, |y|^{-1}, x)\end{aligned} \tag{1.4}$$

with smooth functions F_1 and F_{ij}. Next we observe that the covariant derivative ∇ can be explicitly written and that we can replace X_i by w_i in (1.3). Since the mean curvature of $\partial B_{1/r_0}(\tau_0)$ is given by $\sum_{i=1}^{n} g(\nabla_{w_i} Z, w_i)$, we deduce form (1.3) that

$$\begin{aligned} H(r,\tau,0) &= n\left(1 - \frac{\sigma}{2|y|^{n-1}}\right) + \frac{1}{2}(g_{ii,k}x^k - g_{ij,k}x^i x^j x^k)\left(1 - \frac{\sigma}{2|y|^{n-1}}\right) \\ &+ \frac{1}{r|y|^{n+1}}\widetilde{F}_1(g_{ij}, |y|g_{ij,k}, h_{ij}, |y|h_{ij,k}, |y|^{-1}, |y|^{-1}y, x) \\ &+ \frac{1}{|y|^n}\widetilde{F}_2(g_{ij}, h_{ij}, |y|^{-1}, x) + \frac{1}{r}h_{ij,k}F_{ijk}(g_{\ell m}, h_{\ell m}, |y|^{-1}, x) \\ &+ \frac{1}{r|y|}h_{ij}\widetilde{F}_{ij}(g_{k\ell}, |y|g_{k\ell}, h_{k\ell}, |y|h_{k\ell}, |y|^{-1}, |y|^{-1}y, x) \\ &+ h_{ij}\overline{F}_{ij}(g_{k\ell}, h_{k\ell}, |y|^{-1}, x). \end{aligned} \tag{1.5}$$

Here r_0, τ_0 have been replaced by r and τ, $y = (\tau + x)/r$, the variable for g_{ij}, $g_{ij,k}$ etc. is y, and $\widetilde{F}_1$, $\widetilde{F}_2$, F_{ijk}, $\widetilde{F}_{ij}$, $\overline{F}_{ij}$ are smooth functions. We have

$$g_{ij,k} = \frac{\sigma(1-n)}{|y|^{n+1}}y^k\delta_{ij} + h_{ij,k}, \tag{1.6}$$

$$|y|^{-1} = r|\tau + x|^{-1} = r(1 - x^i\tau^i + \tau^i\tau^j b_{ij}(\tau, x)) \tag{1.7}$$

for smooth functions b_{ij} which are defined for all $x \in S^n$ and $\tau \in \mathbb{R}^{n+1}$ with $|\tau| < 1$. From (1.5), (1.6) and (1.7) we have deduce

$$\begin{aligned} H(r,\tau,0)(x) = n - \frac{\sigma n^2}{2}r^{n-1} &+ \frac{\sigma n(n-1)(n+1)}{2}r^{n-1}\tau^i x^i \\ &+ r^{n-1}\tau^i\tau^j\tilde{b}_{ij}(\tau,x) + r^{n-1}f(r,\tau,x), \end{aligned} \tag{1.8}$$

with smooth functions $\tilde{b}_{ij}$, f for $r > 0$, $|\tau| < 1$ and $x \in S^n$ such that

1. $\|f(r,\tau,\cdot)\|_{C^1(S^n)} \le C(|\tau|)r$,
2. $\|d_\tau f(r,\tau,\cdot)\|_{C^1(S^n)} \le C(|\tau|)r$,
3. $\|\frac{\partial f}{\partial r}(r,\tau,\cdot)\|_{C^1(S^n)} \le C(|\tau|)r$

for a positive continuous function $C(t)$ defined for $|t| < 1$, where $d_\tau f$ denotes the differential of f in τ. We put $\tilde{f} = f + \tilde{b}_{ij}\tau^i\tau^j$.

Our goal is to find solutions τ, φ of the equation

$$H(r,\tau,\varphi) = n - \frac{\sigma n^2}{2}r^{n-1}. \tag{1.9}$$

We consider $H(r,\tau,\cdot)$ as a mapping from $C^{2,1/2}(S^n)$ into $C^{0,1/2}(S^n)$ and let H_φ denote the differential of H w.r.t. φ. Set $g_{r,\tau} = r^2 g$. Then $H(r,\tau,\varphi)$ is the mean curvature of S^n and $H_\varphi(r,\tau,0)$ is just the Jacobi operator $\Delta + \|A\|^2 + Rc(\bar{\nu})$ on S^n relative to the metric $g_{r,\tau}$ where Rc denotes Ricci curvature and $\bar{\nu}$ the inward unit normal of S^n. We indicate the dependence on $g_{r,\tau}$ as follows

$$H_\varphi(r,\tau,0) = \Delta_{r,\tau} + \|A_{r,\tau}\|^2 + Rc_{r,\tau}. \tag{1.10}$$

Note that $g_{r,\tau}$ converges to the Euclidean metric in the C^4 topology as $r \to 0$. It follows that $H_\varphi(0,\tau,0) = L := \Delta_{S^n} + n$, where Δ_{S^n} is the standard Laplace operator on S^n. Let Ker denote the kernel of L which is spanned by the Euclidean coordinate functions. We have the orthogonal L_2-decompositions $C^{2,1/2}(S^n) = \mathrm{Ker} \oplus \mathrm{Ker}^\perp$ and $C^{0,1/2}(S^n) = \mathrm{Ker} \oplus L(\mathrm{Ker}^\perp)$. Let P denote the orthogonal projection from $C^{0,1/2}(S^n)$ onto Ker and T: Ker $\to \mathbb{R}^{n+1}$ the isomorphism sending $x_i|_{S^n}$ to $\mathbf{e}_i$ = the i-th coordinate basis. Put $\widetilde{P} = TP$. Then

$$\widetilde{P}(H(r,\tau,0)) = \frac{1}{2}n(n-1)(n+1)\sigma\omega_{n+1}r^{n-1}\tau + r^{n-1}\widetilde{P}(\tilde{f}(r,\tau,\cdot)), \tag{1.11}$$

where $\omega_{n+1} = \mathrm{vol}(B_1(o))$. To solve (1.9) we introduce the following expansions

$$\begin{aligned} H(r,\tau,\varphi) = H(r,\tau,0) &+ r\int_0^1\int_0^1 H_{\varphi r}(sr,\tau,t\varphi)\varphi\, ds\, dt \\ &+ \int_0^1\int_0^1 H_{\varphi\varphi}(0,\tau,st\varphi)\varphi\varphi\, ds\, dt + L\varphi, \end{aligned} \tag{1.12}$$

where the subscript r means the partial derivative in r and

$$H_{\varphi\varphi}(r,\tau,\psi)\varphi\varphi' = \frac{d}{dt}\, H_\varphi(r,\tau,\psi + t\varphi')\varphi\big|_{t=0}.$$

We first consider the equation

$$\widetilde{P}(H(r,\tau,r^{n-1}\varphi)) = 0. \tag{1.13}$$

Dividing it by r^{n-1} and applying (1.11), (1.12) we can reduce it to the following

$$\frac{1}{2}n(n-1)(n+1)\sigma\omega_{n+1}\tau + \widetilde{P}(\tilde{f}(r,\tau,\cdot)) + r\widetilde{P}(q(r,\tau,\varphi)) = 0, \tag{1.14}$$

where

$$q(r,\tau,\varphi) = \int_0^1\int_0^1 H_{\varphi r}(sr,\tau,tr^{n-1}\varphi)\varphi\, ds\, dt \\ + r^{n-1}\int_0^1\int_0^1 tH_{\varphi\varphi}(0,\tau,str^{n-1}\varphi)\varphi\varphi\, ds\, dt.$$

(Note that $\widetilde{P}L = 0$.) Easy computations show that q has the following properties

1. $\|q(r,\tau,\varphi)\|_{C^{0,1/2}(S^n)} \le \beta(\|\varphi\|)$,
2. $\|d_\tau q(r,\tau,\varphi)\|_{C^0(S^n)} \le \beta(\|\varphi\|)$,
3. $\|q_\varphi(r,\tau,\varphi)\| \le \beta(\|\varphi\|)$, with $q_\varphi\colon C^{2,1/2}(S^n) \to C^{0,1/2}$,
4. $\|\frac{\partial q}{\partial r}(r,\tau,\varphi)\|_{C^{0,1/2}(S^n)} \le \beta(\|\varphi\|)$.

Here and in the sequel, β denotes a positive continuous function on $\mathbb{R}$ and $\|\varphi\| = \|\varphi\|_{C^{2,1/2}(S^n)}$.

Let $Q(r,\tau,\varphi)$ denote the left hand side of (1.14). Then for each bound on φ, $Q(r,0,\varphi)$ approaches zero uniformly as $r \to 0$. This follows from the properties of q and $\tilde{f}$. Also by virtue of these properties, the differential $d_\tau Q$ is nearly the identity for small r and τ. Hence we can apply the implicit function theorem to obtain a unique solution $\tau(r,\varphi)$ of the equation (1.14) and hence (1.13) for small r which lies near zero. There holds $\tau(r,\varphi) \to 0$ as $r \to 0$. We also have the following estimates

$$\|\tau_\varphi\| \le \beta(\|\varphi\|)r, \qquad \left|\frac{\partial\tau}{\partial r}\right| \le \beta(\|\varphi\|).$$

These estimates are easy consequences of the properties of q and $\tilde{f}$. It follows that

$$|\tau(r,\varphi)| \le \beta(\|\varphi\|)r.$$

Now we replace τ in (1.9) by $\tau(r,\varphi)$, φ by $r^{n-1}\varphi$, and divide (1.9) by r^{n-1}. Then (1.9) is reduced to

$$L\varphi + \frac{\sigma n(n-1)(n+1)}{2}\tau^i(r,\varphi)x^i + \tilde{f}(r,\tau(r,\varphi),\cdot) + rq(r,\tau(r,\varphi),\varphi) = 0. \tag{1.15}$$

By the above argument for finding $\tau(r,\varphi)$ and the properties of $\tilde{f}$, q we can find a unique solution $\varphi(r)$ of (1.15) for small r which lies near zero. There holds $\varphi(r) \to 0$ as $r \to 0$. Moreover, we have the following estimates:

$$\|\varphi(r)\| \le Cr, \quad \left\|\frac{d\varphi}{dr}\right\| \le C$$

for a constant $C > 0$.

The uniqueness property of $\tau(r,\varphi)$ and $\varphi(r)$ can be stated precisely as follows

Proposition 1.1. *For each positive number C there are positive numbers $R_1 = R_1(C)$ and $R_2 = R_2(C)$ with the following properties:*

1) *if $Q(r,\tau,\varphi) = 0$, $0 < r \le R_1$, $|\tau| \le R_2$ and $\|\varphi\| \le C$, then $\tau = \tau(r,\varphi)$;*

2) *if $H(r,\tau(r,\varphi),r^n\varphi) = n - \frac{\sigma n^2}{2} r^{n-1}$, $0 < r \le R_1$ and $\|\varphi\| \le C$, then $\varphi = \varphi(r)$.*

We omit the easy proof. (For part 2) one utilizes the equation (1.15) to show that φ is small.)

Now we consider the family of surfaces

$$\mathcal{F}_0 = \{\, \widetilde{S}_r = S_{1/r,\tau(r,\varphi(r)),r^{n-1}\varphi(r)} : 0 < r \le r_0 \,\},$$

where r_0 is chosen as follows. Set $C = \sup_{r \le R_1(1)} \|\varphi(r)\| + 1$, $r_1 = \max\{ r : |\tau(r,\varphi)| \le R_2(C)$ with $\|\varphi\| \le C \}$ and then $r_0 = \min\{r_1, R_1(1), R_1(C)\}$. These surfaces are diffeomorphic to S^n. By construction, $\widetilde{S}_r$ has constant mean curvature $nr - \frac{\sigma n^2}{2} r^n$. Since all the equations and known functions in the above construction are smooth, we conclude that $\mathcal{F}_0$ is a smooth family of constant mean curvature spheres. Geometrically, we obtained this family by moving the center (the origin) of the Euclidean spheres $\partial\mathbb{B}_r(o)$ to $\tau(r,\varphi(r))/r$ and then performing the Euclidean normal perturbation $r^{n-2}\varphi(r)$. It remains to show that $\mathcal{F}_0$ is a foliation. Put $\psi(r,x) = r^{-1}\tau(r,\varphi(r)) + r^{-1}x + r^{n-2}\varphi(r)(x)$, where $x \in S^n$. By the estimates for τ and φ we easily see that the maps $v(r,\cdot) := \psi(r,\cdot)/|\psi(r,\cdot)|$ approach the identity map from S^n to S^n in the C^2 norm as $r \to 0$. Hence $v(r,\cdot)$ is a smooth diffeomorphism for small r. We set $\tilde{\varphi}(r,x) = |\psi(r, v^{-1}(r,x))|$ where $v^{-1}(r,\cdot)$ denotes the inverse of $v(r,\cdot)$. Then $\widetilde{S}_r$ is the Euclidean normal graph of $\tilde{\varphi}(r,\cdot)$ over S^n. Employing the estimates for τ and φ we compute

$$\begin{aligned}\frac{\partial\psi}{\partial r} &= -\frac{x}{r^2} + \frac{1}{r}\frac{\partial\tau}{\partial r} + \frac{1}{r}\tau_\varphi\left(\frac{d\varphi}{dr}\right) - \frac{\tau}{r^2} + (n-2)r^{n-3}\varphi(r) + r^{n-1}\frac{d\varphi}{dr}\\ &= -\frac{1}{r^2}(x + O(r)),\\ d_x\psi &= \frac{1}{r}(I + O(r)), \quad d_x v = I + O(r),\end{aligned}$$

and finally

$$\frac{\partial\tilde{\varphi}}{\partial r} = \frac{1}{|\psi|}\psi\cdot\left(\frac{\partial\psi}{\partial r} + (d_x\psi)\left(\frac{\partial v^{-1}}{\partial r}\right)\right) = -\frac{1}{r^2}(1 + O(r)).$$

We conclude that $\tilde{\varphi}(r,x)$ is strictly decreasing w.r.t. r for r small (or better, $\tilde{\varphi}(\rho^{-1},x)$ is strictly increasing for ρ large). Hence $\widetilde{S}_r$, $\widetilde{S}_{r'}$ are disjoint for small r, r' with $r \neq r'$. We replace r_0 by a smaller positive number if necessary. Then $\mathcal{F}_0$ constitutes a smooth foliation. It is easy to see that $\mathcal{F}_0$ is balanced and regular at ∞. Thus we have proven the existence part of the Main Theorem.

2. Uniqueness.

Let Ω be the asymptotically flat end considered above and $\mathcal{F}_0$ the constructed foliation. Let Ω_1 be the support of $\mathcal{F}_0$, i.e. the subdomain of Ω foliated by $\mathcal{F}_0$. Note that $\Omega\setminus\Omega_1$ is compact.

Definition 6. A *constant mean curvature foliation* is a codimension one C^2 foliation with closed leaves of constant mean curvature.

We have

Uniqueness Theorem. *Let $\mathcal{F}$ be a weakly balanced and regular constant mean curvature foliation on a closed subdomain Ω' of Ω such that $\Omega\setminus\Omega'$ is compact. Then either $\mathcal{F}_0$ is a restriction of $\mathcal{F}$ or $\mathcal{F}$ is a restriction of $\mathcal{F}_0$. In other words, $\mathcal{F}_0$ is the unique maximal weakly balanced and regular constant mean curvature foliation in Ω_1.*

Proof. Let $\mathcal{F}$ be as described in the theorem. It is easy to see that the leaves of $\mathcal{F}$ can be parametrized as a C^2 family S_t, $0 < t \leq 1$ with $S_t \neq S_{t'}$

if $t \neq t'$ and $\lim_{t\to 0} \operatorname{diam} S_t = \infty$ (See Lemma 2.1 in [Ye1]). Then S_t lies on the interior side of $S_{t'}$, provided that $t > t'$. We put $\ell(t) = (\operatorname{diam} S_t)^{-1}$ and $S_t^* = \alpha_{\ell(t)}(S_t)$, i.e. S_t^* is the dilation of S_t by the factor $\ell(t)$. Let $g_t = \ell(t)^2 \alpha^*_{1/\ell(t)} g$. Then g_t converges to the Euclidean metric in the C^4 topology away from the origin. Each S_t^* has constant mean curvature in the metric g_t. Because $\mathcal{F}$ is regular, the second fundamental form of S_t^* in g_t is bounded above by a constant independent of t.

We claim that $\operatorname{dist}(o, S^*_{t_k})$ is uniformly bounded away from zero. Assume the contrary. Then we can find a sequence $S^*_{t_k}$ with $t_k \to 0$ which converges to an immersed surface S_∞ in $\mathbb{R}^{n+1} \setminus \{o\}$ of constant mean curvature. S_∞ has uniformly bounded second fundamental form and diameter 1. The only possible self-intersections of S_∞ are of the type that an embedded piece of S_∞ meets another from one side. A priori, S_∞ may have many components. Let S'_∞ be one component. Then $\overline{S}'_\infty \setminus S'_\infty = \{o\}$. Since S_∞ has uniformly bounded second fundamental form and constant mean curvature, the origin is a removable singularity, i.e. $\overline{S}'_\infty$ is a smooth immersed surface of constant mean curvature. Applying the classical Alexandrov reflection principle we deduce that $\overline{S}'_\infty$ is a round sphere. It follows that $\overline{S}_\infty$ is either a round sphere containing the origin or the union of two round spheres meeting tangentially at the origin. But the construction of weak balance does not allow such limits as one readily sees. Thus we arrive at a contradiction and the claim is proven.

Consider an arbitrary sequence $S^*_{t_k}$ with $t_k \to 0$. By the above claim and the above arguments, a subsequence converges to a round sphere S_∞ of diameter 1 such that the origin lies in the open ball bounded by S_∞. We conclude that

$$\widetilde{H}(t) \operatorname{diam} S_t = 2n + h(t) \tag{2.1}$$

for a function $h(t)$ which converges to zero as $t \to 0$, where $\widetilde{H}(t)$ denotes the mean curvature of S_t (in the metric g). In particular there is for small t a unique solution $r(t)$ of the equation

$$\widetilde{H}(t) = nr - \frac{\sigma n^2}{2} r^n. \tag{2.2}$$

Now we set $\widehat{S}_t = \alpha_{r(t)}(\widetilde{S}_t)$ and $\widetilde{g}_t = r(t)^2 \alpha^*_{1/r(t)} g$. Then the mean curvature $H(t)$ of $\widehat{S}_t$ in $\widehat{g}_t$ is $n - \frac{\sigma n^2}{2} r(t)^{n-1}$. By the above arguments concerning S_t^* we deduce that every sequence $\widehat{S}_{t_k}$ with $t_k \to 0$ contains a subsequence converging to a round sphere S_∞ such that the origin lies in the open ball bounded by S_∞. Then

$$|a_\infty| \leq b(\mathcal{F}) < 1.$$

We conclude that for small t every $\widehat{S}_t$ is the normal graph of a smooth function φ_t over a round sphere of radius 1 and center $a(t)$ such that $|a(t)| \leq b(\mathcal{F})$ and φ_t converges to zero in the C^4 norm as $t \to 0$. By the arguments in the proof of Lemma 2.3 in [Ye1] we can find (for small t) $v(t) \in \mathbb{R}^{n+1}$ such that $\limsup_{t\to 0} |v(t)| \leq b(\mathcal{F})$, $\widehat{S}_t = S_{1,v(t),\widehat{\varphi}(t)}$ for a smooth function $\widehat{\varphi}(t)$ over S^n with $\widehat{\varphi}(t) \to 0$ in the C^4 norm as $t \to 0$. Moreover, the projection $P(\widehat{\varphi}(t))$ vanishes. Roughly speaking, we can move center $a(t)$ slightly to achieve that the defining function of $\widehat{S}_t$ as a graph has zero kernel component. Note that $S_t = S_{1/r(t),v(t),\widehat{\varphi}(t)}$ and

$$H(r(t), v(t), \widehat{\varphi}(t)) = n - \frac{\sigma n^2}{2} r(t)^{n-1}. \tag{2.3}$$

Applying (2.3), (1.8), (1.12) and the inequality $\limsup_{t\to 0} |v(t)| \leq b(\mathcal{F}) < 1$ we obtain for small t

$$\|\widehat{\varphi}(t)\| \leq C r(t)^{n-1} \tag{2.4}$$

for a constant $C > 0$. Next we estimate $v(t)$. To this end we apply (1.5) and calculate $H(r, \tau, 0)$ in a way somewhat different from (1.8). We have

$$\begin{aligned} H(r,\tau,0)(x) = n &- \frac{n\sigma}{2} r^{n-1} \frac{1}{|x+\tau|^{n-1}} - \frac{n(n-1)\sigma}{2} r^{n-1} \frac{1}{|x+\tau|^{n+1}} \\ &- \frac{n(n-1)\sigma}{2} r^{n-1} \frac{\tau^k x^k}{|x+\tau|^{n+1}} + r^{n-1} f(r,\tau,x), \end{aligned} \tag{2.5}$$

where f is the same function as appearing in (1.8). It follows from (2.3), (2.4), (2.5) and (1.12) that

$$\left| P\left(\frac{1}{|x+v(t)|^{n-1}} + (n-1)\frac{1}{|x+v(t)|^{n+1}} + (n-1)\frac{v(t)\cdot x}{|x+v(t)|^{n+1}} \right) \right| \leq C r(t) \tag{2.6}$$

for a positive constant C. For a fixed small t we change coordinates such that $v(t) = \ell \mathbf{e}_1$ for some $\ell \in (0,1)$. Then

$$- \int_{S^n} \frac{x^1}{|x+v(t)|^{n-1}} d\,\mathrm{vol} - (n-1) \int_{S^n} \frac{x^1}{|x+v(t)|^{n+1}} d\,\mathrm{vol}$$

$$
\begin{aligned}
&- (n-1) \int_{S^n} \frac{(v(t)\cdot x)x^1}{|x+v(t)|^{n-1}} d\,\mathrm{vol} \\
&= -\int_{S^n} \frac{x^1}{|1+\ell x^1|^{n-1}} d\,\mathrm{vol} - (n-1)\int_{S^n} \frac{x^1}{|1+\ell x^1|^n} d\,\mathrm{vol} \\
&\geq (n-1) \int_{x^1>0} x^1 \left(\frac{1}{|1-\ell x^1|^n} - \frac{1}{|1+\ell x^1|^n} \right) d\,\mathrm{vol} \\
&\geq 2n(n-1) \int_{x^1>0} \frac{(x^1)^2}{(1+\ell x^1)^n (1-\ell x^1)^n} d\,\mathrm{vol} \\
&\geq (n-1)(n+1)\omega_{n+1}\ell,
\end{aligned}
$$

whence

$$
|v(t)| \leq C r(t) \tag{2.7}
$$

for a constant $C > 0$.

Setting $\varphi^* = r(t)^{1-n}\widehat{\varphi}(t)$ we have $Q(r(t), v(t), \varphi^*(t)) = 0$. Applying (2.4), (2.7) and Proposition 1 we deduce that $v(t) = \tau(r(t), \varphi^*(t))$ for small t. Applying (2.3) and Proposition 1 we then conclude that $\varphi^*(t) = \varphi(r(t))$ for small t. Consequently $S_t = \widetilde{S}_{r(t)}$ for small t. A continuity argument then shows that either $\mathcal{F}_0$ is a restriction of $\mathcal{F}$ or $\mathcal{F}$ is a restriction of $\mathcal{F}_0$. □

References.

[B] R. Bartnik, *The mass of an asymptotically flat manifold*, Comm. Pure Appl. Math. **34** (1986), 661–693.

[HY1] G. Huisken and S. T. Yau, *Foliation by constant mean curvature spheres*, speech by Huisken at "Heat Flow in Geometry", Hawaii, Summer 1989.

[HY2] G. Huisken and S. T. Yau, *Foliation by constant mean curvature spheres*, speech by Huisken at "Calculus of Variation", Oberwolfach, July 1992.

[LP] J. M. Lee and T. H. Parker, *The Yamabe problem*, Bull. Amer. Math. Soc. **17** (1987), 37–91.

[SY] R. Schoen and S. T. Yau, *Positivity of the total mass of a geneal space time*, Phys. Rev. Lett. **43** (1979), 1457–1459.

[Ya] S. T. Yau, *Problem Section, Seminar of Differential Geometry*, Ann. Math. Studies 102, Princeton University Press 1982.

[Ye1] R. Ye, *Foliations by constant mean curvature spheres*, Pacific J. Math. **147** (1991), 381–396.

[Ye2] R. Ye, *Foliations by constant mean curvature spheres: existence, uniqueness and structure*, preprint.

RECEIVED JULY 5, 1995.

MATHEMATICS INSTITUTE
RUHR-UNIVERSITY BOCHUM

AND

DEPARTAMENT OF MATHEMATICS
UNIVERSITY OF CALIFORNIA
SANTA BARBARA, CA 93106

Publications of Stefan Hildebrandt

1. Über die Identität der Sobolewschen und der Calkin-Morreyschen Räume. Math. Ann. **148**, 226–237 (1962)
2. Rand- und Eigenwertaufgaben bei stark elliptischen Systemen linearer Differentialgleichungen. Math. Ann. **148**, 411–429 (1962)
3. Einige konstruktive Methoden bei Randwertaufgaben für lineare partielle Differentialgleichungssysteme und in der Theorie harmonischer Differentialformen. I. J. Reine Angew. Math. **213**, 66–88 (1963)
4. Über das alternierende Verfahren von Schwarz bei positive definiten selbatadjungierten Differentialgleichingssystemen. Ann. Acad. Sci. Fenniae, Ser. A1. **336/2** (1963)
5. Constructive proofs of representation theorems in separable Hilbert space. Commun. Pure Appl. Math. **17**, 369–373 (1964) (with E. Wienholtz)
6. The closure of the numerical range of an operator as spectral set. Commun. Pure Appl. Math. **17**, 415–421 (1964)
7. Numerischen Wertebereich und normale Dilationen. Acta. Sci. math. **26**, 187–190 (1965)
8. Über den numerischen Wertebereich eines Operators. Math. Ann. **163**, 230–247 (1966)
9. Über das Randverbalten von Minimalflächen. Math. Ann. **165**, 1–18 (1966)
10. Entwicklungslinien der Variationsrechnung seit Weierstraß. Wiss. Abh. Arbeitsgemeinschaft Nordrein-Westfalen **33**, Festschr. Gedächtnisfeier Karl Weierstraß, 183–240 (1966) (with E. Hölder, R. Klötzer, S. Gähler)
11. Über Minimalflčhen mit der Rand. Math. Z. **95**, 1–19 (1967)
12. Über die Lösung nichtlinearer Eigenwertaufgaben mit dem Galerkinverfahren. Math. Z. **101**, 255–264 (1967)
13. Über die alternierenden Verfahren von H.A. Schwartz und O. Neumann. J. Reine Angew. Math. **232**, 136–155 (1968)
14. Zur Konvergenz von Operatorprodukten im Hilbertraum. Math. Z. **105**, 62–71 (1968) (with D. Schmidt)
15. Boundary behavior of minimal surfaces. Arch. Ration. Mech. Anal. **35**, 47–82 (1969)
16. Über Flächen konstanter mittlerer Krümmung. Math. Z. **112**, 107–144 (1969)
17. Randwertprobleme für Flächen mit vorgeschriebenser mittlerer Krümmung und Anwendungen auf die Kapilaritätstheorie, I; Fest vorgegebener Rand. Math. Z. **112**, 205–213 (1969)
18. Randwertprobleme für Flächen mit vorgeschriebenser mittlerer Krümmung und Anwendungen auf die Kapilaritätstheorie, I; Freie Ränder. Arch. Ration. Mech. Anal. **39**, 275–293 (1970)
19. On the Plateau problem for surfaces of constant mean curvature. Commun. Pure. Appl. Math. **23**, 97–114 (1970)

20. Einige Bemerkungen über Flächen beschränkter mittlerer Krümmung. Math. Z. **115**, 169–178 (1970)

21. Zum Randverhalten der Lösungen dewisser zweidimensionaler Variationsprobleme mit freien Randbedingungen. Math. Z. **118**, 241–253 (1970) (with K. Goldhorn)

22. Some remarks on minimal surfaces in Riemannian manifolds. Commun. Pure Appl. Math. **23**, 371–377 (1970) (with E. Heinz)

23. On the number of branch points of surfaces of bounded mean curvature. J. Differ. Geom. **4**, 227–235 (1970) (with E. Heinz)

24. On the regularity of sufaces with prescribed mean curvature at a free boundary. Math. Z. **118**, (289–308 (1970) (with W. Jäger)

25. Unendliche Produkte von Kontraktionen. Math. J., Indiana Univ. **20**, 909–911 (1971)

26. Some recent contributions to Plateau's problem. Diff.-Geom. im Großen, Ber. Tagung Math. Forsch.-Inst. Oberwolfach 1969, Ber. Math. Forsch.-Inst. Oberwolfach **4**, 163–170 (1971)

27. Ein einfacher Beweis für die Regularität der Lösungen gewisser zweidimensionaler Variationsprobleme unter freien Randbedingungen. Math. Ann. **184**, 315–331 (1971)

28. Über einen neuen Existenzsatz für Flächen vorgoschriebener mittlerer Krümmung. Math. Z. **119**, 267–272 (1971)

29. Maximum principle for minimal surfaces of continuous mean curvature. Math. Z. **128**, 253–269 (1972)

30. Two-dimensional variational problems with obstructions, and Plateau's problem for H-surfaces in a Riemannian manifold. Commun. Pure. Appl. Math. **25**, 187–223 (1972) (with H. Kaul)

31. On the regularity of solutions of two-dimensional variational problems with obstructions. Commun. Pure. Appl. Math. **25**, 479–496 (1972)

32. Interior $C^{1+\alpha}$-regularity of solutions of two-dimensional variational problems with obstacles. Math. Z. **131**, 235–240 (1973)

33. Variational problems with obstacles and a volume constraint. Math. Z. **135**, 55–68 (1973) (with H.C. Wente)

34. Regularity of solutions of variational problems with obstacles and volume constraints. Symp. Math. **14**, Geom. Simplett. Fis. Mat., Teor. Geom. Integr. Var. Minim., Convegni 1973, 493–497 (1974)

35. Some regularity results for quasilinear elliptic systems of second order. Math. Z. **142**, 67–85 (1975) (with K.O. Widman)

36. Dirichlet's boundary values problems for harmonic mappings of Riemannian manifolds. Math. Z. **147**, 255-236 (1976) (with H. Kaul, K.O. Widman)

37. Harmonic mappings into Riemannian manifolds with non-positive sectional curvature. Math. Scandinav. **37**, 257–263 (1970) (with H. Kaul, K.O. Widman)

38. On the Hölder continuity of weak solutions of quasilinear elliptic systems of second order. Ann. Sc. norm. super. Pisa, Cl. Sci., IV. Ser. 4, 144–178 (1977) (with K.O. Widman)

39. An existense theorem for harmonic mappings of Riemannian manifolds. Acta Math. **138**, 1–16 (1977) (with H. Kaul, K.O. Widman)

40. Mathematical aspects of 't Hofft's eigenvalue problem in two-dimensional quantum chromodynamics. I: A variational approach, and nodal properties of the eigenfunctions. Manuscr. Math. **24**, 45–79 (1978)

41. Mathematical aspects of 't Hofft's eigenvalue problem in two-dimensional quantum chromodynamics. II. Ark. Mat. **17**, 29–38 (1979)

42. Variational inequalities for vectorvalued functions. J. Reine Angew. Math. **309**, 191–220 (1979) (with K.O. Widman)

43. Satse vom Liouvilleschen Typ für quasilineare elliptische Gleichungen und Systeme. Nachr. Akad. Wiss. Gött., II. Math.-Phys. Kl. 1979, 41–59 (1979) 9with K.O. Widman)

44. On variational problems with obstacles and integral constraints for vector-valued functions. Manuscr. Math. **28**, 185–206 (1979) (with M. Meier)

45. Mathematical aspects of 't Hooft's eigenvalue problems in two-dimensional quantum chromodynamics. Part III. Math. Z. **168**, 233–240 (1979)

46. Minimal surfaces with free boundaries. Acta. Math. **145**, 251–272 (1979) (with J.C.C. Nitsche)

47. The two-dimensional analogue of the catenary. Pac. J. Math. **88**, 247–278 (1980) (with R. Böhme, E. Tausch)

48. Harmonic mappings and minimal submanifolds. Invent. Math. **62**, 269–298 (1980) (with J. Jost, K.O. Widman)

49. Boundary behavior of minimal surfaces with free boundaries. Recent contributions to nonlinear partial differential equations, Lect. Paris 1978/79, Res. Notes Math. **50**, 77–102 (1981)

50. Entire solutions of nonlinear elliptic systems and an approach to Bernstein. Nonlinear partial differential equations and their applications, Coll. de France Semin., Vol. 1, Res. Notes Math. **53**, 251–292 (1981)

51. Estimation a priori des solutions faibles de certains systemes non lineaires elliptiques. Semin. Goulaoute-Meyer-Schwartz, Equations Deriv. Partielles 1980-1981, Exposë No. **17** (1981) (with M. Giaquinta)

52. On the boundary behavior of minimal surfaces with a free boundary which are not minima of the area. Manuscr. Math. **35**, 387–410 (1981) (with M. Grüter, J.C.C. Nitsche)

53. Optimal boundary regularity for minimal surfaces with a free boundary. Manuscr. Math. **33**, 357–364 (1981) (with J.C.C. Nitsche)

54. A priori estimates for harmonic mappings. J. Reine Angew. Math. **336**, 123–164 (1982) (with M. Giaquinta)

55. Liouville's theorem for harmonic mappings, and an approach to Bernstein theorems. Semin. differential geometry, Ann. Math. Stud. **102**, 107–131 (1982)

56. Nonlinear elliptic systems and harmonic mappings. Proc. of the 1980 Beijing Symposium on Differential Geometry and Differential Equations, Vol. 1, 481–515, Science Press, Beijing, China 1982

57. A uniqueness theorem for surfaces of least area with partially free boundaries on obstacles. Arch. Ration. Mech. Anal. **79**, 189–218 (1982) (with J.C.C. Nitsche)
58. Partielle Differentialgleichungen und Differentialgeometrie. Jährcaber Dtsch. Math.-Ver. **35**, 129–149 (1983)
59. Quasilinear elliptic systems in diagonal form. Systems of nonlinear partial differential equations. Proc. NATO Adv. Study Inst., Oxford/U.K. 1982, NATO ASI Ser. C **111**, 173–217 (1983)
60. Geometric properties of minimal surfaces with free boundaries. Math. Z. **184**, 497–509 (1983) (with J.C.C. Nitsche)
61. Variationsrechnung - ein dreihundertjähriges mathematisches Forschungsgebiet. in: Forschung in der Bundesrepublik Deutschland. herausgegeben von der DFG, Verlag Chemie, Weinheim, 481–488 (1983)
62. Euler und die Variationsrechnung. Zum Werk Leonhard Eulers. Vortr. Euler-Kolloq., Berlin 1983, 21–33 (1984)
63. Minimal surfaces with free boundaries and related problems. Variational mathods for equilibrium problems of fluids, Meet. Tronto/Italy 1983, Asterisque **118**, 69–88 (1984)
64. Calculus of Variations Today, Reflected in the Oberwolfach Meetings. in: Perspectives in Mathematics, Anniversary of Oberwolfach, Birkhäuser, Basel, 321–336 (1984)
65. Minimsl surfaces with free boundaries. Miniconference on P.D.E., Canberra, C.M.A., A.N.U. Preprint, August 1985
66. Harmonic mappings of Riemannian manifolds. In: E. Giusti (Ed.), Harmonic mappings and minimal immersions. Lect. Notes Math. **1161**, Springer, Berlin Heidelberg New York 1985, 1–117
67. Mathematics and optimal form. Scientific American Library, W.H. Freeman, New York 1985. (with A.J. Tromba)
68. Panoptimum. Mathematische Grundmuster des Vollkommenen. Spektrum-der-Wissenschaft Verlagsgesellschaft, Heidelberg (with A.J. Tromba)
69. Free boundary problems for minimal surfaces and related questions. Commun. Pure Appl. Math. **39**, Suppl., S 111- S 138 (1986)
70. Regularity for stationary surfaces of constant mean curvature with free boundaries. Acta Math. **156**, 119–152 (1986) (with M. Grüter, J.C.C. Nitsche)
71. Boundary configurations spanning continua of minimal surfaces. Manuscr. Math. **54**, 323–347 (1986) (with R. Gulliver)
72. Remarks on some isoperimetric problem. in: Meeting 100th birthday Leonida Tonelli, Pitman, London (1986)
73. On two isoperimetric problems with free boundary conditions. Variational methods for free surface interfaces. Proc. Conf., Menlo Park/Calif. 1985, 43–51 (1987)
74. On the analyticity of minimal surfaces at movable boundaries of prescribed length. J. Reine Angew. Math. **379**, 100–114 (1987) (with U. Dierkes, H. Lewy)
75. Calculus of variations and partial differential equations. Proc. Trento/Italy 1985, Lect. Notes Math. **1340**, Springer (1988) (with D. Kinderlehrer, M. Miranda (Ed.))

76. Partial differential equations and calculus of variations. Lect. Notes Math. **1357**, Springer (1988) (with R. Leis (Ed.))

77. The calculus of variations today. Math. Intell. **11**, No. 4, 50–60 (1989)

78. Partielle Differentialgleichungen und Variationrechnung. Festschrift zum Jübilaum der Dtsch. Math.-Ver. (1991) (with J. Bemelmans, W. v. Wahl)

79. Embeddedness and uniqueness of minimal surfaces solving a partially free boundary value problem. J. Reine Angew. Math. **422**, 69–89 (1991) (with F. Sauvigny)

80. On one-to-one harmonic mappings and minimal surfaces. SFB 256, Univ. Bonn, Preprint (with F. Sauvigny)

81. Minimal surfaces in a wedge. Preprint (with F. Sauvigny)

82. Uniqueness of stable minimal surfaces with partially free boundaries. J. Math. Soc. Japan (with F. Sauvigny)

83. Minimal surfaces I: Boundary value problems. Grundlehren math. Wiss. 295 (1992), Springer (with U. Dierkes, A. Küster, O. Wohlrab)

84. Minimal surfaces II: Boundary regularity. Grundlehren math. Wiss. 296 (1992), Springer (with U. Dierkes, A. Küster, O. Wohlrab)

85. Laudatio auf Fritz John. In: Sitzungeberichte der Berliner Mathematischen Gesellschaft. Herausgegeben vom Vorstand der Gesellschaft. Jahrgänge 1988–1992, 193–198 (1992)

86. Contact transformations, Huygens's principle, and calculus of variations. Calc. Var. **2**, 249–281 (1994)

87. On Hölder's transformation. to appear in: J. Fac. Sci. Univ. Tokyo Sect. IA (1994)

88. Von mathematischen Kultur. Vortrag anläßlich der Auszeichnung mit dem von Staudt-Preis, Mitteil. der Dtsch. Math.-Ver. **4** (1994)

89. Calculus of Variations I. The Lagrangian formalism. Grundlehren math. Wiss. 310 (1995), Springer (with M. Giaquinta)

90. Calculus of Variations I. The Hamiltonian formalism. Grundlehren math. Wiss. 311 (1995), Springer (with M. Giaquinta)